Immersed Tunnels

Immersed Tunnels

Richard Lunniss and Jonathan Baber

CRC Press
Taylor & Francis Group
Boca Raton London New York

CRC Press is an imprint of the
Taylor & Francis Group, an **informa** business
A SPON BOOK

CRC Press
Taylor & Francis Group
6000 Broken Sound Parkway NW, Suite 300
Boca Raton, FL 33487-2742

First issued in paperback 2017

Version Date: 20130123

ISBN 13: 978-1-138-07618-1 (pbk)
ISBN 13: 978-0-415-45986-0 (hbk)

Visit the Taylor & Francis Web site at
http://www.taylorandfrancis.com

and the CRC Press Web site at
http://www.crcpress.com

Contents

List of figures

List of tables

Foreword

Faced with having to cross a waterway, an engineer must decide whether to use a bridge above it or a tunnel underneath. If for whatever reason a bridge is not feasible or desirable, the next decision to be made is what type of tunnel should be used. The key parameter is the depth required for a safe tunnel. This is key, because the depth largely determines the total length of the tunnel and, thereby to a large degree, the cost. There are three basic alternatives of depth, listed here in ascending order: (1) bored, (2) immersed, and (3) submerged floating.

The bored tunnel is most likely the deepest as it generally requires an overhead cover of a full diameter or more below the sound bottom of the waterway. The immersed tunnel requires only a 2–3 m cover below the natural waterway bottom to protect it from dragging anchors or sinking vessels. The submerged floating tunnel is the shallowest alternative as it only needs to be located at a safe navigation depth. The latter has great potential advantage in crossing the deep, wide fjords of Norway. The submerged floating tunnel concept has been studied extensively in that country and others, but no project of that kind has actually been built as of yet.

What this all leads to is the fact that the immersed tunnel has emerged more and more as the most efficient way of crossing under a waterway. In addition, this method offers other very real advantages over bored tunnels. The factory-style production in a dry dock of prefabricated elements that can be floated to the site, lowered into a trench, and joined with watertight immersion joints to very accurate alignment is a construction method with little of the inherent risk or contingency of a bored tunnel.

The first immersed tunnel ever constructed was a sewer tunnel under Boston Harbor in 1893, and the first transport tunnel was a two-track rail tunnel between Detroit, Michigan, and Windsor, Canada, in 1910. The first transport immersed tunnel constructed in Europe was the Maastunnel in the Netherlands that was completed in 1943, being built during the time that spanned the occupation of Holland during World War II. The latter tunnel was designed as a concrete box–type tunnel. That configuration started immersed tunnel construction in Europe and is still used exclusively to the present day. Immersed tunnel design in the United States, until just recently, was largely that of single and double-shell steel acting compositely with concrete. As immersed tunnel technology grew in acceptance for both service and transport tunnels, its use became worldwide. Other countries, such as Japan and China, adopted both concrete box and steel shell technologies and added some refinements of their own.

In recent years, some truly outstanding projects have been or are currently being constructed involving immersed tunnels. The Øresund Link between Denmark and Sweden with its 4-km-long Drogden Tunnel; the 3.7 km Busan-Geoje immersed tunnel in South Korea; and the Bosphorus rail tunnel, the deepest immersed tunnel in the world at 58 m, are examples of successful milestone projects. Currently, the Hong Kong–Zhuhai– Macao Crossing, with its 5.7 km immersed tunnel, and the 19 km Fehmarn immersed tunnel between Germany and Denmark are the longest immersed tunnel projects proposed to date.

I became involved in the design of my first immersed tunnel in 1954. It was the first Hampton Roads Bridge Tunnel (HRBT). At the time, it seemed like science fiction to me that you could put a tunnel together underwater. Later, in 1970, I went to Virginia to help supervise construction of the parallel second HRBT. Working through the construction of that immersed tunnel was, for me, a marvelous learning experience that led to a lifelong career in this special niche of civil engineering. While in Boston in the 1990s, at work on the Central Artery/Tunnel Project, I joined Working Group 11 (Immersed and Floating Tunnels) of the International Tunnelling Association (ITA). The working group was preparing a "State-of-the-Art Report" on immersed and floating tunnels, and I spent a lot of time at the Massachusetts Institute of Technology Engineering Library researching all existing published papers on all the immersed tunnels constructed to date. From this research, plus much support and help from Dr. Nestor Rasmussen, a world expert in the field, we jointly produced the ITA catalogs of immersed service and transport tunnels.

During this research in Boston, I was surprised to find how little was available in the way of papers on immersed tunnel projects. These were major public works using very innovative technologies still in development, but little seemed to be available that documented them. Beyond this, I found that there also was little available in the way of manuals or technical books to help engineers undertake the design of immersed tunnels. There were a few books that devoted a section, or a chapter or two, to the subject, but they tended to reflect a rather narrow range of experience. The ITA State-of-the-Art Report came closest to a useful design guide with its second edition in 1997, but this is now 14 years old and the technology has moved on considerably since then.

Thus, this comprehensive book on all aspects of developing an immersed tunnel project is a very much needed resource and should be welcomed by planners and designers worldwide. It is written by two fine engineers whom I got to know while I helped to prepare a proposal for the construction of the Marmara project in Istanbul, Turkey. I enjoyed working under their direction and was greatly impressed by their breadth of knowledge and experience, which included key roles in the design and construction of the Drogden Tunnel of the Øresund Link and other important immersed tunnel projects. Although ours was not the winning proposal, it happened that I was later asked by the Turkish Ministry to oversee the construction of the immersed tunnel portion. During the period I worked on the project (2005–2007), I sometimes found myself wishing that some of the construction methods proposed earlier by Richard and Jon had been used by the contractor on the job.

This book covers all the aspects of planning and designing an immersed tunnel project. Early in the project, one must come to grips with the environmental impacts of dredging and dredge spoil disposal, impacts on fishes and their migration, casting basins, water table drawdown, and so on. The basic cross section of the tunnel is determined by the traffic envelope, to be sure, but just as important are the space requirements for ventilation, fire and smoke control, emergency evacuation, service piping, and so on. These considerations, and many others, not purely the mathematics of structural design, are carefully described in this book.

I therefore believe that this book will be found to be a very useful reference for civil engineers in the process of developing an immersed subsea tunnel project for the first time or even for those engineers already experienced in some of the immersed tunnel methods, but who are facing a new project with unusual site conditions or constraints.

Walter C. Grantz, PE
Immersed Tunnel Consultant, Virginia Beach
Virginia, USA

Acknowledgements

First and most importantly, our heartfelt thanks go to Margaret, Louise, Imogen, and Fred for their forbearance during the writing, and particularly to Louise for giving up her dining table for the final drafting period.

We thank Capita Symonds Ltd., for whom both the authors worked for a great many years, for allowing us to draw on various project materials and experiences gained while under their employment.

Particular thanks to Walter Grantz for providing a great many photographs, providing inspiration, and kindly agreeing to write the Foreword.

We also show our appreciation to colleagues at the ITA Working Group 11 for sharing their experiences and expertise over the years in the interests of furthering knowledge about this fascinating construction technique. This book has been written in the same spirit of sharing knowledge.

Thanks also to Brendan Moore who kindly assisted with the preparation of a number of figures. Photographs have been included with kind permission from many colleagues, individuals, and organizations, to whom the authors are most grateful.

We would also like to thank our colleagues... Maureen... manuscript... or for... during the writing, and you... setting up the dining table for the final cleanup...

We thank... for whom help... always welcomed and... grateful... to draw on various... materials and... to each other's achievements.

Thanks to Walter Oracle for providing... from... inspiration, and kindly agreed to... was the purpose...

We wish to express our gratitude to the FDA, Working Group of all experiences and expertise over the years in particular the knowledge from the remaining sources a result of this work, in between 1973 and... have... share...

... Elsevier... Moore who kindly assisted... in the copyright... In a number of cases, people have been included who... had permission... credited... blueprints, and organizations, somewhere the...

Chapter 1

Introduction

An immersed tunnel is an ideal way of constructing a tunnel across a waterway, as the method relies on water to transport and place the tunnel. The first immersed tunnel was built in 1893, and there are now approximately 180 tunnels of this type that have been built worldwide. This is a relatively low number compared to other forms of tunnel construction, which suggests the field is still in its infancy. In fact, the construction methods draw from many other areas of established engineering practice; it is more the combination of techniques and the marine construction operations that are unusual. With only one or two tunnels opening each year, the knowledge of the technique is restricted to a small community of designers and constructors. With this book, we hope to demystify some of the techniques and broaden the construction industry's awareness of immersed tunnels.

To the uninitiated, it may seem that immersed tunneling is quite a narrow field of engineering, whereas, in fact, it is anything but that. The range of skills and disciplines required for an immersed tunnel project is extremely broad, encompassing structural engineering; complex geotechnical engineering; transport planning; highway and railway engineering; environmental engineering; maritime engineering; mechanical and electrical engineering; building structures; the complex issues of tunnel operation and safety; and of course, a few specialist items specific to immersed tunnel methods. This book attempts to describe how these can all be brought together to create an immersed tunnel crossing of a waterway, recently described by a Dutch colleague as "a most beautiful way of building a tunnel."

Although there are several different types of immersed tunnels, the following basic principles of their construction are always the same:

- The tunnel is constructed from a series of prefabricated sections, or elements, each typically 100–200 m long.
- The elements are designed to float so that they can be transported to the tunnel location.

1

- The elements are lowered into a trench that has been dredged in the bottom of the waterway and joined together.
- The trench is backfilled and the tunnel finishes are completed from the inside.

The layout of a typical immersed tunnel is shown in Figure 1.1. It consists of a central section with several immersed tunnel elements that are placed under the waterway. At the banks of the waterway, the type of construction changes to cut and cover tunnel as the alignment rises from beneath the waterway. When the alignment has reached about 5 or 6 m below the bank level, then the cut and cover tunnels end in portal structures, and open approach ramps continue to ground level. There are variations to this general layout; for example, the immersed tunnel could connect directly to a bored tunnel, but in general, these three parts—immersed tunnel elements, cut and cover tunnels, and open approaches—are common to nearly all immersed tunnel schemes.

Immersed tunnels tend to be concentrated in certain geographical areas as a result of the geological and transport infrastructure requirements of various countries. The main areas of construction are the United States, Europe, Japan, and China, with many other countries having one or two such tunnels. Most immersed tunnels accommodate roads or railways but they have also been built as utility tunnels carrying water and power supplies and also as cooling water intakes and outlets for power generation stations. Indeed, the first immersed tunnel ever constructed was a sewage tunnel. Because of this, their dimensions vary enormously, but a typical immersed tunnel carrying a dual carriageway road is about 30 m wide, 9 m high, and generally, approximately 1–2 km long.

The intention of this book is to provide guidance for designers and constructors and explain the wide variety of techniques available to them. It is not to promote immersed tunnels as being intrinsically better than any other form of water crossing, such as bridges or bored tunnels, but rather

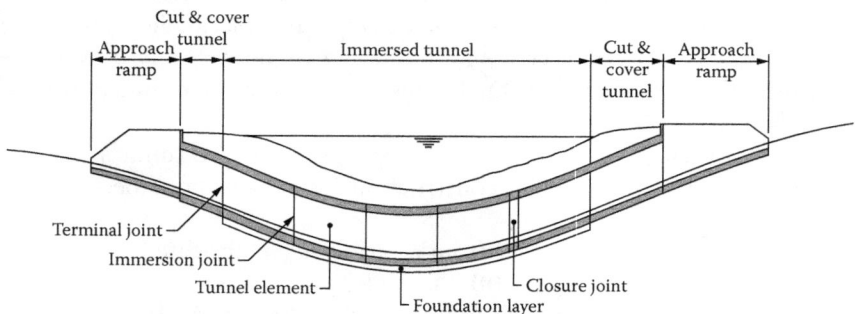

Figure 1.1 General immersed tunnel layout.

to show the merits of immersed tunnels and explain where and how they could be constructed. It gives a general introduction to immersed tunnels from initial planning, through design and construction, to operation and maintenance. We hope that it will promote the use of a technique that is frequently overlooked or misunderstood. Many papers and symposia have been dedicated to immersed tunnels and these cover particular aspects or projects, but this is the first time all aspects of design and construction have been drawn together in a single volume.

There are many different types of immersed tunnel, and practices differ across the world; we have endeavored to bring together the many aspects of current practice. Throughout the text, examples of projects are given where specific techniques have been used or where there are particular issues of which the designer or constructor should be aware. The authors are based in Europe and the majority of their experience is with the concrete immersed tunnels that are used in Europe, so inevitably the book tends to focus on those. We have, however, included sections on steel tunnels to give a complete overview of immersed tunnels. The book provides a general guide for the construction industry, to those engaged in the planning, feasibility, design, construction, or operation of an immersed tunnel, and also to those who have a passing interest in such tunnels. It also provides a broad knowledge base suitable for use in universities.

The book starts by tracing the historical development of the immersed tunnel, leading to the types of tunnel in use today. It then describes the factors that have to be considered when developing an immersed tunnel scheme and in particular, the environmental impact, and the mechanical and electrical systems that are installed, which are a vital part of any transportation tunnel. The second half of the book goes into the details of the design and construction of an immersed tunnel before finishing with operation and maintenance aspects and a review of important contractual issues.

Chapter 2

Development of the immersed tunnel

The idea of the immersed tunnel arrived some time before a project was actually realized. The first concepts were developed in England in the early 1800s, at the time Brunel was starting out on his Thames Tunnel in London. The birth of immersed tubes and shield-driven tunnels therefore occurred at around the same time, even though immersed tubes were much slower to be implemented.

In 1803, a British engineer, Henry Tessier du Mottray, proposed linking England and France by an immersed tunnel constructed from cast iron tunnel elements laid on the bed of the English Channel. This was one of a number of similar schemes proposed at the time, but the imminent threat of a French invasion by Napoleon meant that none of these ever progressed. In 1808, another British engineer, Richard Trevithick, proposed a method of construction for a crossing of the river Thames that involved building sections of tunnel within dewatered cofferdams formed of timber piles. Once completed, the brick tunnel sections would be backfilled to the original riverbed level, and the cofferdam removed and reconstructed 50 ft further along the tunnel alignment. By progressing the cofferdam across the river, the tunnel would be formed. Although this was essentially a cut and cover method of construction, it featured many elements of the techniques now employed for immersed tunnels and was an important stepping stone toward the development of the first ideas for building them.

The tunnel was proposed to be of brick construction, although he later suggested the tunnel sections could be cast iron. Trevithick's proposals were submitted to the Thames Archway Company, which was trying to build the first tunnel under the Thames, but were not adopted, and in 1809, the company launched a competition for a new crossing of the Thames. They received 54 proposals, and in 1810, accepted the one from Charles Wyatt. This was to become the first true immersed tunnel concept. Wyatt's idea was to excavate a trench and immerse 50 ft long brick cylinders into it. The ends of the cylinders would be sealed with temporary spherical brickwork bulkheads to enable them to be watertight and to float. Each would have a simple ballasting arrangement for sinking.

Wyatt's scheme was well engineered; for example, he had considered the possible impact of ships' anchors damaging the tunnel and ensured the trench would be deep enough so that once placed and backfilled, there would be 6 ft of earth covering the tunnel. The Thames Archway Company decided to trial the new technology to test the methods and outcomes, in particular, the method of forming the tunnel joints, the strength of the cylinders, the accuracy of placement that could be achieved, and the disruption to river traffic that would be caused. John Isaac Hawkins was appointed to construct two 25 ft long cylinders with an internal diameter of 9 ft. The trial was carried out in shallow water so that the tops of the cylinders could be inspected at low water, and manhole access was provided to enable internal inspections. The wall thickness of the tubes was 13½ in and each cylinder weighed 52 t, requiring 8–10 t of water ballast for immersion. The cylinders were built on submersible barges and scaffolding was constructed in the river to lower and position the cylinders. Because of the heavy river traffic, there was a frequent need to repair the scaffold following numerous collisions. The cylinders were transported by tying them alongside a barge. Once they were maneuvered into the scaffolding, lowering lines were attached to the cylinders along with masts to control positioning. After immersion, gravel backfill was placed manually around the cylinder to lock it in position. Hawkins's scheme is shown in Figure 2.1.

When the second element was placed, a mixture of mud and gravel was placed around the joint and the tunnel dewatered. Although some leakage of the joint occurred, it was considered that it would be possible to seal the joints with puddled clay. Although the concept was considered technically feasible, undoubtedly the methods of sealing the joints may have proved problematic in the full tunnel construction and would have needed some further engineering development. Sadly, because of the cost of the trials in 1811, the Thames Archway Company decided to abandon the project, but it was the first full-scale use of the technique and was groundbreaking engineering for its time.

The development of ideas continued in the United Kingdom after this, through to the mid-nineteenth century, by engineers such as John de la Haye, who published extensive discourse on the possible applications and construction methods for submerged tunnels in *The Mechanics' Magazine, Museum, Register, Journal, and Gazette* in 1845. He considered the use of cast iron submerged elements to construct tunnels in a number of locations around the United Kingdom and for a Dover to Calais crossing to France. He proposed external ballasting methods and looked closely at the safety benefits and cost benefits the technique would have compared to the new shield tunneling techniques being used beneath the Thames. In fact, a number of new projects were proposed in the mid-nineteenth

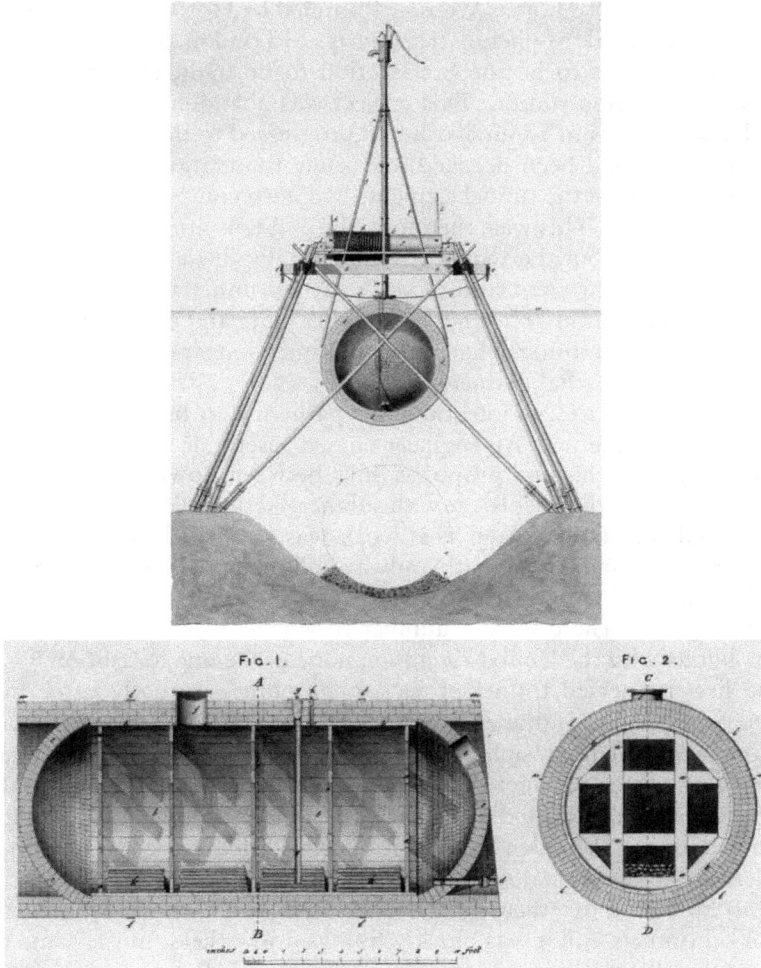

Figure 2.1 Charles Wyatt's immersed tunnel proposals. (Courtesy of Institution of Civil
Engineers Library.)

century that used the immersed tunnel idea. These included further proposals for crossing the English Channel by French engineers, but they were
not progressed due to continued national security concerns. There were
also a number of immersed tunnels proposed on railway projects in various western European countries. At the same time, ideas were beginning
to emerge in the United States. However, the next attempt at construction
was back in the United Kingdom, when a new immersed tunnel beneath the
Thames in London was proposed in 1865 for the Waterloo and Whitehall

Pneumatic Railway. Thomas Webster Rammell had formed a company that envisaged a network of pneumatic subways in London, and the Waterloo–Whitehall line was to be the first section to be built, crossing the river Thames at Waterloo station. This project was abandoned in 1868 following a banking crisis in 1866. Work had progressed to the extent that parts of the riverbed had been dredged and some foundation works had been constructed. One of the tunnel elements had been completed and two more partly constructed. This was, therefore, the first true attempt at building an immersed tunnel, but the works were eventually dismantled. The type of construction was different from Wyatt's earlier tunnel; the tunnel elements were constructed from ¾ in thick iron boiler plate and had an internal and external brickwork lining. They were also much longer cylinders, each 235 ft long, with an internal diameter of 10 ft.

In the United States, momentum was gathering to build transportation tunnels with the method. An engineer named Joseph de Sendzimir was particularly active in this and proposed links between Lower Manhattan and Brooklyn crossing the East River with submerged tunnel elements fabricated from bolted iron boiler plate. This work was published in 1857. He also looked into building a pneumatic railway, following in Rammell's footsteps in 1866, using submerged iron tubes to connect Manhattan with Brooklyn and New Jersey. There were a number of similar ideas across the United States, but as with the United Kingdom, none ever came to fruition.

The first immersed tunnel project to be built eventually came in the United States right at the end of the century in 1893. It was a rather unglamorous scheme, a siphon beneath a 60 m wide tidal sea inlet called Shirley Gut, which conducted sewage from Boston to Deer Island station. The tunnel was constructed from brick and concrete; it was 100 m long and 2.7 m in diameter. Wooden bulkheads were installed at the end of each tunnel element, and external steel flanges at the ends of the elements enabled them to be bolted together after placing. Its opening marked the birth of immersed tunnels and it was quickly followed by others. In the same year, a twin-tube 200 m sewer scheme was constructed in Paris beneath the river Seine using the same technique, and 7 years later in 1900, a 185 m long culvert and a 43 m long sewer were constructed in Denmark. From these humble beginnings, the construction technique was able to be developed and applied to larger-scale transportation tunnels. The first of these to be constructed was the Detroit River Tunnel that opened in 1910. This was a border crossing railway tunnel built under the St. Clair river between Detroit, Michigan, in the United States and Windsor in Canada, for the Michigan Central Railway. The tunnel was designed by the American engineer W.J. Wilgus and comprised twin watertight steel tubes placed in a dredged trench, which were then surrounded by concrete. The tunnel was made up of 10 elements, each 80 m long with one shorter closure element of 20 m. The floating element can be seen in Figure 2.2.

Figure 2.2 Detroit River Tunnel. (Courtesy of Library of Congress, Prints and Photographs Division, Detroit Publishing Company Collection.)

There was a considerable steel shipbuilding industry in the United States, and it made sense to make use of that expertise in the construction of the prefabricated units that formed the tunnel. The tubes were built by the Great Lakes Engineering Works in Detroit. They were used to building freighters for the Great Lakes and so they used a form of construction that was familiar to them. The tubes were constructed in pairs and each tube was designed to carry one railway track. The tubes were circular, 23 ft 4 in in diameter, and built from 3/8 in steel plate riveted together with lap joints that were caulked similar to boiler construction. They were strengthened by circumferential stiffener angles (4 × 3 × 3/8 in) riveted to the inside of the plate at 12 ft centers. To provide additional temporary support during the placing operation, internal steel rods, similar to bicycle wheel spokes, were placed at each stiffener. The tubes were fitted with external steel diaphragms, again at 12 ft centers, which in turn supported wooden shuttering on the vertical sides that formed an open box around the tubes. This external box contained the permanent ballast concrete that was poured around the tubes once they had been placed in position in the trench. The minimum thickness of concrete around the steel tube was 3 ft. The ends of the tubes were sealed with wooden bulkheads, and rubber gaskets were used to seal the joints between the elements. All the tube sections were built within 20 months.

The trench was dredged across the river using a clamshell dredger. The material excavated was clay with layers of sand and gravel and the water depth varied between 20 and 50 ft. The bottom of the trench was checked for depth and width by dragging a steel beam along it. Once the trench had been excavated, a steel grillage was placed on the bottom under each joint between tubes to provide support when they were placed. The grillages placed under the joints were large enough to engage with the end diaphragms of each tube and shims were placed to achieve the required levels. This was an interesting, if perhaps expensive, method of overcoming the problem of obtaining the correct relative vertical position of adjacent tunnel elements, which is still a challenge using modern techniques.

To overcome the negative buoyancy of the tubes, water was introduced through the bulkheads with each of the end bulkheads being fitted with a valve that could be opened from the outside. The inside of the tubes were compartmentalized into three sections by two interior bulkheads, and 60 ft long air cylinders were placed on top of the elements. These external air cylinders were also divided into three compartments. This arrangement enabled the trim of the element to be controlled very accurately during the placing operation. The submerged weight of the whole element during placing was about 500 t. The sinking operation itself was very similar to those of today. Large anchored floating barges fitted with derricks were used, and the time taken to submerge an element was approximately 2 hours. The elements were joined together using bolted flanges at the end of each unit, and these connections were made by divers.

Concrete was then tremied into the open box around the steel shells. At the time, there was no experience of placing such large volumes of concrete underwater and the contractor had to undertake numerous experiments to develop a successful concrete mix and a method of placing it. This operation was controlled by divers who ensured that the concrete surrounded the shells and provided sufficient cover to the top. The trench was then backfilled, first with granular material to about half the tunnel height and then the remainder was filled with clay that had been excavated from the trench. The water in the elements was then pumped out through the previously placed units. This enabled work to continue inside the elements with the removal of the wooden bulkheads and the placing of the inner reinforced concrete lining. The successful completion of the tunnel in 1910 demonstrated the feasibility of large-scale immersed tunnel construction and found practical solutions to the technical challenges it presented. The steel shell immersed tunnel had been born, and many of the techniques used for this first tunnel were carried forward, reused, and developed for the many tunnels that were to follow in the next decades.

Figure 2.3 LaSalle Street Tunnel.

STEEL SHELL TUNNELS

Following the success of the Detroit River Tunnel, steel shell immersed tunnels became an established method of building tunnels under waterways in the United States. Two more rail tunnels were to follow in quick succession. The LaSalle Street rail tunnel in Chicago was opened in 1912, although this was only one element long. This was a single-shell tunnel and the first of its kind. The steel shell can be seen under construction in the Goose Island dry dock in Figure 2.3. Shortly after this, the Harlem River railway tunnel in New York opened in 1914. The first road tunnel in the world was the Posey Street Tunnel between Oakland and Alameda in California, which was completed in 1928, although this was, in fact, a cylindrical reinforced concrete tunnel.

The first three tunnels were all slightly different in how they were constructed, but from thereon, designs for steel tunnels in the United States developed subsequently into two clear types:

- Double steel shell
- Single steel shell

The original Detroit River Tunnel has been described as being both of these. In fact, it was a kind of hybrid. It only featured one steel shell, but it used the construction methodology that would come to be used for the double-shell tunnel and, in fact, the tunnel is closest to this form. The only difference was that timber shuttering was used to form the outer shell. The first true double-shell tunnel with a steel outer shell was the Harlem River Tunnel. Figure 2.4

Figure 2.4 Harlem River Tunnel. (Photo Courtesy of New York State Public Service Commission.)

shows one of the tunnel elements for the Harlem River Tunnel supported on a series of flatboats prior to its immersion.

In 1930, the second tunnel beneath the Detroit River, the Detroit–Windsor Tunnel, was completed. The design of this second tunnel developed the hexagonal external shaped double shell section that became the most popular shape for tunnels in the United States. Figure 2.5 shows this geometry and is based on the Second Hampton Roads Tunnel that opened in 1976. The stiffened circular steel shell is fabricated for the whole length of the element. External diaphragms are added at intervals along the length, to which the external steel shell, also known as a form plate, is fitted. Concrete is placed around the outside of the shell between it and the external form plates. This concrete provides the main ballast to prevent uplift and also acts as corrosion protection for the circular structural steel plate. Inside the circular steel shell, an internal reinforced concrete lining is placed, and this in conjunction with the circular steel shell provides the primary structural strength of the tunnel.

The single steel section also developed in the United States. This also has a stiffened outer steel shell that is normally about 1/2 in thick. It is stiffened internally both transversely and longitudinally with the longitudinal stiffening being continuous over the full length of the element. Inside the steel shell, a reinforced concrete lining is placed, which acts compositely with the steel. This design does not use the tremie concrete as ballast but typically uses a ballast container on the roof that is filled with stone. In this type of design, the structural steel shell remains on the outside and exposed to the waterlogged ground. It is therefore necessary to provide some form of external corrosion protection such as a cathodic protection system. A typical single shell design is shown in Figure 2.6, which was used for the Bay Area Rapid Transit (BART) Tunnel in San Francisco, California.

A similar method of construction is used for both forms of steel shells, and they both make use of shipyard production techniques. First, the steel shells are fabricated as a series of regular modular subassemblies. Then, several of these subassemblies are welded together to form the continuous element. This is often done directly on a slipway adjacent to the waterway,

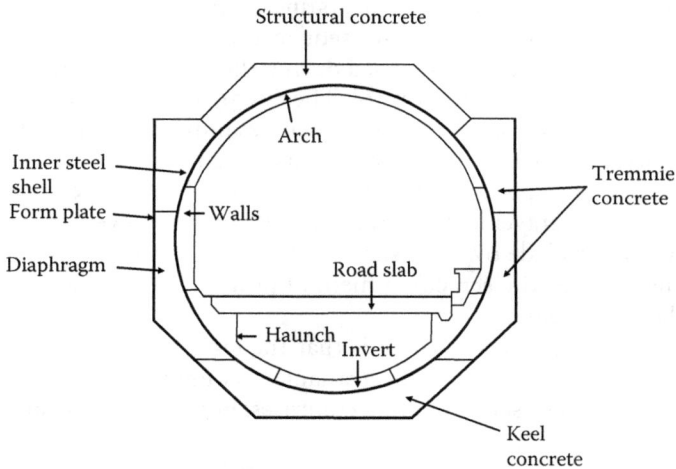

Figure 2.5 Second Hampton Roads double steel shell arrangement.

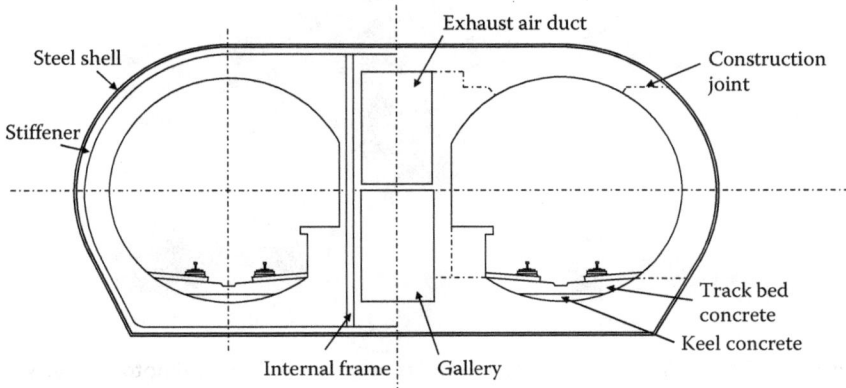

Figure 2.6 BART Tunnel single steel shell section.

to facilitate launching the element. The elements are preferably launched sideways to minimize the stresses imposed, but they can be designed for end launching if required. Before launching, the reinforcement for the inner concrete lining is placed together with other internal equipment and a certain amount of keel concrete is also placed to increase the draft of the element and to give it stability while afloat. The ends of the tubes are sealed with watertight bulkheads and the element launched similar to a ship launching.

It is then towed to an outfitting jetty where the internal concrete is placed while the element is afloat. The outfitting jetty can be some considerable distance from the shipyard and some steel shells have been transported on semisubmersible barges. This concreting operation is shown in Figure 2.7, which shows the draft of the elements increasing as more of the internal concrete is placed. The concreting sequence is carefully programmed so that the steel shell is not overstressed during the operation and also to keep the trim of the element under control. The designer has to take these temporary loading conditions into account when designing the unit and set limits for the contractor to adhere to during construction.

The steel shells are generally founded directly onto screeded gravel foundations. The gravel bed is screeded to the correct levels by dragging a screeding grid or blade over the surface. The base of a steel element section is fairly narrow, so the screeding method produces a sufficiently accurate bed to place the element on.

The advantage of the steel shell is that the fabrication facilities required are not extensive and are generally available in countries with a shipbuilding industry. They can be set up in other countries, but if there is no shipbuilding

Figure 2.7 Ted Williams steel tunnel elements at concreting jetty. (Photo courtesy of W. Grantz.)

tradition, the materials and fabrication skills may not be available and will have to be imported. The shells are also relatively lightweight with a shallow draught of about 2 ft, so they can easily be towed to the outfitting facility without the need for deep navigation channels. It is also possible to tow them considerable distances. For example, the elements for the 63rd Street Tunnel in New York were fabricated in Port Deposit, Maryland, and towed 300 km to Norfolk, Virginia, where the concrete was placed and then towed a further 480 km to the site in New York. Even longer journeys have been undertaken with the steel shells carried on semisubmersible barges. The Ted Williams Tunnel elements were transported from Baltimore to Boston, a distance of about 960 km along the Atlantic coast. In theory, there is no limit to the distance that a steel shell tunnel can be transported on a semi-submersible vessel, which means that a very wide range of fabrication facilities can be considered during the planning stage of the tunnel.

Some steel shell designs have taken slightly different shapes, but the principles remain the same. In general, steel shells have remained popular in the United States, with most being the double shell type, but not so much in other parts of the world, where concrete tunnels have generally been constructed. There have been exceptions, such as the first Cross-Harbor Tunnel in Hong Kong, but in Europe, concrete tunnels have been used exclusively. This is not due to a lack of shipbuilding facilities, but rather different economic drivers.

CONCRETE TUNNELS

In Europe, even though the immersed tunnel concept had originated in the United Kingdom, the abandonment of the early Thames crossing trials led to the development of shield tunneling, which became the preferred tunneling method there until the 1980s. A number of small utility immersed tunnels were constructed in the early 1900s in Germany, France, and Denmark. These were generally steel pipeline tunnels and culverts. The first concrete immersed tunnel of any significance to be constructed was the Friedrichshagen Tunnel in Germany, a rectangular pedestrian tunnel that was completed in 1927. The first concrete transportation tunnel was finished a year later in 1928 in the United States, the Posey Street Tunnel between Oakland and Alameda in California. Just over a decade later, the immersed tunnel technique was adopted by the Dutch. In response to a growing traffic congestion problem across the bridges over the Maas in Rotterdam, three engineers from the Rotterdam Department of Public Works traveled to the United States in 1929 to study the immersed tunnel construction methods being used there. The result was the letting of a construction contract for the Maastunnel in 1937.

To satisfy an expanding road and railway infrastructure network, the Dutch had to provide many relatively short waterway crossings in their low-lying delta country with its many natural and man-made waterways. The height of the bridges needed to allow ships and barges to pass underneath would require long approach structures. These would be intrusive in such a flat landscape as well as adding to the overall cost of the crossing. In these circumstances, it became economic to construct an immersed tunnel due to its shorter overall length. The soft alluvial sandy soils and high water table present in the Netherlands lent themselves to the adoption of the immersed tunnel technique. The option of using a tunnel shield that had been successfully developed in the United Kingdom would have been very difficult with the high water table and soft permeable ground. The Dutch also had a history and capability in marine construction that could quickly adapt to building an immersed tunnel.

The Dutch, however, did not adopt the steel shell design pioneered in the United States. Steel prices were relatively much higher in Europe than the United States, so the Dutch developed a reinforced concrete section. The overall principle of the tunnel construction is the same, but the tunnel elements were built out of reinforced concrete rather than steel. This is a very adaptable form of construction, more so than the steel shell, and this enabled a rectangular cross section to be developed. The Dutch waterways are not very deep, so the rectangular section was easily capable of resisting the hydrostatic pressures. The rectangular section also matched better the rectangular traffic envelope required for road tunnels.

The use of a rectangular cross section was made possible by the development by the Danish contractor Christiani & Nielsen of a method of injecting a sand foundation into a space below the tunnel element. At the time, the techniques were not available to screed a gravel bed accurately enough to allow a stiff wide concrete box to be placed on it without the possibility of inaccuracies in the bed levels causing unacceptable stresses in the box. The Maastunnel in Rotterdam (Figure 2.8), which was completed in 1942 despite wartime conditions, was the first Dutch tunnel and was the first large transportation immersed tunnel constructed outside the United States.

Subsequently, a few concrete immersed tunnels were built around the world, but the real impetus for concrete immersed tunnels came in the Netherlands in the 1960s as the need for improved transport links grew. The first was the Coen Tunnel, in Amsterdam, built between 1961 and 1966 for the Rijkswaterstaat, the Dutch Ministry of Transport and Water Management, which again involved Christiani & Nielsen both as contractor and designer. Several such tunnels were then built in the area (the Netherlands, Belgium, and Denmark) in the 1960s. There was a similar large building program in the Netherlands in the 1970s.

Figure 2.8 The Maastunnel under construction in 1941. (Photo courtesy of A. Scheel.)

The rectangular concrete section became the standard construction technique for these many European road and rail tunnels. The Dutch and the Danish contractors and designers involved developed considerable expertise and experience, which they subsequently managed to export around the world. Concrete immersed tunnels were also built in Germany and Sweden. Christiani & Nielsen also built two concrete tunnels in Canada in the 1960s with Per Hall as the designer. At this time, the Japanese also showed interest in the technique. Their first immersed tunnel, at Haneda, adopted a different approach with a prestressed concrete design, albeit with a steel outer layer, and was opened in 1964.

The technique spread around the world, and in general, the reinforced concrete section became the preferred option. The United States favored their steel shells, Europe universally adopted reinforced concrete, while the Japanese developed both. The most concentrated development of immersed tunnels is across Hong Kong Harbor, where five immersed tunnel crossings have been built (two road, two rail, and one road/rail combined), and at least two more are proposed. The shallow depth of immersed tunnels is particularly appropriate to somewhere like Hong Kong, where urban development presses in along the sides of the harbor and the approaches to the tunnel have to be as short as possible. In addition, the underlying granite could make it very difficult to bore a tunnel.

As more tunnels were built, the construction methods were refined. For example, initially, the sand foundation material under the tunnel was jetted in place using external plant. A rig was supported on the top of the tunnel and moved along the length of the tunnel. A sand–water mixture was piped down the outside of the tunnel and injected in underneath the tunnel. The use of such plant in the waterway was an obstruction to navigation that would be better avoided. The Dutch developed a method of placing the foundation by injecting it through ports cast through the floor of the tunnel elements. This enabled the operation to be carried out without obstructing the waterway.

Another major development was the introduction of the segmental concrete tunnel element in the 1960s and 1970s. Before this, the tunnel elements, which were about 100 m long, were built as monolithic reinforced concrete elements. Building such large concrete sections led to early thermal shrinkage cracking, and such cracks go right through the concrete section and provide a path for water to leak into the tunnel. As well as being unsightly and affecting the internal tunnel finishes, the leakage could cause chlorides to penetrate the concrete, jeopardizing the long-term durability of the reinforced concrete. Thus, the early concrete tunnel elements had an external waterproofing membrane to make them watertight. To avoid the need to apply the membrane, which was time-consuming as well as expensive, the Dutch developed the technique of dividing the 100 m long element into a number of individual segments each about 20–25 m long. These segments could be cast without any early thermal shrinkage cracks, so the element did not need an external water-proofing membrane. The effort was put into making the concrete itself watertight rather than surrounding potentially cracked concrete with a watertight membrane. The segments have to be temporarily prestressed together to form a continuous element while it is being towed and placed, but this prestress is then cut after the element has been placed.

The technique was developed in stages; segmental construction was used for land tunnels initially on projects such as the Schiphol airport tunnel that was completed in 1966. Short segments were constructed with a waterproof articulation joint between them to prevent ground-water entering the tunnel. The concrete tunnel structure still had an external waterproofing membrane applied because cracking could not be entirely eliminated. This type of construction was first transferred to the immersed tunnel method for the Heinenoordtunnel that opened 3 years later in 1969. The method of temporarily prestressing segments together to create tunnel elements was established for the first time on this project. The final step in the development of the segmental construction method was made for the Vlaketunnel that was opened in 1975. The technique of cooling the concrete during curing meant the cracking could be prevented and the need for a waterproof membrane was eliminated. The technique was successful, and since the 1980s, it has been adopted on all Dutch tunnels and for the majority of the tunnels built in western Europe.

Thereafter both segmental and monolithic concrete designs were used around the world, the choice often being dependent on the client's view of which was the more watertight. Some, like the Dutch, were satisfied that the measures taken to produce crack-free concrete in the segmental method were sufficient. Others were not so convinced and preferred the security of an external waterproofing system. As discussed in Chapter 11, external waterproofing systems are not problem free and do not always give the watertightness envisaged.

It is worth noting that in the early 1970s, the Dutch were also pioneering techniques for full-section casting of concrete tunnels for their smaller utility tunnels. By constructing short tunnel segments on end, they effectively removed construction joints from the segment structure and prevented cracking from occurring. Segments could then be assembled together to form tunnel elements. This was successfully achieved for the Amsterdam–Rhine canal culvert in 1971 and the Hollandsche Diep and Oude Maas pipeline tunnels in the mid-1970s. This was the precursor to utility tunnels built in Asia, but more importantly fed into the development of full-section casting techniques used for the much larger-scale Øresund and Busan Tunnels.

Concrete tunnels have remained little used in the United States. The two Posey Street Tunnels, built in 1928 and 1962, and the 2002 Boston Fort Point Channel Tunnel are the only concrete tunnels to have been built there. Apart from economic arguments, another reason was that U.S. authorities favored transverse or semi-transverse ventilation. They did not believe that the longitudinal ventilation system used in Europe and elsewhere could provide clean enough tunnels. They also considered their ventilation systems inherently safer in the event of a fire. Transverse ventilation systems are easy to accommodate in the basically circular cross section used in steel shell tunnels because of the spare space above and below the rectangular traffic envelope. This provides space to duct air into and out of the tunnel, so the United States was happy with their approach. In the relatively short Dutch tunnels, which are typically up to 1 km long, longitudinal ventilation was used, with jet fans in the traffic bore assisting the natural piston effect of the moving traffic. Space is at a premium and not having to provide wider or deeper cross sections to accommodate the air ducts was a big advantage.

Worldwide experience and analysis and testing has eventually satisfied U.S. authorities that the longitudinal system is safe in the event of fire and so now concrete immersed tunnels are more acceptable in the United States. Additionally, with the decline of the steel industry, it is becoming more attractive to contractors to construct in concrete as it gives greater flexibility in how they approach a project and a greater pool of labor resources to draw from. The Second Midtown Tunnel project in Virginia is progressing as a concrete tunnel for these reasons, and it is likely that very few, if any, steel tunnels will be constructed in the United States in the future.

COMPOSITE SANDWICH TUNNELS

The use of steel–concrete composite sandwich construction is a more recent development that has mostly been promoted in Japan, although a lot of research and testing has also been carried out in the United Kingdom.

Construction consists of a sandwich of concrete between two steel plates. Typically the steel plates are about 300 mm apart and connected with shear studs. The concrete is placed between the steel plates, so a very fluid self-compacting mix is required. Placing this concrete and ensuring sufficient compaction and complete filling of the void between the plates is one of the main challenges of this method. Because the steel concrete sandwich is very strong structurally, the composite section is a very elegant structural solution. However, in any immersed tunnel, the air space to carry the road or rail traffic requires a fixed amount of ballast to hold it down. So, although the composite design can give thinner structural members, the weight still has to be provided by ballast boxes on the roof or internal ballast under the roadway. The internal steel skin also does not lend itself to the many box outs and openings that are needed in a modern road tunnel.

While research and one-third scale testing of sandwich sections have been carried out in the United Kingdom by the Steel Construction Institute to the extent that design rules have been developed (SCI Publication 132, 1997), the concept has not yet been used on a full-scale tunnel in Europe. It has been left to the Japanese to pioneer the use of steel/concrete composite sandwich construction on the Kobe and Naha Tunnels.

TUNNELS IN JAPAN

Apart from the Netherlands and the United States, Japan has also been instrumental in the development of immersed tunnels and warrants specific mention. The first Japanese immersed tunnel was finished in 1944. This was the Aji River Tunnel and carried both road and pedestrians, but it seems the first tunnel of great significance is considered to be the Haneda Ebitori River Tunnel, completed in 1964 for the Haneda monorail. It is a rectangular single shell steel box 7.5 m high and 11 m wide, and the tunnel consists of one 56 m long element. The first tunnels in Japan were all quite short and mostly single shell type construction. Many were built more as underwater bridges than what we would now consider as a conventional immersed tunnel. The first multiple element tunnel was the Haneda Tama River Tunnel built in 1970 beneath the Tama River and the Keihin Channel in Tokyo.

Since then, a number of tunnels have been constructed, generally using the single steel shell construction method or reinforced concrete monolithic tunnel elements. The majority are rectangular in cross section. In recent years, the Japanese have been more innovative in their immersed tunnel designs than any other nation. They have built steel, concrete, and sandwich composite tunnels, developed novel closure joints, and developed seismic-resistant tunnel element joints.

Rate of construction of immersed tube tunnels worldwide

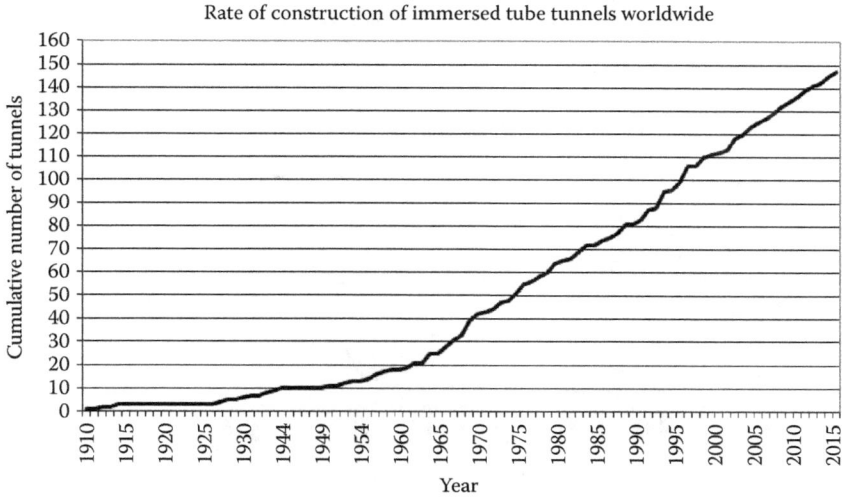

Figure 2.9 Rate of immersed tunnel construction worldwide for transportation tunnels.

GROWTH IN IMMERSED TUNNELS WORLDWIDE

After the first transportation tunnel was built in 1910, it took a further 35 years for the number of projects to reach double figures. Having reached that landmark, there was a brief slowdown during the period of World War II, but afterwards, there was a gradual increase in the number of tunnels built over the next 20 years. In this time, there were a similar number of projects being built in the United States and in Europe. However, from the 1960s, the rate of construction in Europe and to a slightly lesser degree in Asia increased significantly and has stayed at an average of around one project opening each year ever since. This is shown in Figure 2.9.

The geographical distribution of immersed tunnels illustrates the dramatic increase in numbers in Europe from the 1960s as transportation links became more developed. This rate has been maintained subsequently, as can be seen in Figure 2.10.

There has been a general trend toward longer immersed tunnels as construction techniques have improved, although many tunnels in the 1–2 km range are still being built. The distribution of tunnel length with time can be seen in Figure 2.11. Longer tunnels such as the planned 19 km road and rail Fehmarn Crossing between Denmark and Germany will redefine the graph.

The majority of tunnels built to date have been constructed in just three countries: the United States, the Netherlands, and Japan. However, the

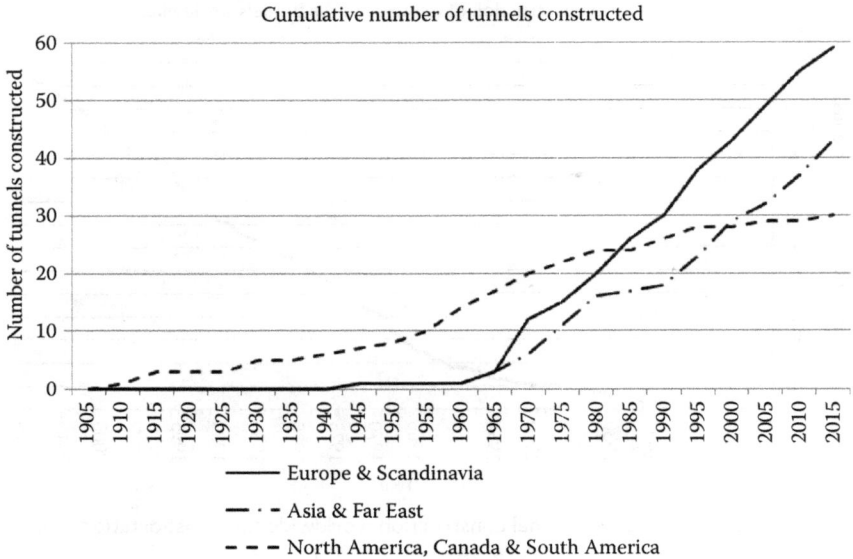

Figure 2.10 Rate of construction of transportation tunnels by region.

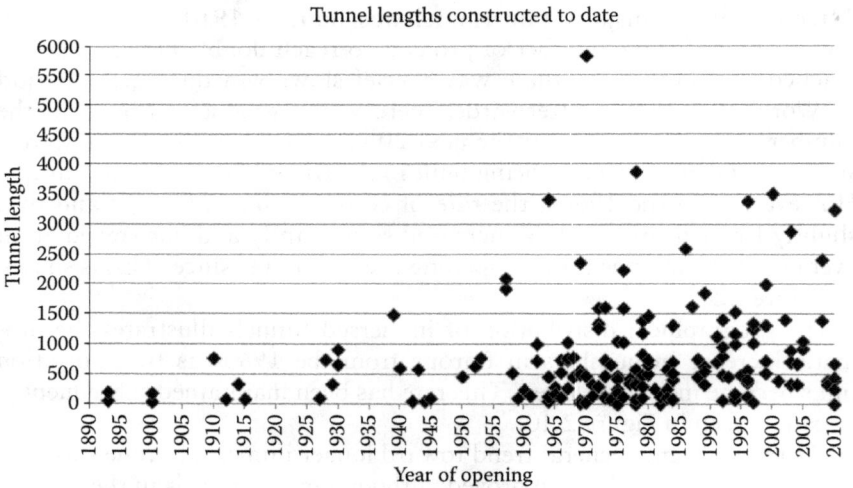

Figure 2.11 Distribution of immersed tunnel lengths with time.

People's Republic of China, including Hong Kong, has seen the greatest increase in the rate of immersed tunnels being built, and since the year 2000, has constructed as many as the total number built in the Netherlands. Mainland China has many low-lying regions with major cities built on estuaries and rivers that particularly lend themselves to using the immersed

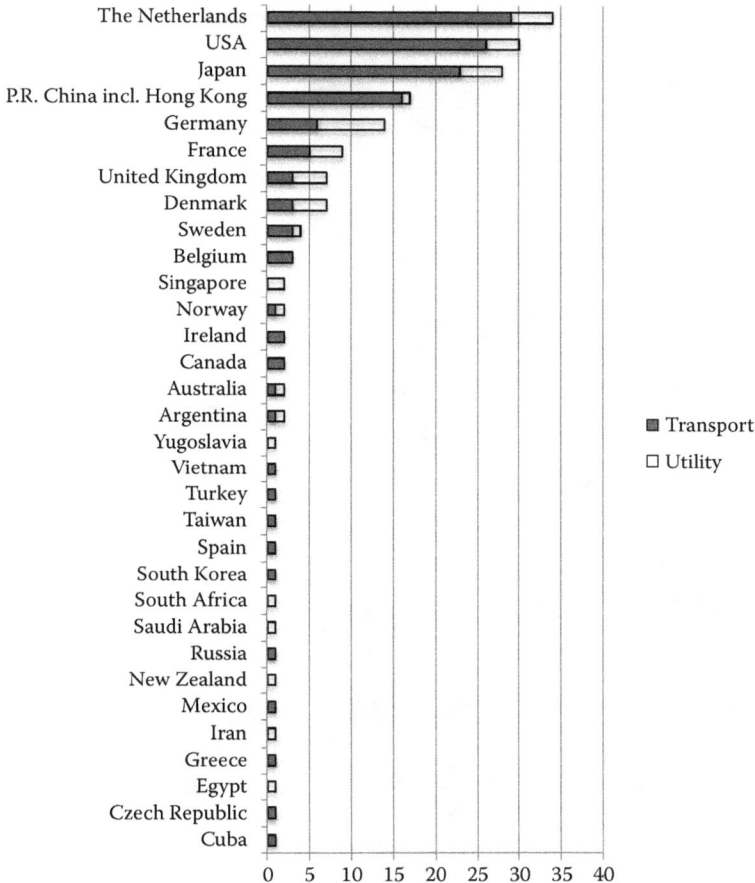

Figure 2.12 Number of tunnels constructed by country.

tunnel construction method. With the rapid expansion of the country's infrastructure as a result of high levels of national investment, it is perhaps of no surprise that China is building so many immersed tunnels. It is likely to remain the country with the greatest number of new tunnel projects for many years to come. Figure 2.12 shows the number of immersed tunnels constructed by country as of 2011, including those under construction at the time. Current trends favor the use of concrete tunnels, either monolithic or segmental, as shown in Figure 2.13.

A full chronological list of tunnels constructed to date and currently under construction is given in Table 2.1. Details of many of these tunnels are provided in the International Tunneling Association's catalog of immersed tunnels, which is published on their website (www.ita-aites.org).

Cumulative number of tunnels constructed, by type

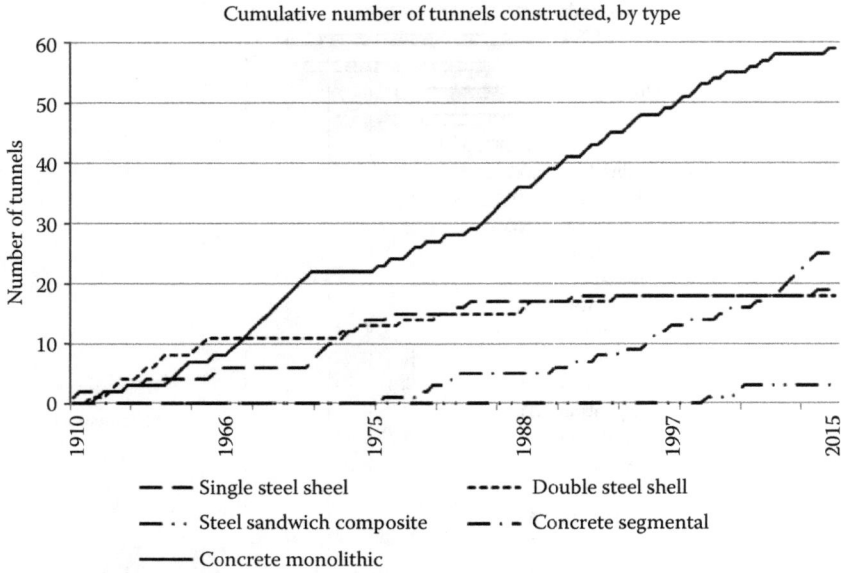

Figure 2.13 Trends in structural form.

Table 2.1 Listing of immersed tunnels constructed by date

Date	Country	Immersed tunnel	Type
1893	United States	Shirley Gut, Boston	Sewer
1893	France	Paris sewer	Sewer
1900	Denmark	Nyhavn Culvert	Sewer
1900	Denmark	Langebro sewer	Sewer
1910	United States	Detroit River	Rail
1912	United States	LaSalle Street	Rail
1914	United States	Harlem River	Rail
1915	France	Paris sewer	Sewer
1915	Germany	Karlsruhe sewer	Sewer
1927	Germany	Friedrichshagen	Pedestrian
1928	United States	Oakland–Alameda (Posey)	Road
1929	Germany	Cologne sewer	Sewer
1930	Argentina	Buenos Aires outfall sewer	Sewer
1930	United States	Detroit–Windsor	Road
1939	Denmark	Gentofte sewer outlet	Sewer
1940	United States	Bankhead, Mobile, Alabama	Road
1942	United States	State Street, Chicago, Illinois	Rail
1943	The Netherlands	Maastunnel	Road and pedestrian/cycle

Table 2.1 Listing of immersed tunnels constructed by date (Continued)

Date	Country	Immersed tunnel	Type
1944	Japan	Aji River, Osaka	Road and pedestrian
1945	The Netherlands	Scheveningen sewer	Sewer
1950	United States	Washburn, Texas	Road
1952	United States	Elizabeth River 1, Virginia	Road
1953	United States	Baytown, Texas	Road
1957	United States	Baltimore Harbor	Road
1957	United States	Hampton Roads 1, Virginia	Road
1958	Cuba	Havana	Road
1959	Canada	Deas Island, Vancouver	Road
1960	South Africa	Durban Harbor sewer	Sewer
1961	Germany	Rendsburg	Road
1962	United States	Webster Street, Oakland, California	Road
1962	United States	Elizabeth River 2, Virginia	Road
1964	Egypt	Cairo Syphon	Water
1964	United States	Chesapeake Bay	Road
1964	Sweden	Liljeholmsviken	Rail
1964	Japan	Ebitori River, Tokyo	Road
1964	Japan	Haneda monorail, Tokyo	Rail
1965	Sweden	Stockholm cable tunnel	Utility
1965	France	Siphon de la Mutatiere, Lyon	Water
1966	The Netherlands	Coen	Road
1966	Germany	Wolfsburg	Pedestrian
1966	The Netherlands	Rotterdam metro	Rail
1967	New Zealand	Marsden power station water intake	Water
1967	The Netherlands	Benelux	Road
1967	Canada	Lafontaine, Montreal	Road
1967	France	Vieux-Port, Marseille	Road
1968	Sweden	Tingstad, Gothenburg	Road
1968	The Netherlands	IJ-tunnel, Amsterdam	Road
1969	Belgium	J.F. Kennedy tunnel, Antwerp	Road and rail
1969	The Netherlands	Heinenoord	Road
1969	Denmark	Limfjord	Road
1969	Argentina	Parana Tunnel	Road
1969	Japan	Dojima River, subway, Osaka	Rail
1969	Japan	Dohtonbori River, subway, Osaka	Rail
1970	Japan	Atsumi cooling water, Aichi	Water
1970	Japan	Haneda (Tama River), Tokyo	Rail
1970	United States	Bay Area Rapid Transit, California	Rail
1971	Japan	Haneda (Keihin Channel), Tokyo	Rail

(Continued)

Table 2.1 Listing of immersed tunnels constructed by date (Continued)

Date	Country	Immersed tunnel	Type
1971	United States	Charles River, Boston	Rail
1972	Japan	Dokai Bay conveyor, Kitakyushu	Conveyor
1972	United States	Calvert Cliffs outfall	Water
1972	Hong Kong	Cross-Harbor	Road
1973	The Netherlands	Amsterdam canal tunnel	Water
1973	The Netherlands	Hollandsch Diep service tunnel	Utility
1973	United States	63rd Street, New York	Rail
1973	United States	Interstate Route 10, Alabama	Road
1973	Japan	Kinura Port, Aichi	Road
1974	Denmark	Odense heating tunnel	Utility
1974	Japan	Ohgishima, Kawasaki	Road
1975	The Netherlands	Oude Maas service tunnel	Utility
1975	Germany	Unterwasser outfall	Water
1975	Germany	Elbe	Road
1975	The Netherlands	Vlake	Road
1975	Russia	Kanonerski Tunnel, St. Petersburg	Road
1975	Japan	Sumida River, subway, Tokyo	Rail
1976	United States	Cove Point LNG	Utility
1976	United States	Hampton Roads 2	Road
1976	France	Paris metro	Rail
1976	Japan	Tokyo Port	Road
1977	Japan	Dokai Bay gas pipeline, Kitakyushu	Utility
1977	The Netherlands	Drecht	Road
1978	France	Blayais power station outfall	Water
1978	United Kingdom	Kilroot power station outfall	Water
1978	Yugoslavia	Bakar Golf conveyor	Conveyor
1978	Germany	Bramsche Syphon	Water
1978	United States	Lake Piru outfall	Water
1978	The Netherlands	Prinses Margriet	Road
1978	The Netherlands	Kil	Road
1979	Iran	Halileh power station outfall	Water
1979	United Kingdom	Portsmouth sewer outfall	Sewer
1979	Japan	Keihin Conveyor Tunnel, Tokyo	Conveyor
1979	United States	WMATA, Washington, DC	Rail
1979	Hong Kong	Hong Kong Mass Transit	Rail
1979	Japan	Kawasaki Port, Kawasaki	Road
1980	The Netherlands	Hemspoor	Rail
1980	The Netherlands	Botlek	Road
1980	Japan	Daiba, Tokyo	Rail
1980	Japan	Tokyo Port Dainikoro, Tokyo	Road
1981	Australia	Dora Creek power station intake	Water

Table 2.1 Listing of immersed tunnels constructed by date (Continued)

Date	Country	Immersed tunnel	Type
1982	Saudi Arabia	Al Khobar water intake	Water
1982	Belgium	Rupel	Road
1983	Norway	Karmoey Tunnel	Utility
1983	Germany	Main metro	Rail
1983	France	Bastia Old Harbor, Corsica	Road
1983	Germany	S-bahn Rhein-Main	Rail
1984	Germany	Bramsche Canal Syphon, Hase Tunnel	Water
1984	Germany	Brokdorf power station intake	Water
1984	The Netherlands	Coolhaven, metro Rotterdam	Rail
1984	Taiwan	Kaohsiung Harbor	Road
1984	The Netherlands	Spijkenisse metro	Rail
1986	Singapore	Jurong Utility Tunnel	Utility
1987	United States	Fort McHenry, Baltimore	Road
1988	United States	Second Downtown, Norfolk	Road
1988	Denmark	Guldborgsund	Road
1989	Germany	Ems	Road
1989	France	Marne River	Road
1989	The Netherlands	Zeeburger	Road
1989	Hong Kong	Eastern Harbor Crossing	Road and rail
1991	United Kingdom	Conwy	Road
1991	Belgium	Liefkenshoek	Road
1992	United Kingdom	Sizewell B power station outfall	Water
1992	United States	Monitor Merrimac	Road
1992	Australia	Sydney Harbor	Road
1992	The Netherlands	Grouw	Road
1992	The Netherlands	Noord	Road
1993	P.R. China	Pearl River, N. Huangsha Guangzhou	Road and rail
1994	France	Météor, Paris	Rail
1994	United States	Ted Williams, Boston	Road
1994	The Netherlands	Willemsspoor	Rail
1994	Hong Kong	MTRC (metro rail) advance unit	Rail
1994	Spain	Bilbao metro	Rail
1994	Japan	Tama River, Tokyo	Road
1994	Japan	Kawasaki Fairway, Kawasaki	Road
1995	The Netherlands	Schiphol	Rail
1996	United Kingdom	South Humber power station outfall	Water
1996	Hong Kong	Lamma power station no. 3 intake	Water
1996	United Kingdom	Medway Tunnel	Road
1996	The Netherlands	Wijkertunnel	Road

(Continued)

Table 2.1 Listing of immersed tunnels constructed by date (*Continued*)

Date	Country	Immersed tunnel	Type
1996	P.R. China	Yong Jiang River, Ningbo	Road
1997	The Netherlands	Piet Hein	Road and rail
1997	Hong Kong	Airport Railway Tunnel	Rail
1997	Hong Kong	Western Harbor Crossing	Road
1997	The Netherlands	Aquaduct Alphenad Rijn	Aquaduct
1997	Japan	Sakishima, Osaka	Road and rail
1999	Singapore	Tuas Bay Cable Tunnel	Utility
1999	Ireland	Jack Lynch Tunnel, River Lee, Cork	Road
1999	Japan	Tokyo Port, Rinkai, Tokyo	Road
1999	Japan	Minatojima, Kobe	Road
2000	Denmark	Drogden Channel, Øresund	Road and rail
2001	P.R. China	Chang Hong, Ningbo	Road
2002	Japan	Niigata Minato, Niigata	Road
2003	Greece	Aktion Preveza	Road
2003	United States	Fort Point Channel, Boston	Road
2003	Germany	Warnowtunnel, Rostock	Road
2003	P.R. China	Shanghai ring road, Huangpu (Yangtze)	Road
2003	The Netherlands	2nd Benelux Tunnel	Road, rail, and cycle
2004	Czech Republic	Prague metro line C, Vltava Tunnel	Rail
2004	The Netherlands	Caland Tunnel	Road
2005	The Netherlands	Dordtsche Kil (HSL)	Rail
2005	The Netherlands	Oude Maas (HSL)	Rail
2008	The Netherlands	A73 Roermond Tunnel	Road
2008	Turkey	Marmaray, Bosphorus Crossing	Rail
2009	Vietnam	Thu Thiem, Saigon River, Ho Chi Minh City	Road
2010	Norway	Bjørvika, Oslo	Road
2010	Ireland	River Shannon, Limerick	Road
2010	South Korea	Busan–Geoje	Road
2011	United Kingdom	New Tyne Tunnel, Newcastle	Road
u/c	Japan	Second Kinuura, Aichi	Road
u/c	Japan	Yumeshima, Osaka	Road and rail
u/c	Japan	Shin-Wakato, Kitakyushu	Road
u/c	Mexico	Coatzacoalcos	Road
u/c	The Netherlands	Amsterdam North-South Line metro	Rail
u/c	The Netherlands	Second Coen, Amsterdam	Road
u/c	The Netherlands	N57 Middelburg	Road
u/c	P.R. China	Hong Kong Zhuhai Macao Bridge-Tunnel	Road

Table 2.1 Listing of immersed tunnels constructed by date (*Continued*)

Date	Country	Immersed tunnel	Type
u/c	P.R. China	Zhoutouzi Tunnel, Guangzhou	Road
u/c	P.R. China	Fenjiang Road Tunnel	Road and Rail
u/c	P.R. China	Haihe Tunnel	Road
u/c	P.R. China	Lunto Bio Island Tunnel No. 1	Road
u/c	P.R. China	Lunto Bio Island Tunnel No. 2	Road
u/c	Sweden	Söderstrom	Road

Chapter 3

Current forms of immersed tunnel

We have seen from the historical development of immersed tunnels that there are now two principal choices of immersed tunnel in use today. These are the concrete tunnel and the steel tunnel. Concrete tunnels comprise elements of either monolithic or segmental form, whereas steel tunnels can be formed of elements that are single shell, double shell, or sandwich construction. There are variations of each type, but both provide durable long-life structures that can be constructed safely. It is helpful to elaborate a little further on each type of construction to provide an introduction to current construction practice, and this chapter goes into further broad description by way of this introduction, but reference should be made to all subsequent chapters to understand the full detail of design and construction techniques.

MONOLITHIC CONCRETE ELEMENT CONSTRUCTION

The first, simplest and most straightforward of the concrete tunnel options is the monolithic element. Each tunnel element is a continuous structure that acts as a beam. Measures are taken to control cracking during construction, and it is usual to provide an external waterproofing membrane. Construction is generally of reinforced concrete although prestressed concrete has been used for structural strength and to prevent cracking. Articulation of the tunnel structure is provided at the immersion joints between the tunnel elements.

The tunnel element, which nowadays is typically between 100 and 200 m long, is constructed as a continuous reinforced concrete box; the floor, external walls, and roof being about 1 m thick and the internal walls between 300 and 700 mm. The base, walls, and roof are all rigidly connected together with the reinforcement being continuous throughout the section and across the construction joints. A typical multi-modal tunnel cross section is shown in Figure 3.1.

The monolithic form of construction provides great flexibility to the designer in that it can easily be adapted to different shapes to accommodate

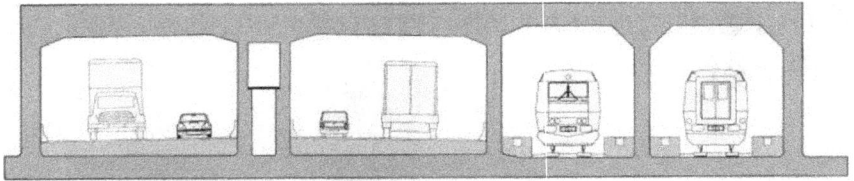

Figure 3.1 Typical concrete immersed tunnel section.

variations in the width or height of the cross section, or the length of the element. This is particularly useful where lane widths vary or emergency lay-bys have to be provided within the immersed tunnel. Such variation in geometry can be provided in segmental tunnels, but the regularly spaced segment joints may constrain locations of particular features.

The continuous box element is strong enough to resist all the temporary forces it will have to carry during towing and placing as well as the permanent loadings applied to it. Shear keys are provided at the joints between the elements to transfer shear across the joints and also to make sure that the elements remain in alignment in the long term. These shear keys can be steel-type knuckle joints or more simple reinforced concrete shear keys. The vertical shear keys are normally placed in the walls with transverse shear keys located in the floor or the roof.

These shear keys are not generally installed immediately after the element is placed. It is usual to place the foundations first, and then release the element onto its permanent foundation before installing the keys. That way a lot of differential movement is allowed to take place before the elements are locked together, which in turn reduces the forces to be carried by the shear keys.

Historically, the construction joints between the concrete pours often received no special treatment to make them watertight. The laitance was removed before the fresh concrete was poured, but apart from that, no other watertightness measures were taken. Some designers required more reassurance of watertightness and placed 150 mm wide steel strips centrally along the construction joint to act as an internal waterstop. Nowadays, it is more common to require a combination of measures, and a waterbar is usually combined with either a grout injectable tube or a hydrophilic sealant placed in the joint to give protection against leakage. However, it is still possible that the joint might not be completely watertight. Equally, it is accepted that there is a higher risk of early thermal shrinkage cracks as, historically at least, only limited measures are taken to control the heat of hydration in the setting concrete. Most usually, these limit the maximum temperature in the concrete, as well as the temperature differential between the core and the surface of a section, and the mean temperature compared to a previously cast section. As understanding of concrete behavior in its early ages increases, it is becoming more usual to see full thermal analysis and stress analysis being carried out. This used to be carried out just for the segmental forms of the tunnel, but it is now also applied to the monolithic tunnel element to gain the best possible understanding of the behavior and to minimize the risks of cracking.

Watertightness in a monolithic section is achieved by a combination of controls during construction to reduce cracking and the application of an external waterproofing membrane. For a monolithic concrete tunnel, this is considered to provide the best long-term solution to watertightness and hence durability. The early monolithic concrete tunnels were provided with a steel membrane formed from a 6 mm thick steel plate. Steel panels are welded together and used as permanent external formwork for the floor and the outer walls. Shear studs attached to the inner surface of the steel plate ensure attachment to the concrete. Only a nominal number of studs are used, so no composite structural action occurs between the steel and the concrete, and the steelwork acts solely as a waterproofing membrane.

As the steel membrane provides watertightness, it is vital that there be no gaps between the steel plates. All the welds joining the plates together have to be tested for watertightness during fabrication. This is a considerable undertaking because of the many kilometers of welds needed to join all the plates together.

The overlapping details between the membrane used underneath the tunnel and the membrane used on the roof and the sides of the tunnel need some careful consideration. If a spray-applied membrane is used for the roof, this is a relatively easy matter to address and there is a simple overlapping distance that the spray membrane should achieve over the steel/plastic membrane. If a sheet membrane is used, there is often a clamping arrangement to ensure the edge of the sheet membrane is secured.

Providing an external steel waterproofing membrane raises another complication: the long-term durability of the membrane. The watertightness of the tunnel depends on it, but the steel is exposed to a saline water environment, so it needs some protection to avoid being penetrated by corrosion. On the plus side, there is little oxygen in the tunnel backfill, which slows the general rate of corrosion, but there is still a possibility for localized pitting corrosion, which could penetrate the membrane. This is normally prevented by applying protection in the form of a bituminous coating or a protective layer of epoxy paint. Even this may not be enough to provide the confidence required for the 100-year design life of a piece of major infrastructure like a transportation tunnel. Cathodic protection of the steelwork provides this additional security. This can either be a sacrificial anode protection system or an impressed current cathodic protection system.

The sacrificial anode system, which usually consists of zinc anodes fixed to the outside of the steel membrane, is designed to make up the difference between the expected life of the protective coating and the design life of the tunnel, with an additional allowance made for a possible percentage of defects in the coating. An impressed current system can be installed at the outset, or provision can be made to retrofit one later if subsequent monitoring of the steelwork corrosion shows that it is needed.

Another possible option for corrosion protection is to provide an additional sacrificial thickness to the steel membrane. This is designed so that the required thickness of steel still exists at the end of the design life of the tunnel. However, the drawback is that opinions vary on the degree of corrosion that occurs in the long term. Experience from extracted sheet pile walls in a marine environment would indicate very low rates of corrosion, mainly due to the lack of oxygen in the soil surrounding the pile. The piles though would have been driven into the ground, whereas in constructing a tunnel the fill material around the tunnel is disturbed, which could lead to a more aerated material with entrapped oxygen resulting in greater corrosion rates. Additionally, on long structures such as tunnels, differences in electrical potential can cause longline effects that can result in greater corrosion rates in certain parts of the tunnel. Local corrosion rates can also be increased by electrical potential changes in the backfill regime, such as an intrusion of freshwater. Because of these uncertainties, a sacrificial thickness is not appropriate and is not used today.

Apart from being somewhat problematic to install during construction, the external steel membrane has a major problem in use. Despite all the care that is taken during construction, it is very unlikely that the membrane will be in continuous contact with the concrete over its entire area. Thus, if the membrane is defective and leaks somewhere, it is likely that the water penetrating it will spread under the membrane in the small space between it and the concrete. Eventually, the water will find a defect in the concrete and damp patches will be noticed on the inside of the tunnel. The problem is that these damp patches are probably not opposite the defect in the membrane, so identifying the location to repair is not possible. Grout can be injected around the damp patch, but grouting pressures must be carefully controlled or they will cause further separation of the membrane from the concrete, making the problem worse rather than solving it.

These difficulties with steel membranes were eliminated by the development of spray-on waterproofing membranes, which are robust enough to resist the placing and backfilling processes. Steel plate is still provided to the floor as this has to be robust enough to stand up to the construction activities of placing the reinforcement and concrete. Clearly, it is also not possible to apply an exterior coating to the base of the concrete after it has been constructed. The spray-on membrane is therefore only applied to the walls and the roof.

A few tunnels have used plastic membranes in place of steel, which offers a lower-cost solution. While this is relatively new, it is likely to become more commonplace. Tough plastic materials are available that are highly resistant to puncture and abrasion, and preformed sheets that have a T-lock system to connect the sheet to the concrete are used. The T-lock also serves to compartmentalize the membrane and limit the distance any penetrating water can travel.

SEGMENTAL CONCRETE ELEMENT CONSTRUCTION

The segmental form concrete tunnel element was developed from the original monolithic tunnel element to avoid the need for an external waterproofing membrane. In a monolithic element, the main risk of leakage is through shrinkage cracks in the concrete. These cracks go right through the thickness of the concrete and create a water path from the outside to the inside of the tunnel. They are caused, for example, when the freshly poured wall concrete cools and contracts, but is restrained by the hardened concrete of the base on which it has been cast. Measures were taken on monolithic sections to minimize this by limiting the temperature differentials between the freshly placed concrete and the previously placed hardened concrete. These measures could not, however, be guaranteed to be completely successful and so an external membrane was provided to ensure the watertightness of the completed section.

The Dutch found that if the length of the concrete tunnel section was limited to about 22–25 m, it could be cast without any shrinkage cracks developing. The section was not cast in one pour, but typically, the base was cast first and then the walls and the roof were added in one further concrete pour. This second pour was carefully controlled so that the temperature differentials between it and the base were not sufficient to cause the concrete to crack as it cooled. The stresses developed as the concrete cooled were kept below the developing tensile strength of the concrete. This was achieved by a combination of concrete mix design, controlling the temperature of the setting concrete with water pipes cast into the concrete, and insulating the formwork.

Cement replacement mixes were designed, using pulverized fuel ash or blast furnace slag cement to replace ordinary Portland cement in the mix. This reduced the heat of hydration within the concrete and therefore the temperature that developed as the concrete hardened. Cooling water pipes were cast into the walls and the roof. These controlled the temperature gradient in the boundary region between the existing and the fresh concrete. External insulation was applied to the shutters to prevent the concrete from cooling too quickly. With these measures, the development of stresses in the section could be controlled so that they never exceeded the tensile strength of the setting concrete. Typically, the design was such that the tensile stresses did not exceed 70% of the tensile strength of the maturing concrete at the time.

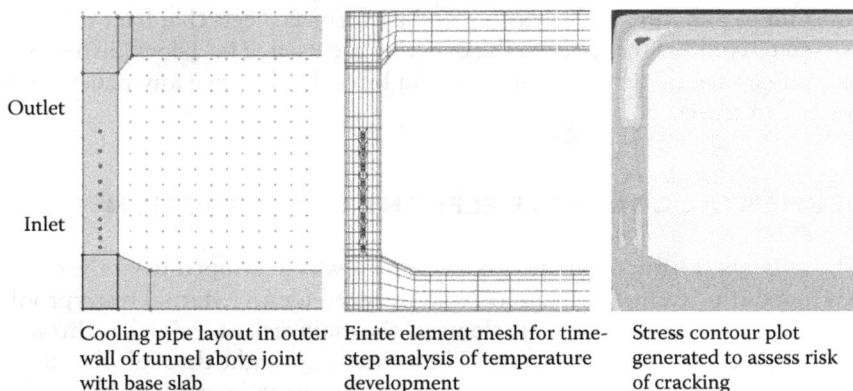

| Cooling pipe layout in outer wall of tunnel above joint with base slab | Finite element mesh for time-step analysis of temperature development | Stress contour plot generated to assess risk of cracking |

Figure 3.2 Cooling arrangement and stress analysis.

A detailed thermal analysis of the section was carried out to identify the stress contours in the cross section. Figure 3.2 shows some typical analysis output with the arrangement of cooling water pipes in the wall, the finite element model used, and the resulting stresses. The darker areas indicate the highest longitudinal tensile stresses in the concrete and hence the areas most likely to crack. The figure shows no dark areas close to the construction joint between the base slab and the wall and hence the effectiveness of the cooling pipes. Note that there are some dark areas toward the top of the section that may require further curing measures if there is a risk of cracking.

Through the application of these construction techniques, it became possible to cast the cross section for an immersed tunnel without any cracks. The concrete itself became the watertight barrier and there was no need for an external membrane. However, that was not the end of the development, because elements of 25 m are not very practical for immersed tunnels.

A complete tunnel element was built up of several such segments, for example, five 25 m long segments, to form a typical 125 m long tunnel element. The segments are match cast against each other with the joints between segments being discontinuous and debonded. These segments are then clamped together with longitudinal prestress so that they behave as a single homogeneous tunnel element for transporting and placing. The prestressing cables were regarded as temporary and were later cut at each of the joints between the segments after the element had been placed. The result is a reinforced concrete tunnel with regularly spaced articulation joints and hence very little longitudinal bending. Cracking due to shrinkage and early thermal effects has been eliminated, so no external waterproofing membrane is required. Articulation occurs not only at the joints between the elements but also at the segment joints. Figure 3.3 illustrates the segmental arrangement of the Øresund Tunnel, where each 175 m long tunnel element was made up of eight 22 m long segments.

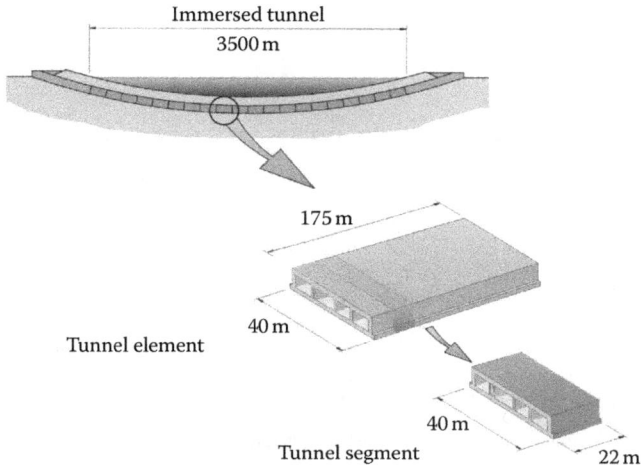

Figure 3.3 Example of segmental construction of the Øresund Tunnel.

Although watertight concrete had been produced, there was a weakness in the design in that several more joints had been introduced between the segments and each of these was a potential source of leakage. The joints between the segments were kept as simple as possible to minimize construction problems. They were simple match cast half-joints that acted as shear keys between the segments and prevented any vertical or transverse differential movement. They did allow some axial movement so that as the concrete aged and the temperature changed, it would not develop any tensile stresses, which might cause cracking right through the section in the long term. They also allowed some rotation to accommodate any differential settlement that might occur. To make these segment joints watertight, a continuous waterstop was introduced within the joint. These waterstops were developed so that grout could be injected around the ends so that any porosity in the concrete near the tips of the waterstop could be sealed. The arrangement is shown in Figure 3.4.

With these techniques, the segmental form tunnel element was realized and quickly gained acceptance. The omission of the external waterproofing saved both cost and time, although the cost saving was partially reduced by the additional measures required to produce the crack-free concrete. However, it did not and has not received universal acceptance for concrete tunnels, as some owners still prefer what they consider the additional security of an external watertight membrane.

A variation of the segmental construction method can be used on smaller tunnels that are used to carry utilities. This is similar to bridge deck construction where the element is made up of segments that are glued together

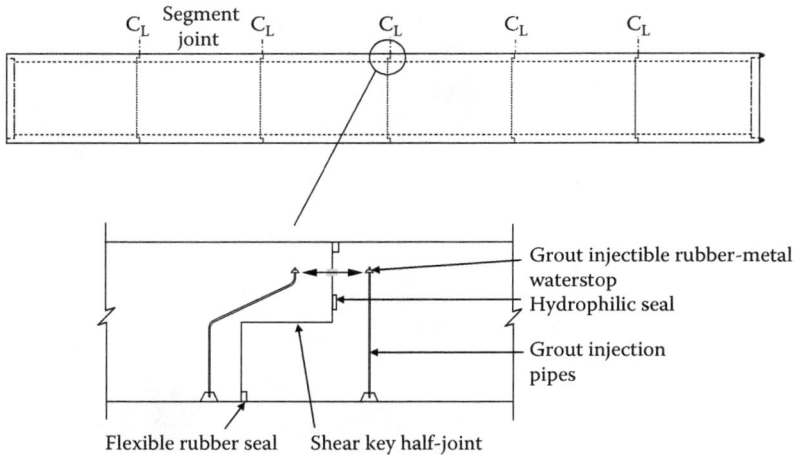

Figure 3.4 Typical segment joint detail.

and permanently prestressed to form a monolithic tunnel element. Whether an external waterproofing membrane is needed depends on the precise method of construction, but it is not normally provided.

PRESTRESSED CONCRETE

A variation of the monolithic reinforced concrete element is to prestress it with permanent longitudinal prestress. This form of tunnel element can have advantages in reducing the amount of longitudinal reinforcement and also the overall compressive stress it provides tends to close any cracks in the concrete, reducing the likelihood of leakage. Despite this, however, it is common to apply an external membrane to the outside of such tunnels. The disadvantage of the method is that an extra activity is required during construction, and unless great care is taken with the grouting and with the detailing around the anchorages, the prestress can compromise the long-term durability of the tunnel.

There is also sometimes a need to consider transverse prestress. As the capacity requirements of tunnels increase, particularly to accommodate more road traffic, the widths of the internal compartments will also increase. These larger internal bores increase the roof and floor spans and thus a reinforced concrete section becomes uneconomic and transverse prestress might be needed in the floor and the roof. The transverse prestress does, however, increase the risk of durability problems arising as the prestress anchorages are sometimes positioned on the outside of the element and must be well protected against the external environment or made from nonferrous material.

Transversely prestressed concrete tunnels are not common but do exist; for example, the western element of the Bjørvika Tunnel in Oslo was prestressed transversely to enable the roof slab to span over the three main running lanes and a slip road to come into the tunnel. Such designs are relatively rare as immersed tunnel designers usually prefer to avoid permanent prestressing if at all possible.

SINGLE STEEL SHELL

Both steel and concrete tunnels are technically viable options for nearly all immersed tunnels. Each has its advantages and disadvantages. Steel tunnels are perhaps easier to fabricate and need smaller facilities, but set against this is the generally cheaper material cost of concrete tunnels in most parts of the world. Steel shells developed originally in the United States and have been predominant there but have been built throughout the world. Steel tunnel elements are monolithic. Articulation can take place at the joints between the elements, although often these joints have been welded up to provide a continuous tunnel structure.

A single steel shell element has an external steel shell fabricated typically from 10 mm steel plate. This does not have to be the traditional circular steel tunnel shape as can be seen in Figure 3.5. The steel shell provides strength and watertightness. A reinforced concrete lining approximately 700 mm thick is placed inside the steel shell and is usually designed to act compositely with it. As the steel shell provides a watertight barrier, it is usual to protect it against corrosion in some way, generally by a cathodic protection system.

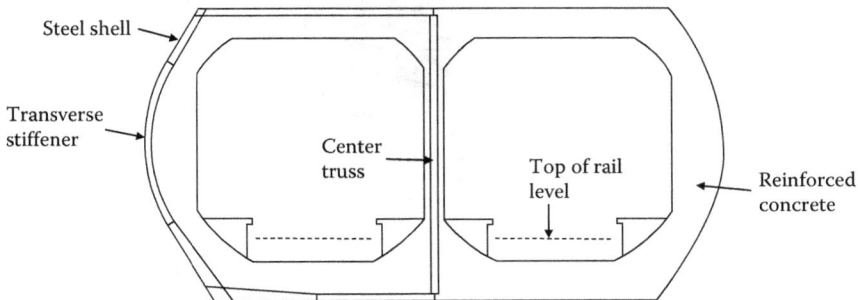

Figure 3.5 Single shell tunnel cross section—Cross-Harbor Tunnel, Hong Kong.

The steelwork for the shell can be fabricated in a comparatively small facility or in an existing shipyard. One big advantage of the technique is that the steel shells are relatively easy to build and do not require the large fabrication facility needed for a concrete tunnel. This can also be an advantage if there are strong environmental objections to the construction of a large casting basin needed for a concrete tunnel element.

It is possible to build a rectangular single steel shell tunnel element and then pour the internal concrete while it is afloat. Such a section is shown in Figure 3.6. To enable the steel shell to be robust enough to float, it has to be stiffened with longitudinal and transverse stiffeners. In fact, the detailing of such a section uses shipbuilding practices. Shear studs are attached to the inner face of the steel plate to mobilize the composite action between

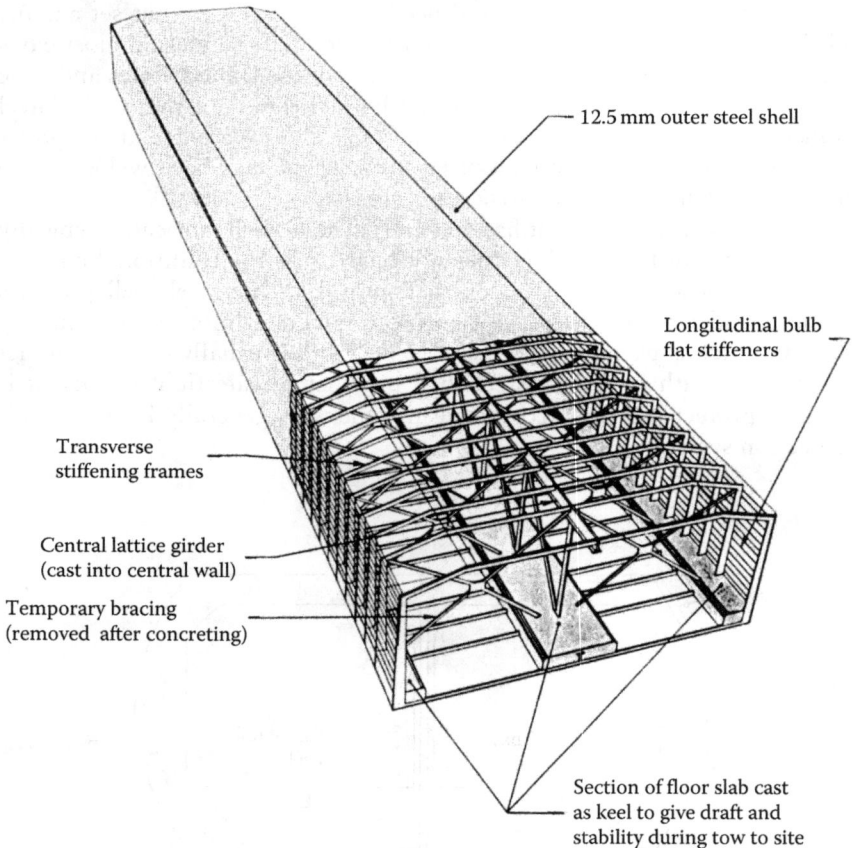

Figure 3.6 Single steel shell concept developed for the Conwy Tunnel.

the steel and the concrete. The single steel shell arrangement shown in Figure 3.6 was developed as a design option for the Conwy Tunnel in the United Kingdom, a dual two-lane road tunnel. The rectangular shell is 25 m wide and 9 m high, and the length of the tunnel elements was chosen as 110 m.

DOUBLE STEEL SHELL

The cross section of a double steel shell has two steel skins. There is an inner steel shell, which is thinner than the steel shell of a single shell tunnel and typically 8 mm thick. This provides strength and watertightness and is stiffened and constructed in a similar way to the single steel shell form. This inner steel shell is reinforced by an internal concrete lining that is designed to act compositely with the steel shell. Around this inner shell is an outer steel box, referred to as a form plate that forms a void for the ballast concrete. This outer form plate is slightly thinner than the inner shell at typically 6 mm. The inner shell and form plate are connected by a series of transverse diaphragm plates that compartmentalize the void between the shells and add stiffness to the structure. The space between the two shells is filled with concrete, which acts as ballast and also protects the inner steel shell against corrosion. A typical double steel shell is shown in Figure 3.7 and can be seen under construction in Figure 3.8.

The method of construction is similar to that of the single steel section. The steel shell is fabricated from a series of modules, then assembled on a slipway, and launched. It is then towed to an outfitting site and the concrete placed while the element is afloat. This tunnel form is the direct descendant of the original Detroit River Tunnel where the outer "shell" was formed out of timber.

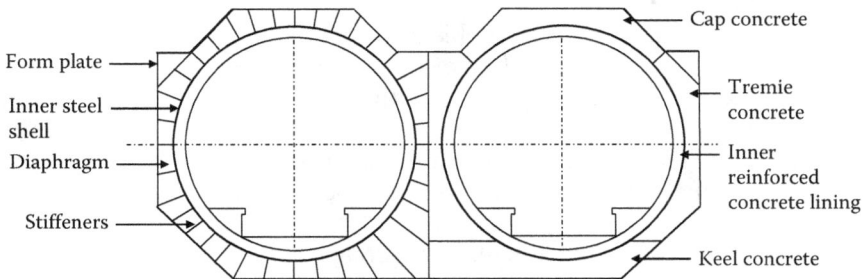

Figure 3.7 Double steel shell.

Figure 3.8 Ted Williams Tunnel under construction. (Photo courtesy of W. Grantz.)

COMPOSITE CONCRETE STEEL SANDWICH

A development of both the concrete and the steel shell sections is the fully composite steel concrete sandwich section. The inner concrete lining of the original single and double steel circular shell sections acts compositely with the steel, but the sandwich design has taken the concept further. The section consists of two thin steel plates acting compositely with an unreinforced layer of concrete between them. To achieve the composite action, shear connectors are attached to the inner faces of both the steel plates. The form of the structure can be seen in Figure 3.9. This produces a very strong and ductile section that in the ultimate condition can accommodate large deflections without rupturing. This is an attractive property where large dynamic loads such as those from severe earthquakes can occur.

The development of modern sophisticated methods of structural analysis has led to a greater understanding of the structural behavior of such composite sections. This, in conjunction with improvements in materials and construction techniques, has made the development of an efficient composite shell possible. The composite section has great structural strength, so the thickness of the section required to resist the structural loadings can be reduced.

The structural action of a double skin composite section is complicated. The spacing of the shear studs that connect the steel plate to the concrete has to be limited so that the plate does not buckle under compression between the shear studs. The studs also resist the shear loading in the section, which

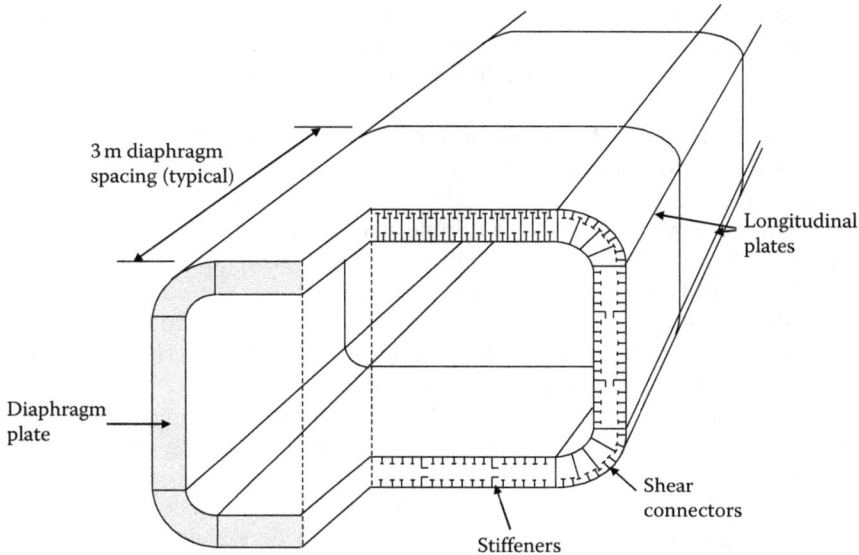

Figure 3.9 Sandwich construction (developed from SCI publication 132).

in a concrete section is the function of shear reinforcement. To be effective at resisting shear, these studs have to extend through the full depth of the section and must be anchored in the compression zone.

The analysis of such a section under varying loads is a specialist subject and the reader should refer to detailed papers on the subject published by the Steel Construction Institute in the United Kingdom (SCI Publication 132, 1997). The section is subject to both compression and bending forces and the composite action is complex. The stiffness of the section varies as the concrete cracks. This variation is normally neglected in the analysis of a concrete section but is significant in this double skin composite. A considerable amount of physical model testing has been undertaken on these sections to validate the analytical methods used for design.

In design terms, such a composite section is an elegant solution to a structural problem. Both the steel and the concrete are used efficiently and produce a very strong section. There are, however, difficulties in transforming this elegant structural concept into a practical immersed tunnel.

The thickness of the walls is less because it is such an efficient structural section. This leads to an overall tunnel cross section that is lighter and therefore has a shallower draught that can make towing to the tunnel site easier. However, this advantage of a thinner section thickness is offset because the weight of the concrete also serves a ballast function. If it is not in the structural section as part of the load carrying members, it has to be applied elsewhere, either as internal ballast or on the roof. To place this extra ballast internally requires constructing larger internal compartments to contain it, so some of the structural advantage is lost.

Added to this, the section is more complicated to construct. Ensuring that the concrete completely fills the section so that the full composite action is obtained is not straightforward. If there is only an outer steel shell, the problem is mainly confined to the roof. Here, ports can be left through the steel shell to let the air out as the concrete is poured so that it can fill the section completely. With the double steel shell, the floor also presents a similar difficulty, further exacerbated by all the stud connectors. Careful mix design and placing techniques are needed to get the concrete to flow through the section between the plates and through and around the shear connectors. The concrete must also be relatively self-compacting, as normal internal vibration techniques are not possible.

These difficulties have been solved and the sections can be constructed satisfactorily. However, there are other aspects to creating an immersed tunnel than just the structure itself. A modern tunnel has to accommodate a large number of facilities. In the traditional circular steel section, there are spaces above and below the rectangular traffic envelope that can be used for these. In a more rectangular section, there are multiple holes and box outs in the walls to hold emergency panels, emergency exits, and electrical control and distribution panels. These add considerably to the difficulty of fabricating the inner steel shell and concreting the section.

Fire in the tunnel is a greater hazard in the sandwich construction because of the significance of the inner steel plate to the strength of the section. Damage to the inner steel plate would lead to a loss of strength, although a layer of fire protection applied to the inner steel skin would guard against this.

These additional practical factors weigh against the double skin composite section, although two such tunnels have been built in Japan: the Kobe and Naha Tunnels. The form of construction lends itself to being fabricated on a slipway or in a shipyard, but equally could be constructed in a dry dock or casting basin. Construction is similar to the conventional steel shell design in that the shell is built up from modular components. The fabricated shell can be made stiff enough to be slid onto a submersible barge or launched and towed to its outfitting station. Concrete can then be placed while the element is afloat and ballast can be placed either internally or externally as with a single steel shell tunnel.

None have been built outside Japan, although the form has been considered at the feasibility stage of various immersed tunnels, for example, the 50 m deep Marmaray Tunnel in Istanbul, where the strength of the composite section had advantages in resisting the hydrostatic pressures as well as providing good ultimate strength in the face of the very high seismic forces expected at the site. Despite this, however, the section was not adopted due to the additional cost and the difficulties of construction.

Chapter 4

Developing an immersed tunnel scheme

Right at the start of the project life cycle, at the planning stage, there are a number of important decisions to be made that will define the characteristics of a crossing. This chapter discusses the important elements of early planning that will be needed to develop an immersed tunnel project to a preliminary stage that can confirm its viability. It touches on a number of subjects that are covered in more detail throughout the book but does so in order to illustrate the broad thinking that is required at the start of an immersed tunnel project.

Some of the many aspects that have to be considered when planning or even just considering whether an immersed tunnel is appropriate for a particular crossing include

- Should it be a tunnel rather than a bridge?
- What is the crossing intended to accommodate?
- What form of tunnel (bored/immersed/cut and cover)?
- Environmental impacts of construction.
- Operational safety and maintenance.

If an immersed tunnel is appropriate, then further questions arise, including

- What form of immersed tunnel (concrete/steel)?
- Where will the tunnel elements be built?
- Effects of construction on third parties.

Each of these involves many issues, many of which are interlinked. This chapter gives a brief overview of the main considerations.

SELECTION OF BRIDGE OR TUNNEL

New crossings are required for many reasons. They can replace existing crossings that are life expired, supplement existing crossings that cannot cope with the volume of traffic, or be part of entirely new routes or developments. Each will have its own particular set of constraints that will have to be overcome, whether they are the logistics of building a new crossing adjacent to an existing one that has to remain in continuous use, or the environmental and social issues associated with an entirely new crossing.

Whether to choose a bridge or a tunnel often depends on environmental considerations and the operation of the waterway. A tunnel is usually, but not always, more expensive than a bridge, so for a tunnel to become the preferred option, there often has to be another reason where the benefits outweigh the additional cost. For example, a tunnel rather than a bridge was chosen at Conwy in the United Kingdom to preserve the views of the estuary and the setting of the World Heritage Site of the walled town of Conwy. Another example is the Øresund Tunnel, part of the fixed link between Denmark and Sweden, which was chosen over a bridge as high bridge piers that would have been needed to span the navigation channels were unacceptable so close to the Copenhagen airport. If the crossing is to carry rail, particularly freight, the long approach spans needed for a bridge may constitute unacceptable visual intrusion in a valuable landscape or, in an urban situation, may just be unachievable because there is no space available, or they would blight the land over which they pass and reduce the development potential of the land.

On busy waterways, access for shipping is also an important factor. In general, when crossing a modern major waterway, the air draught required for a bridge is often 50–60 m, to allow the passage of large freight and cruise ships. In some major ports this clearance may even rise to 80 m. Authorities will often not want to compromise any future development upstream of the crossing, so they will insist on these large clearances under bridges even if no ships of that size currently use the waterway. This air draught is considerably more than the water draught of such ships, which is generally around 15–20 m. So, in general, a bridge has to be higher above the water level than a tunnel does below it. The carriageway on a bridge would be about 65 m above high water level for a 60 m air draught, whereas for a tunnel the carriageway would be closer to 30 m below high water level if the channel is about 20 m deep. This difference makes the approaches to the bridge longer than those required for a tunnel, which in turn increases the cost and also makes a bigger intrusion into the adjacent landscape.

This is illustrated in Figure 4.1, which, as well as showing the difference between a tunnel and a bridge, also includes a comparison between a bored and an immersed tunnel. Bored tunnels are constructed deeper below

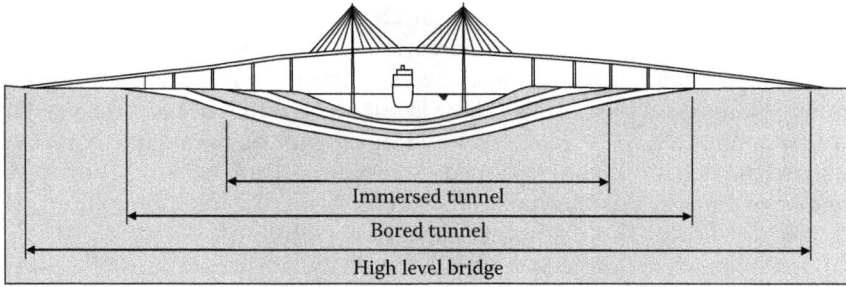

Figure 4.1 Relative length of immersed tunnel, bored tunnel, and high level bridge crossing.

the bed of the waterway than immersed tunnels because tunnel boring machines need competent ground in which to operate. The beds of waterways usually have a layer of soft material at the surface, so the bored tunnel has to go deeper to find better ground, whereas immersed tunnels can be built within these soft bed materials. In addition, the shape of the bored tunnel is broadly circular, resulting in the level of the road or railway being further below the top of the tunnel than in an immersed tunnel, which is often more rectangular in shape. The circular cross-sectional shape of a traditional steel shell immersed tunnel negates the shape advantage against bored tunnels, but the construction depth advantage still stands. Thus, as Figure 4.1 shows, the immersed tunnel gives the shortest overall crossing length, followed by the bored tunnel, and the bridge gives the longest overall crossing length. This depth advantage of immersed tunnels over bored tunnels is being reduced as advances in bored tunneling techniques enable bored tunnels to be constructed with smaller amounts of cover and therefore closer to the surface.

Other factors may come into play in making this decision. Bridges in exposed locations often have to be closed due to adverse weather conditions, such as wind or ice falling from parts of the superstructure, and such interruptions in service may be undesirable or even unacceptable.

TUNNEL CROSS SECTION

When compared to a bored tunnel, the immersed tunnel offers greater flexibility as to the arrangement of bores carrying roads, railways, and pedestrians. Additional lanes and bores can be added to an immersed tunnel relatively easily, whereas bored tunnels will be limited by the diameter that it is currently possible to construct. When considering multimodal crossings, an immersed tunnel can usually provide the crossing as a single tunnel, whereas bored tunnels may need multiple tunnels to provide the same functionality. This could potentially make the bored tunnel more expensive, depending on project-specific circumstances. Comparing the immersed tunnel with a bridge, there is not much difference as bridge decks can be configured to be multilevel and be as wide as necessary to accommodate the desired number of road lanes and rail tracks. For a high-level long-span bridge, this can make the bridge structure very heavy and the costs of the bridge and the tunnel may be very similar.

If an immersed tunnel is selected for study, then care must be taken at an early stage to size the structure correctly. Space is at a premium in immersed tube tunnels. Any internal space required to accommodate whatever the tunnel is carrying has to be weighed down by the concrete and steel of the tunnel structure to keep it in position and stop it floating. Put simply, the greater the airspace inside the tunnel to accommodate whatever it is carrying, the more surrounding concrete has to be provided to keep it down, and hence, the more expensive the tunnel will be.

The cross-sectional designs are based on the necessary clearance envelopes for the road and rail traffic, the space for the mechanical and electrical (M&E) systems required to be installed in the tunnel, the requirements for ventilation (both air supply and smoke extract), and the provision for safe emergency access and escape. When developing the cross section, all the various requirements for space must be rigorously challenged to minimize the airspace in the tunnel. The number of traffic lanes, rail tracks, and their respective envelopes will be defined by the relevant authorities, but apart from these, there are several aspects that leave scope for the designer. Verge widths, hard shoulders, marginal strips, and railway emergency walkways may also be client-defined, but if there is an opportunity to make reductions, then the designer should look for these, always considering the safety of the users and the maintenance staff. The structure will be designed around these requirements to generate a number of potential cross-section arrangements. It is more cost-effective to add width to an immersed tunnel than increase its depth because of the increased dredging requirements and the increased length of approaches associated with going deeper. This is a simple matter of how the additional area of the tunnel affects the dredging volume, as can be seen in Figure 4.2.

Some authorities require lay-bys in their tunnels. They should be avoided, if possible, in an immersed tunnel because of their impact on the element geometry, but in longer tunnels, this may not be possible and special wider elements may be required, such as those used on the Bjørvika Tunnel in Oslo, shown in Figure 4.3. Although a short immersed tunnel, it linked the existing rock tunnels to form a longer network of tunnels beneath Oslo, and as the new link, it was the only location at which a lay-by could be introduced.

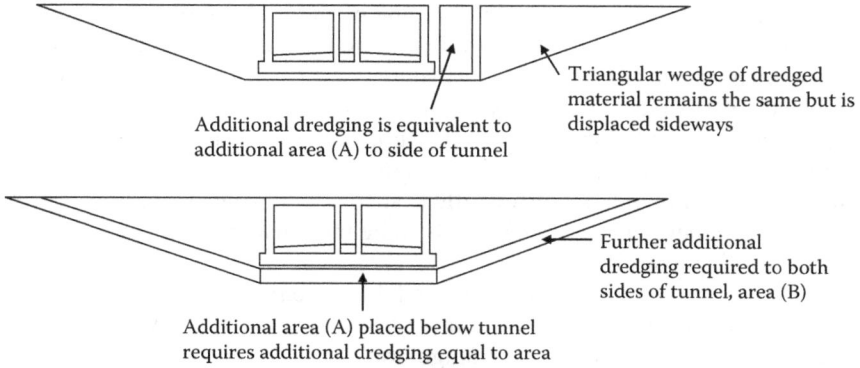

Triangular wedge of dredged material remains the same but is displaced sideways

Additional dredging is equivalent to additional area (A) to side of tunnel

Further additional dredging required to both sides of tunnel, area (B)

Additional area (A) placed below tunnel requires additional dredging equal to area

Figure 4.2 Impact of widening or deepening tunnel on dredging.

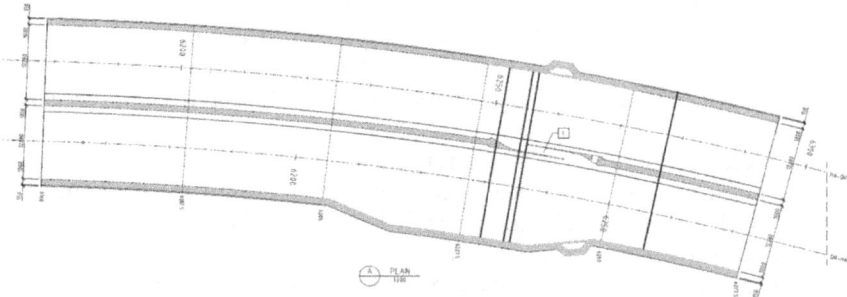

Figure 4.3 Plan and impression of lay-by in tunnel element. (Photo courtesy of Statens Vegvesen.)

In a rail tunnel, the kinematic envelope of the trains will depend on the type of the rolling stock being used. Provision for escape should be in accordance with international standards, such as NFPA 130. Space will also be required for the electrification system, particularly the overhead catenary, if third rail is not being used. Low-height ballast-free trackform systems, such as booted blocks, should also be considered to reduce space, and the amount of equipment to be installed, such as lighting, signaling, power supply, and emergency systems, needs to be included from the outset in the space proofing exercise.

The type of ventilation system used has a particular bearing on the cross section of a road tunnel and must be considered at a very early stage. Underprovision of space at this stage could lead to costly redesign later. Whatever type of ventilation system is chosen, a considerable amount of space is needed to accommodate it and all the other mechanical and electrical equipment, signs and barriers, and internal finishings. Provision must also be included for tunnel users to escape from the tunnel in the event of an incident. There are also tolerances that have to be included for construction. All these must be considered when developing the internal space requirements of the tunnel.

The rectangular shape of a concrete immersed tunnel matches well with the rectangular shape and clearance envelopes required for road and rail traffic. With a steel tunnel that generally has a more circular shape, the arcs above and below the rectangular traffic envelope are dead airspace, which requires a greater weight of concrete to hold it down. This disadvantage is alleviated somewhat as the space above and below the traffic envelope in a steel tunnel can be used for ventilation, drainage, and carrying services, but the generality that the concrete tunnel section better matches the required traffic or rail envelope holds true.

With multilane road, rail, or multimodal crossings, this shape advantage of rectangular concrete immersed tunnels increases. Steel tunnels have often been duplicated to provide the required crossing capacity, whereas the same capacity could be provided within a single rectangular concrete section. Multiple tubes can be provided for road and rail use and elements up to almost 50 m wide have been built, such as the Second Benelux Tunnel in the Netherlands, shown in Figure 4.4, that opened in 2001. However, there are limitations. The structural action of a concrete element becomes more complicated as it gets wider and it behaves more like a two-way spanning slab than a beam. The marine equipment to place the element also becomes larger and more sophisticated.

In road tunnels, bidirectional traffic in a single bore should be avoided on safety grounds, and indeed, this is precluded by many national authorities. The possibility of accidents and the consequences of them are too great. Two- or three-lane carriageways can be readily accommodated in a single

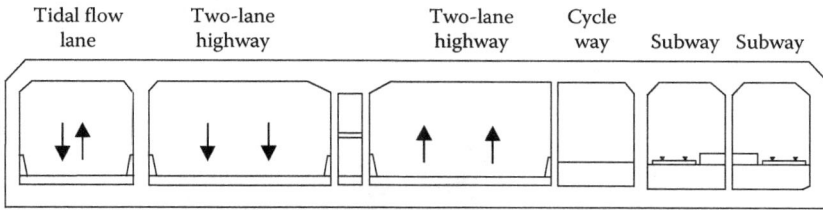

Figure 4.4 Multilane Second Benelux Tunnel.

immersed tunnel compartment, so dual carriageway roads need the conventional two-cell cross section. For more than three lanes, structural issues come into play. As the cells become wider, the roof spans become longer and structural requirements dictate that they become thicker or alternatively require transverse prestressing. Prestressing is possible but is an additional construction complication and requires careful detailing to ensure the durability requirements of a major piece of transportation infrastructure are met. This leads to a four-lane carriageway normally being better accommodated in two bores of two lanes each. The overall area of the cross section will be increased because of the additional verges and hard shoulders, but the cross section will work better structurally and can be more economic overall. This was the layout chosen for the Drecht Tunnel in the Netherlands. The arrangement does, however, have implications for traffic approaching the tunnel from the dual four-lane carriageway. Drivers often make unsafe last-minute lane changes to take the other bore. In this respect, the more costly structural arrangement may be preferred on safety grounds and this risk should be considered in the choice of cross section.

Modern safety requirements call for a separate bore to enable people to safely exit the tunnel in an emergency. In some older tunnels, this was achieved by evacuating people through the central wall into the adjacent bore carrying the other carriageway and may still be acceptable in shorter tunnels, but now a separate escape bore is generally preferred. Typically, these pedestrian escape bores are about 1.2–1.5 m wide; a typical example is shown in Figure 4.5. To be effective, they must be kept clear of equipment, and doors giving access to the bore must not cause an obstruction. The pressure in the escape bore has to be higher than the adjacent traffic bore so that smoke and flames are not drawn into it when a door is opened.

This escape bore can also be used to carry the tunnel services either above or below the pedestrian space. It is not always straightforward to combine the evacuation and services requirements, particularly in long tunnels where large transformer substations are required at intervals along the tunnel and the size of these acts as a blockage to the escape bore. A satisfactory evacuation strategy may be to use the separate central bore for

Figure 4.5 Typical pedestrian escape bore.

immediate evacuation and then route the pedestrians to the other traffic bore when the traffic in it has been stopped.

Ventilation is an important factor in determining the cross section. Ventilation is covered in more detail in Chapter 7, but simplistically in shorter tunnels or those with low vehicle emissions, the tunnel can be ventilated through the traffic bore. This type of ventilation system has been installed in the 4 km long Øresund Tunnel and the 3.2 km long Busan-Geoje Tunnel, and is also planned for the 19 km long Fehmern Tunnel because of the relatively low traffic volumes and low vehicle emissions. So apart from having sufficient space above the carriageway or verge to accommodate the longitudinal jet fans, no additional space is needed. Even this space can be reduced by positioning the fans in niches in the tunnel roof so that the general roof line is lower, rising only where necessary to accommodate a set of fans.

The Saccardo ventilation system is also gaining popularity in immersed tunnels. This uses the traffic bore for ventilation similar to the longitudinal jet fans, but instead of distributing the fans throughout the tunnel, air is blown into the tunnel at the ends through nozzles in the tunnel roof. No additional space is needed within the tunnel above the traffic envelope to accommodate jet fans, so the overall internal height of the tunnel is minimized.

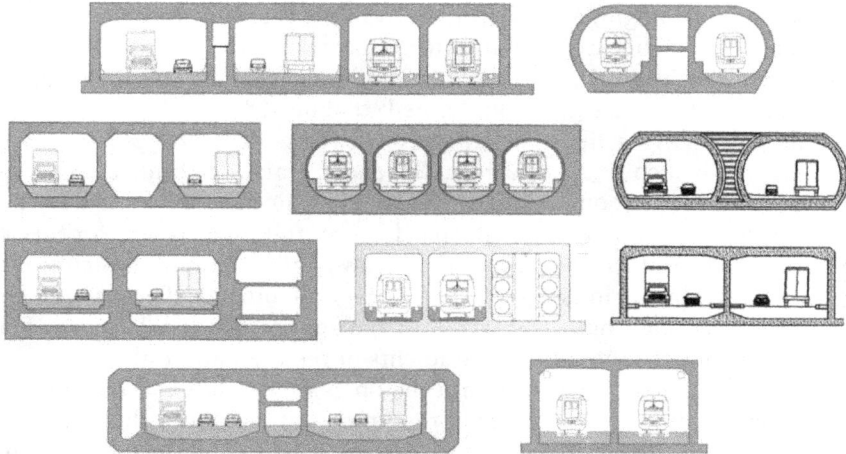

Figure 4.6 Typical tunnel sections.

With longer or busier tunnels, a semi-transverse or a fully transverse ventilation system may be needed. These require additional bores to duct the air either in or out of the tunnel, resulting in a greater overall cross section.

By considering all these parameters, an initial cross section can be developed. Figure 4.6 shows various combinations of traffic and ventilation bores that have been used around the world.

ALIGNMENT

Having determined an initial cross section, the alignment of the road and/ or railway being carried by the tunnel can be developed. The first stage in developing the alignment is to identify all the constraints. These will typically include

- Tying in with the existing infrastructure
- Navigation constraints on width and depth of navigation channel
- Extent of protection to the top of the tunnel, which determines the depth of the top of the tunnel below the bed
- Results of geotechnical studies
- Requirements from environmental assessment
- Requirements/preferences of the local authorities

Some of these will result in rigid constraints that offer no flexibility; others will give the designer the possibility of developing different options that satisfy the constraints.

There will be alignment constraints from the highway and rail authorities. These will differ from country to country. If there is no national standard, then European, U.S., and other standards exist and can be adopted. Highway gradients are generally limited to about 5% or 6%, particularly if the road carries a high percentage of heavy goods vehicles. Rail cannot cope with such steep gradients. Generally, trains are limited to maximum gradients of about 1.6%, although this can increase to 2.5% with the modern rolling stock or high speed trains and about 3.5% for metro systems. Light rail can be steeper than heavy rail and metro systems and is more typically at similar gradients to road alignments. These gradients will determine the length and layout of the approaches to the tunnel. For mixed-use tunnels, the different gradients of the road and rail approaches are generally accommodated within the open approaches and the cut and cover tunnels but come together in the immersed section of the tunnel so that a consistent and uniform tunnel element design is produced. It is usually advantageous to optimize the position of tunnel immersion joints in relation to changes in gradient and areas of curved alignment, to split curvature between tunnel elements, to minimize the impact on the height of the element, and to try to achieve as many uniform element lengths and geometries as possible.

The minimum gradient is normally kept at 0.05%, which is sufficient to maintain longitudinal drainage in the tunnel. A minimum crossfall of 2% is often adopted across the carriageway for drainage purposes. The low point of the tunnel will have a drainage collection sump, which needs to be located away from tunnel element immersion joints. Therefore, the arrangement of elements in relation to the sumps should be considered.

The horizontal alignment should preferably be straight, but this is often not possible and is dictated by the constraints of the approaches to the tunnel. If the alignment is curved, then the cross section must be reviewed to ensure that sufficient allowance for sight lines around the curve has been made. Superelevation of the carriageway may also be required. In practice, it is not much more difficult to build a curved immersed tunnel than a straight one—it just makes setting out the geometry of the tunnel elements a little more complicated.

The other main parameters to be considered for the vertical alignment are the depth of the roof of the tunnel below the bed and the width of the navigation channel. The depth of the tunnel below water level will be largely determined by the navigation authorities. This will be set at a level commensurate with the shipping activities in the waterway, and they may also require a further depth to be included so that future deepening of the navigation channel and development of the waterway are not precluded. The width of the navigation channel will also be specified and this width, combined with the required navigation depth, creates two pinch points on the vertical alignment. For an initial assessment, it can be assumed that the roof of the tunnel is about 1 m thick and the rock protection over the tunnel is 2 m thick for a major waterway and 1 m thick for a minor one. In wider waterways, it may also be necessary to consider the possibility of the location of the navigation channel changing through natural fluvial processes over the lifetime of the tunnel. This may force the alignment to be lower over a longer length of the tunnel.

Dredging can be an expensive and time-consuming part of building an immersed tunnel, especially if it involves contaminated soil, long trips to disposal sites, or dredging periods restricted on environmental grounds. With a circular cross section typical of a steel immersed tunnel, the bottom of the tunnel is further below the carriageway, which requires dredging a deeper trench. Similarly, if the tunnel has to be placed completely below the bed of the waterway, as is generally the case, then the carriageway is further below the bed than for the rectangular cross section. Thus, the bottom of the trench is generally deeper under the bed for the steel cross section than for the rectangular concrete section.

Some dredging cost can be saved if it is possible to leave the tunnel higher than the original bed level of the waterway. It would still be surrounded by fill and protected over the top with a rock protection layer. This is not desirable under the main shipping channel as it would give rise to higher-impact forces on the tunnel should a sunken vessel impact the raised earthworks. However, at the shallow approaches or where there is no significant shipping, it may be possible. The Aktio–Preveza Tunnel in Western Greece followed this approach, as did the Busan Tunnel in South Korea. An important consideration here is whether there are any potential environmental impacts that may preclude this. Some waterways are sensitive to the volume of flow, and port authorities or environmental authorities may not permit long-term blockage of the flow. This was the case for the Øresund Tunnel in Denmark. The impact of current velocities and the potential disturbance to marine species living in the vicinity may have to be considered, as well as the potential for scouring of the bed due to increased velocities, accompanied by deposition of materials elsewhere. Generally, most owners prefer to preserve the original bed level to avoid any problems.

It is possible to accommodate road traffic lane merges and diverges within the tunnel. These are not ideal as they lead to a greater operational safety risk. If they are required, then ideally they should be located in the cut and cover sections and not in the immersed section. For design and construction, the immersed tunnel elements are best kept regular in the cross section. One-off special elements that accommodate diverging and merging traffic have been built successfully, the Fort Point Channel Tunnel in Boston being a good example. This tunnel was in fact two parallel tunnels that carried nine traffic lanes and diverges in tunnel elements that varied between 21 and 47 m in width.

If the tunnel elements are curved horizontally, they are usually built in approximately 20 m long straight bays and the changes in alignment are accommodated by small angular changes at the ends of the bays. If the tunnel is curved both horizontally and vertically, the vertical and horizontal angular deviations are often positioned alternately at the bay joints rather than having both vertical and horizontal angular deviations at the same joint.

There are circumstances when tunnel elements are built with flat planar undersides and cannot be curved to closely follow the vertical curvature of the road or rail alignment. This was the case for the Øresund Tunnel, which was built on skidding beams so that it could be slid within the fabrication facility. In this case, the tunnel element needs to be overheight to accommodate curvature, and the thickness of either the structural slabs or the ballast concrete can be varied to provide the curved alignment.

Some authorities are also now considering the incorporation of road traffic crossovers between carriageways in road tunnels. These would be brought into operation in the event of an incident. Incorporating these into an immersed tunnel section is difficult, although not impossible, but they would require a large opening in the central wall to enable traffic to cross from one carriageway to the other, and would also form a gap in any central escape passage. A special section of tunnel would be required with thickened or stiffened roof slab to deal with the loss of the central supporting wall over a significant length. Similar to any merging lengths, such crossovers should preferably be located at the end of the immersed section in the cut and cover lengths if possible where there is more scope to strengthen the structure to accommodate the central opening.

JUNCTION WITH CUT AND COVER TUNNELS

Having selected the overall alignment, a key decision is where the connections between the immersed elements and the cut and cover sections are located. This is discussed more in Chapter 8, but as a starting point, this can be taken as the point where the tunnel roof meets the low waterline. This ensures that the roof of the immersed tunnel is always below water. A review of the impact of temporary earthworks or cofferdams extending into the waterway is needed to assess if this is acceptable, and a cost assessment of the immersed tunnel construction and the cut and cover tunnel construction is often undertaken to determine the final position of this interface. If the alignment permits it, then the immersed length can be extended landwards to shorten the length of the cut and cover tunnel. This is illustrated in the Thomassentunnel Tunnel shown in Figure 4.7, where the elements were extended in to a temporary water-filled cutting constructed in the riverbank.

Figure 4.7 Extending the immersed tunnel landwards. (Courtesy of Strukton Mergor.)

NUMBER OF TUNNEL ELEMENTS

Having defined the length of the immersed part of the tunnel, the next variable to consider is how many tunnel elements this length is composed of. The element length is again a function of several variables:

- Hydrodynamic loads: As the length of the element increases, the hydrodynamic loads during towing and placing increase. Longer elements are therefore more suited to placing in calmer water, although the cost benefit of longer elements tends to override this in many instances.
- Capacity of the casting facility: The area available may virtually dictate the length and therefore the number of elements. The facility may also place constraints on the width or depth of the elements, particularly if using dry dock facilities.
- Number of placing operations: Each operation is expensive, costing in excess of £1 million, so reducing the number of placing operations reduces the cost as well as the risk associated with placing a

tunnel element. Generally, therefore, the longer the element the better. Reducing the number of placing operations also has program advantages as well as being safer and cheaper.

- Highly curved geometry: It may be desirable to keep the tunnel elements short if there is a high degree of curvature, to avoid highly curved elements that float partially above and partially below water. This is possible to do but is not always preferred. If elements are straight but carry a curved alignment, shorter elements reduce the amount of increased height required to accommodate the curvature.
- Seismic loading: If using monolithic tunnel elements in a highly seismic area, it may be beneficial to have shorter elements to reduce the loading on the elements due to earthquake effects.
- Navigation constraints: If there is a difficult waterway to navigate on the tow route, or locks, or a constrained site that makes maneuvering the element into position difficult, short tunnel elements may be necessary.

Considering all these variables leads to a decision on the element length as well as the length of the immersed part of the tunnel. Generally, it is better to make all the elements the same length as this provides repetition for construction and placing. Exceptions to this equal length do occur. For example, if the ground conditions vary rapidly around the junction of the cut and cover and the immersed tunnel, it may be preferable to introduce one or more shorter elements to cope better with the differential settlement in that area.

GEOTECHNICAL

The geology of the river or seabed is of course one of the main external constraints to the design and construction of an immersed tunnel. At the early planning stages, it is unlikely that the geology will preclude an immersed tunnel solution, unless there is only hard rock beneath the waterway. The main consideration at the planning stage is understanding the conditions sufficiently to determine cost and therefore determine which immersed tunnel solution is the optimum one, and the extent of any foundation works that may need to be considered from an environmental perspective.

In rivers and estuaries, the bed is normally formed from soft fluvial or alluvial deposits that can be readily dredged. An immersed tunnel is a very light structure that usually does not weigh as much as the material it replaces, so they are ideal for use in such soft materials. If there is a rock stratum within the tunnel alignment, this may be able to be ripped out with modern dredging equipment. Alternatively, blasting may be employed although this is often not preferred or even precluded on environmental grounds. This is particularly so in waterways that are important for fisheries.

As important as the founding strata themselves are any rapid changes in strata under the tunnel. This will usually result in increased differential settlement between the tunnel elements and consequent increased shear transfer between the elements. In severe cases, ground treatment or special short tunnel elements may be needed to overcome this. If modest settlements are expected, then the segmental form of construction has benefits, but if large differential settlement is expected, then monolithic elements are more commonly used.

The dredged material has to be disposed of somewhere, and this is an important consideration in the early planning stages. Ideally, as much of the material dredged from the bed as possible will be reused, either directly in the tunnel project as backfilling material or on an associated project that has a requirement for such material. If no such outlet is available, then consideration can be given to using the material as reclamation material for such things as nature reserves, which can make use of such low-grade material. In ports or urban areas, the bed material may be contaminated and require special treatment, which may be very costly.

The most important early activity in relation to all of these issues is making sure enough good data is available to make decisions about what the founding conditions are and what construction activities will be necessary. If ground investigation data is already available for the area, this is a good start, but for most immersed tunnel projects, some data can often be found for the tunnel approaches, whereas there is often little or no data for the marine section. It will be important to instigate ground investigation as soon as possible in the project cycle so that there is adequate data to commence outline design. Planning and costing this work is important, including determining when the investigations are best carried out to manage risk. Typically, a limited investigation is carried out early by the project owners with the intention of gaining enough data to have confidence in the solution, but the detailed investigations are left to the detailed design or design and construct stage of the project, when data gathering can be more targeted.

STRUCTURAL OPTIONS

The appropriate form of tunnel construction will be determined from an options study that considers a number of factors, including its suitability in meeting the employer's requirements, the possible methods of construction, the location and facilities required for construction, and the construction program, construction risk, and cost. At the early planning stage, it may be possible to keep options open and allow the contractor the freedom to choose the form of tunnel. If this is done, then the main factors governing the choice are likely to become material costs and the contractor's preference for the structural form based on the experience of available labor resources.

Often either concrete or steel tunnel solutions can be designed to meet the scheme's requirements, and so from an operational and maintenance viewpoint, there is often no reason to choose one in preference to the other. Steel tunnels require smaller fabrication facilities and are quicker to build. The concrete immersed tunnel form offers more flexibility in the development of the cross section to accommodate the crossing requirements, and thus may quickly be identified as the preferred option for multilane or multimodal crossings. Other factors then come into play to decide the appropriate concrete option to choose.

Local industrial practice is a major contributing factor. Some countries have well-developed steel and concrete industries, and the materials and labor required for either form of construction are readily available. For those countries that do not have both industries, concrete construction would probably be the most appropriate option. Almost every country will have the ability to construct in reinforced concrete, whereas relatively fewer countries have the same capability to construct heavy steelwork. Even if a country has a developed steel industry, such as an established shipbuilding industry, local practice may still come into play. The civil engineering contractors who undertake the tunnel work may not want to take the risk of subcontracting to a shipyard because of fears that they could be held for ransom by the shipyard owners or workers, who may wish to impose their own labor conditions, timescales, and costs. They might prefer to manage that risk themselves by constructing in reinforced concrete.

The choice is further complicated slightly by the fact that the tunnel elements can be transported from one place to another. They could be constructed in one country, say, one that has an established shipbuilding capability, and transported to one that does not. Even this might not be straightforward as the country that is paying for the tunnel might not be prepared to see a major part of the construction cost being spent in a neighboring economic competitor.

The facilities available for construction clearly depend on the local industries. Slipways to build and launch steel tunnel elements will only be available in countries with established shipbuilding or ship repair industries. Although it does not take much land to set up a purpose-built slipway, the construction does need a pool of experienced labor, which would be unlikely to exist in a country without an established heavy steel construction industry.

These local practices may be sufficient on their own to make the decision on the tunnel form, and many of the immersed tunnels built so far tend to reflect local preferences, as shown in Chapter 2. All the European tunnels are reinforced concrete, most American tunnels are steel, Chinese and southeast Asian tunnels are mostly concrete, and Japan uses both forms.

However, it is generally the economics that decide, although this is a somewhat circular argument as the costs will usually reflect local practice and what is available locally.

The physical characteristics of the tunnel site are unlikely to be a factor in choosing the form of tunnel. Both steel and concrete forms of immersed tunnel could be successfully constructed at most sites. However, the buoyant behavior of the element may be significant if the water depth would be restrictive on transporting elements to the site. Concrete tunnel elements have smaller freeboard when floating and require greater water depth. They can be transported partially constructed, for example, with an incomplete roof, and completed at the site, but this raises issues of quality of construction, which could be detrimental to the long-term durability of the structure. The quality issues can be addressed and so this method has been used a number of times, but it comes with an increased risk compared to conventional construction.

The loadings that a tunnel must withstand have to be considered in deciding the appropriate tunnel form. A clear example is the depth of the tunnel, the greater the depth the higher the water pressure it must resist. As these forces increase, the rectangular shape typical of a concrete section becomes less structurally efficient. This leads to rounding or chamfering the corners in concrete sections. This has to be done in such a way that it does not interfere with the required traffic or rail envelope. This clearly favors the rounded shape of the steel shell as it is naturally a more efficient section structurally to resist the hydrostatic pressures. It is not, however, a decisive advantage as reinforced concrete is a very flexible construction material and sections can be designed to withstand the higher pressures. The construction cost of reinforced concrete does increase as the complexity of the cross section increases but is often not sufficient to offset the additional cost of the steel section. For example, the two deepest immersed tunnels built to date, at Istanbul and Busan, which are both 40–50 m deep, are constructed from reinforced concrete.

Watertightness comes into the choice of tunnel structure. The early concrete tunnels were provided with an external steel membrane on the base and walls to make then watertight. There have been advances in the quality of synthetic sprayed on waterproofing systems so that the steel membrane is little used now, except sometimes as a base plate and waterproofing membrane for the base slab. The segmental form of concrete section was developed to avoid the need for an external membrane but does have a weakness in the segment joints. These joints, every 20 or 25 m along the element, have a degree of articulation and therefore rely on rubber waterstops for watertightness. These are generally internal waterstops that are able to be grouted after installation to seal any water paths in the concrete around the waterstop. Although these groutable waterstops do work and can be very effective, they rely on a high standard of workmanship that is not always achieved, even in highly developed countries. For this reason, some clients and some contractors would prefer a monolithic rather than a segmental concrete section. No one can guarantee that one form is more watertight than another; it comes down to individual preferences and experience.

The steel shell in the steel form of construction inherently provides a watertight barrier and no other measures are necessary to make the tunnel watertight, except to provide corrosion protection to any external steel membrane so that it remains in place for the lifetime of the tunnel.

Seismic resistance is often important in the tunnel design. In general, however, the elements themselves are robust enough to withstand the effects of an earthquake and the main issue for design is to ensure that the joints do not spring apart as the seismic waves pass the tunnel and the surrounding ground. This would apply to whatever form of tunnel is chosen, although again the concrete segmental form has more joints to consider, and in seismically active areas, it is more common to use monolithic elements. Despite this, there is a subjective view that the steel shell form has an inherently better seismic performance than the concrete form. If an exceptional event occurs, the additional ductility of the steel shell is likely to enable it to remain intact. This perception is exactly that, a perception and not a reality, and it is possible to design concrete tunnels to perform perfectly satisfactorily in highly active seismic regions. The concrete tunnel at Kobe was only partly constructed when the 1997 earthquake struck and the tunnel was undamaged. The concrete tunnel at Preveza in Greece has performed well despite being subject to several earthquakes since it was built, and a concrete tunnel has been built under the Bosphorus in Istanbul. This is only 25 km away from the North Anatolian Fault with potentially very high seismic forces.

ELEMENT CONSTRUCTION

Where the tunnel elements are to be constructed is an important factor when planning an immersed tunnel. There are many options:

- A purpose built casting basin either adjacent to or within the tunnel site or elsewhere remote from the site
- An existing dry dock or shipbuilding facility
- A floating dry dock

If suitable shipbuilding facilities exist, it does not necessarily follow that the steel form will be adopted. The dry docks used for shipbuilding and maintenance can make suitable places to build concrete tunnel elements. Size, depth, and availability are the key issues. Ship dry docks are often long and narrow and unsuited to the shape of immersed tunnel elements. Some, however, have very large plan areas, which can accommodate the construction of several tunnel elements simultaneously. The width of the dock gate and the water level over the sill have to be suitable to float out the tunnel elements. Even if all these parameters are suitable for tunnel construction, the dock may not be available to meet the construction program. They are often booked up well in advance, which does not suit the stop/start nature of public infrastructure projects where the start dates continually change, making advance leasing of a dry dock an unrealistic prospect. Even if the project start date is fixed, it is unlikely that a thriving dry dock would give up its seasonal and repeat business to be tied up for 6 months to a year for tunnel construction. So although at first sight a dry dock would appear suitable, for one reason or another it very rarely comes about. Disused docks are more frequently used and are a more realistic option.

Once constructed and floating, the elements can then be towed to the site. These tows can be over considerable distances without a problem. The constraints to towing are the sea state and the wave regime that the element has to be designed to withstand during the tow. Insurance companies may occasionally require additional safeguards such as double bulkheads and having pumping equipment aboard for long or exposed sea tows. However, it is normally always possible to find somewhere to build the elements within an economic distance from the site, so long sea tows are the exception. In inland waterways, there may be navigational constraints, and the available width and depth of the waterway for the tow route should always be checked. Extensive dredging of the tow route is likely to be cost-prohibitive and difficult to obtain permissions and permits for.

It is always advisable to make preliminary investigations to identify potential facilities for construction of the elements, either identifying disused docks, existing shipbuilding facilities, or identifying parcels of land that may be used for construction of casting basins. This may be adjacent to the tunnel site or some distance away. The final choice can usually be left to the contractor, but it is helpful if some options have already been identified. In the Netherlands, existing casting basins have been reserved by the promoter at an early stage of planning to ensure the facilities are available for construction of tunnels. The Barendrecht basin has been used repeatedly in this way and has been made available to the contractors.

It is worth noting that environmental legislation is continually evolving, and for some projects, the location where the elements will be built has been determined and fixed during the planning stage to satisfy the environmental authorities of the impacts of construction. This has been done for the Fehmarnbelt Tunnel between Denmark and Germany, for which EU legislation for the environmental approval of large projects required the construction sites to be identified.

ENVIRONMENTAL CONSIDERATIONS

As immersed tunnels cross waterways, there are numerous environmental considerations. First is the waterway itself. River ecology is often fragile and valuable to the community, for example, if the river is a valuable fishing resource. Early discussion is needed with interested parties to identify the issues and reach an agreement on mitigating any effects of constructing the tunnel. This is often possible despite apparent implacable opposition at the outset. Initial resistance can often be overcome by careful explanation of how the tunnel is to be built and taking measures to reduce the temporary inconvenience. Once people know what is proposed, they often find their objections are not as severe as they first thought. For example, dredging and marine operations can be halted at times of the year when the fish migration runs occur. Disposal of dredged material at sea has been halted during whale migrations. In many locations, considerable care must be taken with the marine operations not to deposit fine sediment in the river, as this would interfere with all sorts of marine life. It is particularly important if there are commercial shellfish beds nearby. Compensation measures can be introduced to provide a long-term benefit to offset the short-term inconvenience.

Generally, it is possible to resolve the position with fisheries and reach a solution acceptable to all, and this is not necessarily just financial compensation. The Conwy Tunnel in the United Kingdom, for example, crosses an important salmon river. To compensate for the interference with the salmon runs that was bound to occur during construction, a fish bypass was constructed upstream to allow the fish to get past a waterfall that had previously been impossible for them. This opened up 300 km of additional spawning tributaries for the salmon and so provided a long-term benefit to the river to compensate for the short-term interference during construction. This type of approach to overcome objections may be applicable to other tunnel schemes.

An important early planning task is to identify all of the potential impacts and develop initial mitigation strategies to determine whether the scheme is viable. As with any construction project with sensitive environmental issues, early consultation is key to solving any perceived problems.

Constructing the tunnel will interfere with the current patterns in the waterway. Temporary cofferdams or embankments extending into the waterway from the shore, and the trench itself in the bed will alter the flow patterns. The effect is that the regime of sediment scour and deposition is altered, both in the short term during construction, and possibly permanently, if, for example, the final tunnel cover level is above the original bed level or there is permanent reclamation at the shore. These flow changes in turn could have adverse effects on users of the waterway by silting up channels or moorings or altering the pattern of sandbanks. Equally, the current velocities may be altered in a way that is unacceptable to the navigation authorities and agreement will have to be reached with them over what increases can be tolerated. Hydrographic studies are needed to determine what these changes will be so that the effects may be predicted and mitigation put in place. This may result in amendments to the shape or location of temporary cofferdams or other restrictions that have to be placed on construction. This might include such things as moving moorings temporarily or undertaking maintenance dredging on a quay if the siltation rates are increased. Such studies should be carried out in the early planning phase of a project to determine any constraints for the detailed development that follows.

During the dredging operation for the trench, it is almost inevitable that there will be some spillage that will increase the amount of sediment in the water. It would be unacceptable for this to be carried out without strict limits being imposed and the operation carefully monitored. In very sensitive locations, shrouded grabs can limit the amount of material spilled into the water and silt curtains can be deployed to restrict any sediment to the local area around the dredger and stop it being distributed more widely. The required dredging method is an important consideration for the cost of the project and so this should be assessed early.

Often in busy waterways, the bed material can be contaminated. This is especially so if the river or harbor has been used industrially for a long period when heavy metals and other contaminants may be present in the bed material. This can require the material to be removed so that none is spilled and it may have to be taken to and deposited in special dumping grounds, often under supervision to make sure that none of the contaminated material is lost overboard during transit. Again this can have a big impact on the cost of a project and thus warrants early study and investigation to determine the strategy for handling and disposing of any contaminated materials. Sampling and testing of the bed material should be carried out as early as possible to manage the project risks in this respect.

The construction of a casting basin at the tunnel site is a major activity. The banks of rivers often have special environmental status that would prohibit such a construction. Considerable discussion is needed with the authorities to obtain consent. If this is not possible, then other solutions are available. The tunnel elements can be built elsewhere and towed to the site. They can also be built in the space occupied by the approaches to the tunnel, as shown for the Medway Tunnel in Figure 4.8. If any constraints are to be imposed, they need to be determined at an early stage in the scheme development.

Figure 4.8 Constructing tunnel elements in the approaches. (Courtesy of Kent County Council/BAM Nuttall/Carillion/Philip Lane.)

DISRUPTION TO NAVIGATION

The impact on the waterway navigation must be considered during early planning, particularly the operation of dredgers and the associated barge movements for spoil disposal, and the potential consequences of a sequence of closures of the waterway to immerse the tunnel elements. Consultation with the harbor/port authority or coastguard, as appropriate, should be made early to establish constraints and safeguards that may be necessary to maintain safe navigation in the vicinity of the tunnel crossing during construction. In major shipping lanes, this may involve carrying out simulations of ship navigation through the worksite with narrowed or diverted navigation channels.

Agreement may also be needed for establishing temporary mooring and berthing facilities and for controlling movements of mutlipurpose vessels or work barges and the small craft required to service the larger marine plant, to undertake surveys and support diver operations.

Generally, it is possible to overcome any initial objections or hesitancy on the part of the responsible authorities by careful planning and implementation of procedures to obtain approvals in good time and to coordinate the movements of vessels in the waterway with the relevant authority and to provide notices to shipping and affected commercial enterprises in advance of the works. This is demonstrated by successfully building immersed tunnels across some of the world's busiest harbors, such as Hong Kong and Rotterdam.

COST

Developing a reliable cost estimate is a fundamental aspect of planning an immersed tunnel or indeed any infrastructure scheme. Costs are particular to each country and region but the principles are the same wherever the tunnel is planned. The estimate has to be appropriate to the stage of development of the scheme, and the degree of confidence will increase as the scheme develops, details become clearer, and risks are mitigated. At the initial concept stage, the cost estimate will be a high-level one while there are many options in play, and this will be refined and go into greater detail as the preferred option is identified.

In the initial stages, high-level overall cost estimates can be used to screen the initial options. Broad cost rates are used based on the cost per meter run of tunnel, cost per meter run of approaches, bulk earthworks quantities, and so on. These will give an early indication of the cost and enable comparison against other options, such as different crossing alignments or other forms of crossing, such as a bridge. It will also identify the most critical cost items that need to be identified in greater detail to improve the level of confidence in subsequent estimates.

A cost comparison of the various immersed tunnel types, cross sections, and alignments should always be undertaken. This can be limited to only the major items that differ between the various options. Any aspect that is common to all options need not be considered at this stage. The main elements in an initial high-level cost comparison would be

- Fabrication facility
- Construction of structure
- Transport to site
- Dredging and backfill

Using this simple approach, the preferred option can often be picked out fairly readily in cost terms. A similar exercise will be necessary to consider the environmental implications of each option as some proposals may have fatal environmental flaws. For example, the construction of a large fabrication facility adjacent to the tunnel site may be unacceptable either in an urban situation where the activities would be too intrusive or in a countryside location where such activity would be out of place.

In costing an immersed tunnel scheme, it is necessary to consider both the permanent works and the main temporary works. These are typically

- Fabrication facility for building the elements
- Cost of building the elements
- Dredging the trench and dealing with the resulting material
- Transporting and placing the elements
- Placing the element foundations
- Backfilling the trench around the tunnel
- Ballast exchange and placing internal ballast concrete
- Internal finishings (M&E, cosmetic)
- Cut and cover approach tunnels
- Open approach ramp structures
- Service and ventilation buildings
- Preliminaries and contingency

In developing the initial estimate, the effect of optimism bias needs to be considered. This is a factor increasingly being used on major infrastructure projects to allow for the fact that project outturn costs have historically been greater than the original project estimates, but this is different from contingency, which can be modeled mathematically according to the level of risk or uncertainty that is assigned. There is a demonstrable body of evidence to show that optimism bias exists and in some countries government advice is available on how project sponsors should make explicit, empirically based adjustments to their estimates. This applies, incidentally, not only to a project's costs, but also to the estimate of the construction period and to the estimate of the project's benefits.

For example, U.K. government guidance is that for bridge and tunnel links the median uplift on the estimated capital expenditure could be as high as 55% if the promoters want to be sure that the risk of cost overrun is no more than a 20% probability. If the promoter could accept a 50% probability of the outturn cost exceeding the estimate, then the uplift to account for optimism bias would drop to about 23%. Such figures are constantly under review as the database of outturn scheme costs gets larger over time, so the estimator should seek the latest advice before applying such figures. Optimism bias has resulted from the planning and procedural processes involved in major infrastructure schemes and the strategies adopted by the various parties involved in those processes. Only a few of the parties involved had a direct interest in avoiding optimism bias and it was often in their interests not to avoid it so that costs would be underestimated or the benefits overstated and the progress of the scheme would not be halted.

Over time, however, with the introduction of optimism bias factors, greater emphasis on realistic initial estimates with the corresponding weeding out of overoptimistic initial budgets, and the use of formal cost and risk assessments, optimism bias should be squeezed out of the system.

At every stage in the cost-estimating process, there will be assumptions and uncertainties in the figures. These will reduce as the design develops and more of the cost critical items are studied and uncertainties removed. There will be a level of uncertainty assigned to each component of the cost estimate. These will arise from

- Design risk associated with the complexity of the final design and oversimplification during the preliminary design overlooking or underestimating some effect. Either an overall contingency can be applied or percentages added to the calculated quantities.
- Uncertainty over the unit cost rates used. The local market might react, either positively or negatively, to the scale of a tunnel project, and what other construction is being undertaken in the area at the time.
- Encountering something unforeseen during construction, be it ground conditions or archaeological finds. The likelihood of this depends on and is proportional to the amount of investigation carried out; the more that is known about the ground the less likely it is that something unforeseen will be encountered.

There will, however, be some uncertainty remaining, and a probabilistic analysis such as using a Monte Carlo analysis or something similar can be used to see what the likely variation from the base case will be. Taking a worst case for all these risks will give an unrealistic outcome. With a risk assigned to each component, a mathematical approach can be taken to determine the most likely overall outcome.

The initial estimate can be made from cost comparisons from previous immersed tunnel schemes. These are not always easy to obtain, but published figures are available as a starting point. A more accurate preliminary cost estimate can be obtained from considering the basic elements of construction. From a preliminary estimate of the cross section established from buoyancy principles, the volume of concrete is known and the reinforcement quantity can be estimated. For a segmental concrete element this will typically be between $100-130$ kg/m^3 and for a monolithic element, between $140-180$ kg/m^3. The initial reinforcement quantity can therefore be based on 110 kg/m^3 and 150 kg/m^3, respectively. Detailed design may well be able to reduce this quantity but it is suitable for a first estimate. It will also be a suitable figure for the reinforcement quantity to use for the composite internal lining of a steel shell.

The temporary works involved in placing the elements depend on the equipment favored by the contractor, but for a concrete tunnel with an element length of about 120 m, the placing operation will cost approximately £1 million for each element in an inland or estuarial location, and up to 1.5–2 times this amount in an offshore location. This includes the temporary bulkheads, internal water ballast tanks, as well the provision of the marine plant and the sinking operation itself. The costs of any long tow to the site from the fabrication facility must also be included, although these are relatively small. The hire of four or five tugs for a few days for each element is not expensive in the overall scheme.

The other main cost is that associated with the dredging. The quantity of material to be removed depends on the depth of the tunnel, the alignment, and, importantly, the side slopes of the dredged trench. In very soft material, these could have to be very flat, say 1 in 8, but deeper down as the strength of the in situ material improves, slopes of 1 in 2 to 1 in 4 should be achievable. In weak rock or limestone, steeper slopes of 2 in 1 are achievable. An overall plan for the dredging and backfilling is required. As much of the material as possible should be reused in the scheme as backfill around the tunnel. It is not economical, or environmentally desirable, to dispose of all the material dredged from the trench and then source new material for the backfilling operation. Settling lagoons, temporary stockpiles, and the proportions of dredged material suitable for reuse come into consideration. Thus, for a detailed estimate, an overall materials strategy has to be developed. For the initial estimate, however, bulk estimates of the quantities to be removed and replaced will suffice. The current bulk dredging and backfilling rates for the area in question can then be applied to give the earthworks' cost.

Note that disposal costs for contaminated materials vary widely depending on the project location and applicable environmental legislation. However, in any location, the costs will be considerably greater than the costs for normal material arising from the dredging operations. Care is needed not to underestimate this element of the works.

Simple basic equations can be set up to evaluate the dredging quantities for a particular cross section and depth. These will enable the quantities to be established and compared for different widths and depths of tunnel. Increases in width have significantly less impact on the dredging than increases in depth. Thus, it is generally more desirable to increase the width rather than the height of an immersed tunnel element. For example, a 20% increase in tunnel width can increase the dredged quantity by 10%, whereas a 20% increase in the height of the element will increase the overall dredged quantity by 25–30% for the same element. The cost of the approaches can be estimated from generic rates for cut and cover tunnels and open approach ramps.

Having derived an initial estimate for the overall scheme, the comparative costs of various options for the preferred tunnel route can be evaluated from a comparison of the main items. For example, to compare four tunnel options such as rigid or segmental concrete, composite, and steel, a simple comparative table (Table 4.1) can be drawn up outlining the main quantities.

Table 4.1 Main quantities items for high-level costing

Item
Structural concrete (m³)
Ballast concrete (m³)
Structural steelwork (t)
Reinforcement (t)
Temporary prestress (t)
Groutable waterstops (m)
External membrane (m²)
Formwork (m²)
Dredging (m³)
Backfill (m³)
Casting basin
Outfitting site
Immersion operations (no.)

Applying typical rates to each of these quantities will provide a cost comparison that includes the type of structure, the earthworks, and the major items of temporary works. This will rank the options. This does not give the total cost of each tunnel option, but it identifies the main differences. The cost of the other items, such as M&E systems or the internal finishings, will be similar and will not affect the comparison.

When developing detailed cost estimates in the later stages of a project, it is important not to rely on published all-in rates that are applicable to general construction. An immersed tunnel has a relatively high element of temporary works and method-related cost. Either the temporary works and method-related costs need to be identified separately, or the rates need to be built up from first principles in order to ensure they are correct.

This is not the end of the decision, because the risks and opportunities associated with each option also have to be considered. Does one option offer more or less opportunity for cost reductions or program improvements? For example, the steel options. Even if the costs are similar to the concrete options, they are sensitive to the cost of steel fabrication and to the number of shipyards capable of building the tunnel elements. These aspects can be positive or negative for any option but are important to the overall choice.

One interesting example of cost comparisons between steel and concrete is the Conwy Tunnel in the United Kingdom, which was planned in the early 1980s. This was the first transportation immersed tunnel in the United Kingdom, and comparisons between concrete and steel shell tunnels were made at the preliminary design stage. This identified that a composite steel concrete tunnel element was approximately 18% more expensive than the rigid concrete element. As it was the first such tunnel in the United Kingdom, the client wanted to give all parts of the construction industry equal opportunity, so two detailed designs were developed: one steel/concrete composite and one rigid reinforced concrete. It was left to the market during the tender stage to decide the most economical solution. In this case, no contractor priced the composite option, so although the concrete element proved the most economical, the market difference was not established with any certainty but was thought to be about the original 20%. A similar comparison by the Dutch in the 1980s showed that for a typical Dutch road tunnel that is about 1 km long, the double steel shell solution was about 40% more expensive per meter run than the rectangular concrete section. Added to this are more costly approaches as the carriageway in the double shell design is slightly deeper making the approaches deeper and longer. These two examples are only European experiences and each tunnel must be evaluated on its merits, taking account of the circumstances prevailing at the time. However, there are some generalities that can be concluded from experience.

Typically in a modern road tunnel, the cost of the mechanical and electrical installations has accounted for around 10% of the construction cost of the tunnel. The figure is increasing to around 15% as safety standards improve and systems such as sprinklers are installed. This is not a fixed percentage as it varies with the length of the tunnel and the standard of the installations, but for a typical 1- to 2-km-long tunnel, the figure should be about right. It should be remembered that rule of thumb estimates are not always reliable as circumstances vary so much between particular schemes and different countries.

OPERATION AND MAINTENANCE COSTS

As well as developing a construction cost estimate, the whole life costing of the tunnel must be evaluated. This should be done at the planning stage for budgeting purposes as a project owner may have separate budgets for the new build and the ongoing operation and maintenance. While the ongoing operation costs may not be of so much interest to a design and build contractor, they will certainly be of interest to the project owner and the tunnel operator, and if the tunnel is to be built as part of a concession, then the concessionaire will need to understand these costs in detail to undertake the necessary financial modeling when bidding for the concession.

To develop a whole life cost estimate, the full range of activities for the operation and maintenance of the tunnel must be considered for the mechanical and electrical equipment installed. Energy consumption, routine maintenance, and replacement of life expired equipment have to be considered and there is often a trade-off between the initial capital cost of equipment and the subsequent maintenance or replacement. The better and more expensive the equipment initially, the longer it should last and the less maintenance it should require. Costs for other operational activities such as manning toll booths and dealing with breakdown recovery may also need to be included.

The cost of operating and maintaining the tunnel has to be considered at the outset when deciding between tunnel and bridge options. Tunnels cost more to operate than bridges because of the energy used in lighting and ventilation. Once built the tunnel structure requires very little maintenance and is not subject to the same degradation risks as, say, the cables of suspension bridges. However, the initial cost of the M&E equipment and the building required to house the control equipment need to be factored in as well as the annual operation and maintenance costs. Tunnels are also equipped with comprehensive monitoring and control systems that again cost money to operate. These aspects are discussed in more detail in Chapter 7.

CONSTRUCTION PROGRAM

The construction program needs to be considered during the planning stage as it may affect the choice of crossing option. Construction of an immersed tunnel often takes a relatively long time because of the large number of linear processes that have to take place, but this is equally true for bored tunnels and bridge crossings. The main activities that occur in every immersed tunnel scheme are

- Preparation of the facility to construct the tunnel elements
- Construction of the tunnel elements
- Dredging of the tunnel trench
- Construction of the approach structures
- Placing and founding the tunnel elements
- Internal finishing works

Several of these activities are sequential, for example, the approach structures do not have to be finished before the first tunnel element is placed, but at least one of the approaches should be sufficiently complete to enable the first tunnel element to be placed against it. Similarly, the fabrication facility has to be ready before the tunnel elements can be constructed although in the case of a large open earthworks' casting basin, the construction of the elements can be started before the whole basin has been completed. It only needs a sufficiently large area to be excavated to level and dewatered, and then construction can proceed while the remainder of the excavation is completed. This approach has been followed on several concrete tunnels to save time on the overall program.

In simplistic terms, a typical immersed tunnel construction program is illustrated in Figure 4.9. This general program holds true whether a concrete or steel tunnel is being built. The main difference between the two is that with a steel tunnel, the fabrication facility is often either already in existence or can be set up more quickly. The elements are therefore ready

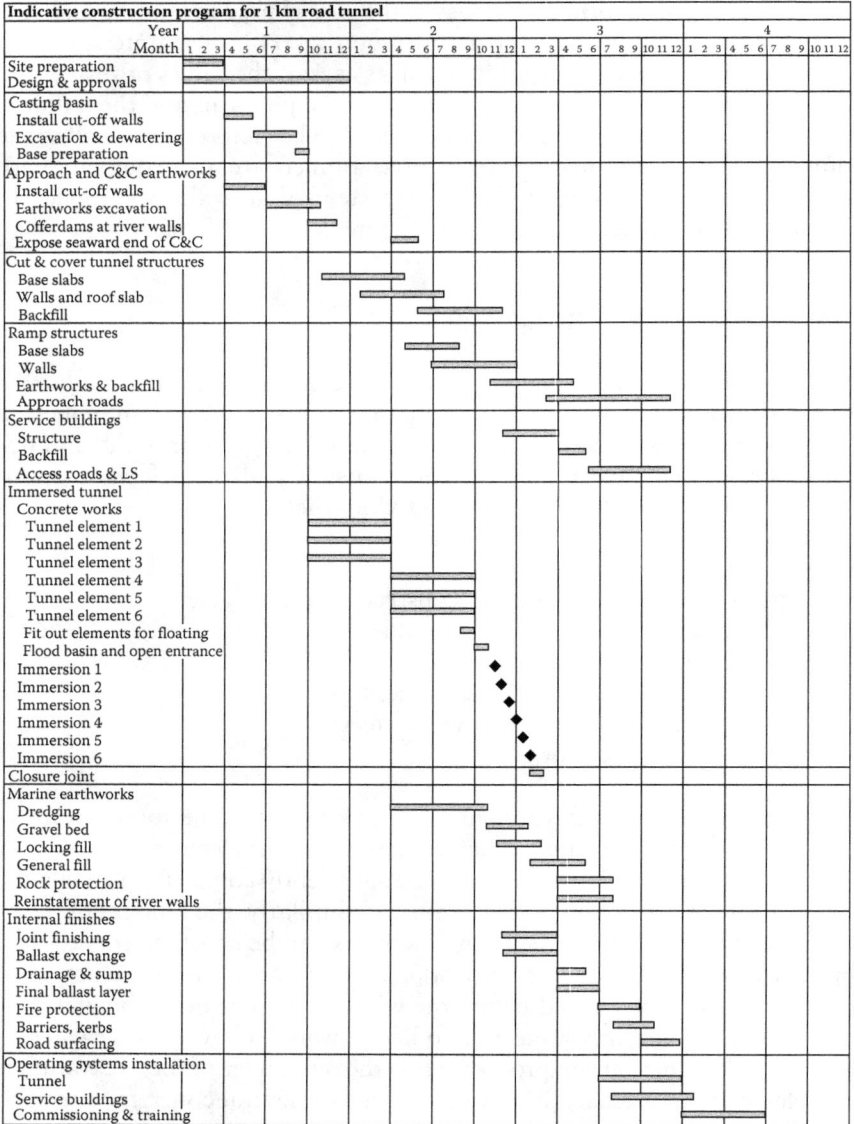

Indicative construction program for 1 km road tunnel				
Year	1	2	3	4
Month	1 2 3 4 5 6 7 8 9 10 11 12	1 2 3 4 5 6 7 8 9 10 11 12	1 2 3 4 5 6 7 8 9 10 11 12	1 2 3 4 5 6 7 8 9 10 11 12
Site preparation				
Design & approvals				
Casting basin				
Install cut-off walls				
Excavation & dewatering				
Base preparation				
Approach and C&C earthworks				
Install cut-off walls				
Earthworks excavation				
Cofferdams at river walls				
Expose seaward end of C&C				
Cut & cover tunnel structures				
Base slabs				
Walls and roof slab				
Backfill				
Ramp structures				
Base slabs				
Walls				
Earthworks & backfill				
Approach roads				
Service buildings				
Structure				
Backfill				
Access roads & LS				
Immersed tunnel				
Concrete works				
Tunnel element 1				
Tunnel element 2				
Tunnel element 3				
Tunnel element 4				
Tunnel element 5				
Tunnel element 6				
Fit out elements for floating				
Flood basin and open entrance				
Immersion 1				
Immersion 2				
Immersion 3				
Immersion 4				
Immersion 5				
Immersion 6				
Closure joint				
Marine earthworks				
Dredging				
Gravel bed				
Locking fill				
General fill				
Rock protection				
Reinstatement of river walls				
Internal finishes				
Joint finishing				
Ballast exchange				
Drainage & sump				
Final ballast layer				
Fire protection				
Barriers, kerbs				
Road surfacing				
Operating systems installation				
Tunnel				
Service buildings				
Commissioning & training				

Figure 4.9 Typical construction program.

earlier and, provided the approach structures can be completed in time, the overall construction period will be slightly less. The steel elements will also be available one at a time and placing can start as soon as one element has been completed rather than having to wait for the completion of a batch of six or eight elements, which is often the case for concrete tunnels. This time advantage for steel tunnels could be helpful in privately financed tunnels where it is important that the toll revenue stream comes on line as quickly as possible. However, this would have to outweigh the likely additional cost of the steel solution.

For a straightforward concrete immersed tunnel up to 2 km in length, the typical construction period is about 3.5–4 years. The preparation of a fabrication, or casting, facility can take up to 9 months before the elements can start to be constructed. Clearly there are program advantages if a suitable casting facility already exists. In the Netherlands, where concrete tunnels have always been used, the authorities did not backfill the casting basin after the construction of the Heinenoordtunnel, but left it mothballed so that it could be used subsequently for the construction of many other tunnels in the course of the next 10 years. This not only provides considerable savings in cost and time for a particular tunnel but also reduces the environmental impact. Similar benefits to program can be obtained if a dry dock facility can be found that is large enough to accommodate element construction, either in batches or singly.

Because it can be difficult to find a suitable casting basin site, it is often worthwhile for the promoter to consider this aspect during the planning stage and make arrangements for a suitable site to be available. Although it can be argued that the contractor should be left to find his own site and obtain the necessary planning approvals, the timescale of the contract may make this difficult, and if this is important to the promoter, he can secure his own interest by making advance arrangements. The contractor then has the option to use the promoter's site or find a better one if he can. In this respect, a contractor can sometimes strike a better commercial arrangement for a site than the promoter who is likely to be a central or local government organization. One caveat to this is the environmental assessment of the basin. The contractor may find a site but be unable to comply with the environmental legislation in time and so will have to use the promoter's site.

The period for constructing the elements clearly depends on the number to be built but typically for six to eight elements would take about a year. The first element is then ready to be placed, so by this time, the trench and any access channels have to be dredged and the approach structure ready to receive the first element. Once the elements have been built, placing them can proceed fairly rapidly. In a tidal estuary, it was common to place them at 14-day intervals to match the tidal cycle so that for eight elements the placing operation would be completed in 4 months. More recently, with improving placing equipment and techniques, the immersion cycle can be much quicker, governed largely by the turnaround time of the equipment.

Once the civil work, such as removing bulkheads and placing the permanent concrete ballast, has been done, the M&E systems have to be installed. Because, at least at first, access is only available from one end of the tunnel, these operations have to be carried out in a linear fashion, requiring careful planning and liaison between the various operations. For this reason, the finishing works do take some time to complete, often taking a year or more after the tunnel elements have been placed. Adding these together gives a total construction period of 3.5–4 years.

The construction period is sensitive to how the tunnel elements are built, at least for concrete tunnels. If they are built in a separate casting basin, then the basic sequence above applies. Sometimes, however, space at the site precludes construction of a separate casting basin and one of the tunnel approaches is used as a basin for building the elements. This procedure was used for the Medway Tunnel in the United Kingdom and the Limerick Tunnel in Ireland. It has a program disadvantage in that one of the approach structures cannot be started until after the elements have been built and floated out of the approaches. It still takes the same time to prepare the approaches for element construction as it would to build a separate casting basin, so there is a program penalty to pay.

The placing sequence of the elements is important in the construction planning process although it does not affect the overall construction time significantly. Which element is to be placed first, and where is the closure joint to be located? It is common to start adjacent to one of the approach structures and proceed to the other approach structure. The closure joint is then located between the last element and the approach structure. However, sometimes, the water depth is insufficient to float the last element into position over the previous one. In such a case, placing would proceed from both approach structures and the closure joint would be somewhere in the middle. It would not usually be right in the middle, as that is likely to be the deepest and busiest part of the waterway, but somewhere toward one side. It is just as important to note that elements are not always placed sequentially from one end to the other. Indeed if there is good reason, the first element placed does not have to be against an approach structure, although doing so would involve an additional closure joint, and there would be programming implications for the internal finishing work as access is not available from one of the approaches.

The construction sequence will have to be discussed and reviewed with the harbor authority responsible for the waterway to obtain their approval and agreement. The environmental authorities may also need to be consulted as they may have requirements that restrict marine activities to certain times of the year. This would be the case if, for example, there were important fish migration periods that required construction to be suspended for a month or two.

If the tunnel elements have to be towed a significant distance across open water to reach the construction site, then the marine climate might affect the construction program. The tows could only take place during suitable weather windows and the whole construction program might have to be set up to accommodate these by including some contingency time. Alternatively, the elements could be designed to withstand greater wave heights, which would increase the periods available for sea tow.

Risks and opportunities

A delay to the project can be caused by something that happens during the construction itself or a delay during the procedural processes beforehand. In the planning stages of the project, it is important to undertake a thorough study of the required and possible construction techniques to identify the risks associated with the various construction options. These may be geotechnical, for example, if the ground is unsuitable for construction of a casting basin, or environmental, such as if large-scale construction of tunnel elements would be inappropriate for a particular location. Consideration, and possible mitigation, of these risks by the promoter of the tunnel will mean that when the contractor is planning his construction he can be reasonably confident that no unexpected difficulty or challenge will arise to delay his construction program. The client does not have to be prescriptive on where or how the tunnel elements are constructed; that is best left to the contractor who will actually build them. However, he does have to set clear parameters, such as the land available for construction within which the contractor will work.

On larger tunnels with more elements, consideration has to be given to the use of multiple casting facilities. These can be used to build batches of six or eight elements at a time. It may well be easier to find two locations for smaller casting basins than to find a suitable site for one large one. Dividing the element construction in this way is also a useful way of reducing the risk in seismically active areas. Damage to a casting basin would only affect the limited number of elements under construction rather than all of them.

The placing sequence is another area where it may be possible to save on construction time. For most tunnels, it is sufficient to place the elements in sequence, starting from one approach and working toward the other. However, on longer tunnels, consideration must be given to placing the elements on two or more fronts. This will attract cost penalties as more marine plant will be needed for placing the elements and the design and locations of the closure joints will be more difficult, but overall, it may be the only way to construct the tunnel within a realistic timescale.

Facilities such as those used for the Øresund Tunnel, described in Chapter 14, also give significant program advantages though they are only cost-effective for the larger tunnels. Significant investment was made in casting facility to establish factory-type construction processes. The factory also permitted continuous working with no interruptions due to adverse weather, a real benefit in Scandinavia, though not so relevant in warmer climates.

All these criteria have to be considered during the development stage of a tunnel scheme. There is no one-size-fits-all solution for an immersed tunnel. An options study is essential to identify the preferred solution although, as has been mentioned, there are certain preferences in certain parts of the world that would enable the preferred solution to be quickly identified.

Chapter 5

Environmental impact

Immersed tunnel proposals are sometimes erroneously dismissed or disregarded in favor of other forms of waterway crossing because the construction techniques and methods to mitigate environmental impact are less familiar or not fully understood. In this chapter, the various environmental impacts that can arise are discussed along with the methods that are usually adopted to mitigate against them. An immersed tunnel is intrinsically no worse or better than any other form of crossing. The issues will be different compared with bridges and bored tunnels and, depending on the specific site and project circumstances, one type of crossing will be favored over another. The key to making the decision as to which is right is dependent on fully understanding the issues for each so that a fair and balanced evaluation of the options can be made.

Environmental legislation is becoming stricter globally as awareness of potential impacts increases. At the same time, methods to manage impacts are becoming more sophisticated, and it is essential to understand the issues and how they can best be dealt with, whether as a planner, designer, constructor, or project sponsor. Having a correct understanding of environmental concerns can open up a greater number of options, rather than close them down. In addition, a greater sense of responsibility is entering the industry with regard to developing in a sustainable manner. The large volumes of materials associated with an immersed tunnel project mean that procurement of materials in a sustainable manner and dealing with waste materials in a responsible way, ideally through the reuse of as much material as possible, are becoming strong drivers in a scheme's development.

The production of an environmental impact assessment (EIA) is a standard requirement in most countries around the world and is used as a tool for ensuring impacts are mitigated as far as possible and that legislation and the requirements of environmental authorities are being adhered to. The objective of any EIA should be to describe the environmental baseline, to fully evaluate the environmental impacts of the scheme, and to explore the potential for mitigation. The output from an EIA should then form project requirements and impose the necessary constraints and conditions for design and construction of the final scheme.

SHORT-TERM VERSUS LONG-TERM IMPACTS

An immersed tunnel, like any other major infrastructure project, will have a considerable environmental impact. There will be a mixture of short-term impacts during construction and long-term impacts during the operation and maintenance of the tunnel. There will be environmental impacts on the waterway, the banks, and the surrounding area through which the approaches will run. All these impacts must be considered, proposals must be developed sensitively, and appropriate mitigation must be taken if the scheme is to gain approval and proceed. People will rightly be concerned about the impact a tunnel and its construction may have on their interests. There will be many discussions and negotiations before a suitable scheme can be agreed upon. Often, the issue is short-term pain for long-term gain. Tunnel construction will inevitably cause disturbance to the surroundings, but if these can be minimized, then the long-term benefits of the tunnel in terms of improved access, reduced congestion, or economic benefit can outweigh them and the scheme becomes acceptable. Immersed tunnel construction can have a greater impact in the short term compared with a bridge, but would generally have less impact in the long term.

TUNNEL APPROACHES

The major impact of an immersed tunnel scheme generally occurs during construction when disturbance of the waterway and the banks is inevitable. Often, the banks are environmentally sensitive, are recreational areas or protected wildlife areas, or are areas of outstanding landscape importance. The tunnel approaches will pass through these, and although the tunnel itself is finally buried, the approaches form a permanent feature of the banks. Keeping the tunnel alignment away from such important sites must be the first aim, but this is often not possible and the case for the tunnel has to be made. This can also be an obstacle for urban tunnels where the waterway banks are built up and the main issue is finding an acceptable clear corridor to approach the river.

In an estuary or tidal river location, constructing the approaches will almost certainly raise environmental issues relating to the intertidal mudflats. They will need excavating with consequent temporary loss of feeding grounds for wading birds or loss of shellfish beds. The construction activity itself may disturb the local bird population and may require, for example, a cessation

Figure 5.1 Construction of Jack Lynch Tunnel. (Photo courtesy of Cork City Council/Robert Bateman.)

of activity during the breeding season. Figure 5.1 shows the construction of the Jack Lynch Tunnel across the River Lee in Ireland, which was in a typical estuarine environment, and illustrates the extent of disturbance to intertidal areas that can occur and that need to be dealt with in a sensitive manner.

As shown in Figure 5.1, the tunnel elements are being constructed in an open dewatered casting basin temporarily reclaimed from the river on the line of the tunnel. The outline of the dredged trench can also be seen. The scale of the construction operation and its local impact on the river bank can be seen from the illustration. However, the disturbance was only temporary and in the final scheme the approaches were incorporated into a realigned bank. On the Conwy Tunnel, a similar casting basin adjacent to the line of the tunnel was left open to the river and developed as a marina. This illustrates how long-term reestablishment can be used to offset temporary disturbance during construction.

Another mitigating environmental measure used at Conwy was that the unsuitable material dredged from the trench was pumped some 1–2 km upstream and deposited on the shoreline to produce a nature reserve and bird sanctuary. The material had to be drained and geotextiles were used to make the pathways safe for public use, but the network of lagoons that was established again provided a long-term benefit to offset the short-term disturbance during construction.

Transplanting flora disturbed by construction is another measure often adopted if flora of sufficient rarity are found on the site. They can then be returned to the site when construction has finished. This is, however, not always as straightforward as it seems. Sometimes, the only habitats suitable to receive the transplanted flora are already home to colonies of the same flora and there is no further room for new plants.

Construction through sensitive shorelines can be minimized by keeping the width of the approaches to a minimum and adopting an approach structure constructed between vertical walls. This "slot" form of approach is not as attractive to the tunnel user as an open cutting would be, but it does considerably reduce the impact of the approaches on the banks of the river. With careful planning of construction activities the worksite can also be restricted to a narrow corridor that is little or no wider than the permanent approach structure. A good example of this was the South Humber Bank Power Station outfall constructed on the River Humber at Immingham in the United Kingdom in 1995. The foreshore mudflats were a designated site of special scientific interest so the worksite was limited to a narrow 15 m wide corridor across the foreshore. The contractor installed a 200 m long combi-pile wall cofferdam progressively from the shoreline, working always within the 15 m wide zone. Decking was installed on top of the cofferdam to give the piling plant a working platform. The decking was also used to excavate the cofferdam and then to support the cranes used to immerse two lines of 6 m wide immersed tunnel elements forming the parallel outfall and the intake tunnels.

MARINE WORKS

The main activity in the waterway itself is the excavation, and subsequent backfilling, of the trench for the tunnel. Placing of tunnel elements is relatively fast and causes no real disturbance to the environment. The earthworks activities can affect the river in many ways, both during construction and afterwards. Dredging, by its very nature, stirs up the bed of the river, resulting in an increased amount of sediment in the river. This could be unacceptable as it can interfere with fish and marine plant life in the river. It can also affect shellfish beds by covering them in unacceptable sediment. However, modern dredging equipment can largely overcome these issues. Cutter suction dredgers can minimize the volume of sediment release into the river during dredging, and other techniques such as shrouded grabs and silt curtains are also available. Generally, the authorities will set a maximum value of sediment allowed and monitor the operation closely to ensure compliance.

The presence of the trench itself will alter the sediment transport regime in the river, affecting the material routinely carried by the river. The sediment falls into two broad types. First are the very small particles, generally less than 0.06 mm diameter. These are carried long distances in suspension and settle out of suspension in still water. They are usually deposited away from the main flows in shallow water. The second category comprises the larger particles greater than 0.06 mm in diameter. These have been picked up by the river flow and are in suspension or are being rolled along the riverbed. The amount of material in suspension depends on the velocity of the flow. The larger-sized gravel and cobbles remain close to the bottom where they are dragged along by the flow. The dynamics of the sediment transport is complicated, particularly in tidal areas where daily flow reversals occur and there is a boundary between the incoming salt water and the fresh water flowing down the river.

As the water crosses the trench, it slows down because of the increased depth of water. The volume of sediment carried in suspension is proportional to the current velocity, so this slowdown means the water can carry less sediment and the excess is deposited. The converse is true as the water leaves the trench and speeds up again. Although the sediment is neither dropped out of suspension nor picked up again immediately as the processes take time to develop, there is a trend for sediment to be deposited on the upstream side of the trench and removed from the downstream side. In a tidal location, this process reverses with the tidal direction so a complex pattern of scour and accretion is set up.

This affects not only the trench but the surrounding area as well. When the effects of any temporary or permanent alterations to the shape of the banks are also considered, the overall sediment transport balance in the river is altered and can become quite complicated. This can lead to new sandbanks being formed and existing ones disappearing. In a straight, uniform stretch of river this is often not an issue but, for example, in a wide, meandering river or in an estuary location the effects could be significant. The effects can be predicted by mathematical modeling of the river, which will identify the changes and any critical areas of scour or accretion. Modeling of this type, however, does need good base data to be of any use. Hydographic surveys are needed to establish the bathymetry, the current velocities, and the volume of sediment. The two types of sediment being transported are known as bed load and wash load. The wash load is the material carried within the water column and the bed load is the material carried along the bed. Measurements of both are needed. If the river is tidal, the surveys must also be carried out over different tidal cycles and also at different seasons as the volume of material being transported downstream varies from season to season.

Such a numerical model will also identify changes in current velocity. This is not only important to sediment transport but, if the river is navigable, even small changes in velocity can be dangerous for small yachts if they are not expected by the crew. For example, if the tunnel trench crosses the river at an angle, then it will deflect the current. Similarly, temporary cofferdams protruding from the banks will alter the flows by forming a constriction to the river that increases the flow speed. This is especially important in a waterway with commercial shipping, and the allowable changes permitted by the navigation authorities can significantly limit the extent of any temporary construction in the river.

The bed material to be excavated must be investigated beforehand. In industrialized areas, the material is often contaminated with heavy metals or the accumulated waste from many years of riverside activity. This contaminated material will have to be disposed of in special dumping grounds. Typically, the top 1–2 m of sediment contain contaminants. An interesting example of this issue is the Bjørvika Tunnel, constructed in Oslo Harbor between 2005 and 2010. The tunnel alignment crossed the site of an old sawmill and there was significant buildup of sawdust in the soils that was highly acidic. In addition, the harbor environment gave rise to a number of contaminants in the sediment. The material was dealt with in many ways: removing it to confined disposal sites, disposing of it at sea in licensed areas, and treating the sawdust for use in compost. This project also gave rise to some interesting studies whereby the clean clay arising from the dredging was used to cap known contaminated sediments elsewhere in the Oslofjord.

Whatever the chosen solution for managing contaminated material, the time to arrange the licenses and permits for disposal needs to be built into the project program, as this can take considerable time. Land-based disposal can be particularly time-consuming and may require extensive testing of the materials to determine compliance with the licensing requirements.

Another consideration in certain parts of the world is the possibility of unexploded ordnance. Maps do exist, although they are obviously very subjective, of where bombs have been dropped, and these can be used to give a preliminary indication of whether the site is likely to hold any unexploded ordnance. This is particularly likely around established port areas that are targets during conflicts. If this is a possibility, then the area of the trench can be investigated with magnetometer-type probes, which will identify any large metal objects on or close to the bed. It is unfortunate that acetylene cylinders are of a very similar shape to some World War II bombs, and this causes numerous false alarms during such surveys.

Archaeological finds during dredging are rare as modern dredgers are powerful machines and unexpected relics may be destroyed before anyone is aware of them. Finds of this sort are more likely when building the approaches through mudflats and banks on each side of the river. Again, if such artifacts are anticipated, surveys and investigations should be carried out before construction starts so that dredging can proceed with due caution or, if necessary, the alignment can be changed to avoid them.

When the tunnel has been placed the trench needs to be backfilled. This can also result in environmental damage from additional unwanted sediment in the river. Placing cohesive material around the tunnel underwater is generally prohibited by the river authorities for this reason. Such material would also generally be unacceptable for engineering reasons. Granular material can be placed carefully by fall-pipe so that it is contained rather than being simply bottom- or side-dumped into the river.

The tunnel trench does not have to be backfilled completely for engineering reasons, although it may be a requirement of the river authority for safety or other reasons. Once the tunnel has been covered and scour protection is placed to protect the backfill, the rest of the trench can be left to fill in naturally. This generally applies only to the hidden part of the excavation that is permanently underwater. The visible excavations through intertidal areas usually have to be returned to their original states.

FISHERIES

Constructing an immersed tunnel can impact the fish and shellfish in the river. The very act of carrying out construction in the river can disturb migrating fish as they are sensitive to the noise and vibration caused by marine plants. For piling works, it is preferable to use push techniques rather than percussive techniques, if ground conditions permit. Mitigation measures are also available and bubble jackets and bubble curtains to prevent the noise arising from piling operations spreading have been developed in the United States. This technique is quite simple and uses perforated pipes laid individually around a pile being driven or around the zone of piling. Air is pumped through the pipework and the curtain of air bubbles that rises through the water column is sufficient to dampen the noise travelling through the water.

In addition, high sediment content introduced into the water by dredging can cause significant oxygen depletion in the water and lead to large-scale fatalities of fish. There is inevitably some short-term disruption during construction, but by careful planning of the operations, this can be minimized so that there is no long-term effect on the fish. For example, if there are significant fish migrations along the waterway, then marine operations can be suspended during the migration season. This practice has been adopted on many immersed tunnels, including those recently built beneath the River Tyne in the United Kingdom and the Bosphorus in Istanbul. For the latter, marine operations were suspended during two fish migrating seasons, both northward and southward each year. Such restrictions will have an impact on the contractor's planning, but generally, the dredging is not on the critical path, so such restrictions can be accommodated within the overall construction program, provided they are clearly defined at the outset.

Other examples include the Bjørvika Tunnel, which crossed the mouth of the Akerselva River that flows into Oslo Harbor. The river is important for the migration of Atlantic salmon and trout. In this instance, no seasonal

Figure 5.2 Silt curtains in use during dredging for the Bjørvika Tunnel. (Photo courtesy of Statens Vegvesen.)

restrictions were applied, but river blockage was carefully controlled and tight control of water quality at the mouth of the river was maintained through the use of silt curtains, as seen in Figure 5.2.

Once a tunnel has been completed, there is no long-term effect on the movement of fish up and down the river. The one caveat to this is that if an impressed current cathodic protection system is installed, there is the potential for electric currents to disturb the fish, and this effect should not be overlooked.

Construction activities can also be detrimental to shell fisheries at or near the site. These can be significantly affected by increased sediment in the river. If necessary, stringent restrictions on the amount of increased sediment that is allowed during construction can be imposed. It may also be necessary to model these changes in sedimentation to satisfy stakeholders that no detrimental effects will occur during construction. If detrimental effects are inevitable, for example to local fishermen and boatmen, then it may be necessary to compensate them for lost income during the construction period.

Environmental authorities will be rightly concerned about the effect of construction on the fisheries. Unfortunately, this often leads to an initial blanket refusal to the marine works associated with an immersed tunnel. It is the experience of the authors, however, that if these issues are fully explained and discussed then it is usually possible to agree on a set of measures that will satisfy the authorities and enable the tunnel to be constructed. This may require several years of monitoring and survey in order to understand the number of different species in the waterway and the seasonal pattern of their behavior. Modeling of flows in the waterway may also be worthwhile to predict the impact of sediment plumes that could arise from earthworks activities in relation to the position of shellfish beds or colonies. Once the baseline conditions are understood, it is possible to enter into meaningful discussions with the relevant authorities about the mitigation measures that are possible and required.

ALGAE

At some tunnel sites there is the possibility of blooms of algae that can hinder construction. These can be a severe handicap to the construction process. For example, on the Øresund Tunnel project, there was the possibility of a severe algal bloom that had the potential to fill the dredged trench. This would have severely disrupted tunnel element placing operations while the trench was cleaned out. Fortunately, such an event did not occur during the construction period, but it highlights another type of risk that a contractor has to be aware of.

WATER QUALITY

As discussed, changes to the water quality during construction can adversely affect the marine biology. Oxygen content and turbidity are factors affecting the quality of marine life and strict limits on the extent of any changes during construction must be imposed. Agitation of sediments that causes them to be mobilized into the water column may deplete oxygen levels. Turbidity is the term used for this condition of the additional sediment being carried. It is possible that this may have an adverse impact on marine flora and fauna, so it is necessary to have an understanding of the marine habitat and species that are present in the watercourse and whether they are sensitive to possible increases in the suspended sediment levels. To do this, baseline surveys must be carried out during the planning phase of the tunnel so that the existing conditions are known and realistic limits can be set for construction.

Water quality monitoring is generally carried out during construction to check that restrictions imposed on the construction activity are being adhered to. This monitoring consists of regular sampling and laboratory analysis of the collected samples. Requirements differ from site to site and the relevant national environmental authorities generally have their own guidelines. Table 5.1 gives typical values to aim for.

Table 5.1 Typical water quality values

Parameter	Typical target values
Dissolved oxygen level to be maintained during construction	5 mg/L (95 percentile) 3 mg/L absolute minimum
Turbidity limit	<30 to 50 FTU
Suspended solids limit	200 mg/L (95 percentile) 500 mg/L absolute maximum

VISUAL ASPECTS

A tunnel imposes far less visual intrusion on a waterway and the surrounding area than a bridge. Once completed it is often almost invisible, except from the air. Nevertheless, there are aspects of a tunnel scheme that do have visual impacts that must be considered. These include service buildings, ventilation inlet and exhaust towers, the architectural appearance of the portals, and the type of approach structures.

A modern tunnel is fitted with a comprehensive environmental control and monitoring system, including lighting, ventilation, communications, and control systems. These systems require a building to house the power supplies, switchgear and controls, and provide a base for the tunnel operation and maintenance staff. The appearance of these buildings requires sensitive treatment, as immersed tunnels are often in areas of natural beauty.

It is possible to locate the building fully underground, as is the case for the Øresund Tunnel. If such a solution is not possible, then it still might be possible to place the building on the tunnel but to shield it visually behind any flood protection bunds that protect the tunnel approaches. This was done for the Limerick Tunnel in Ireland, as seen in Figure 5.3.

An alternative to shielding the building is to make a feature of it. This approach is often adopted when exhaust ventilation towers are required. Generally, there is no way to hide these, so they are best made into architectural features that can act as symbols of the tunnel crossing.

PORTALS

The geometrical shape of the tunnel portals is generally defined by the number of carriageways/rail tracks the tunnel is carrying. How this is treated in architectural terms is often of concern to the tunnel owner who will want some identifying feature to mark the tunnel. For example, at the Medway Tunnel in the United Kingdom, the portals and approach walls are clad in brickwork to match the adjacent historic dockyard setting through which

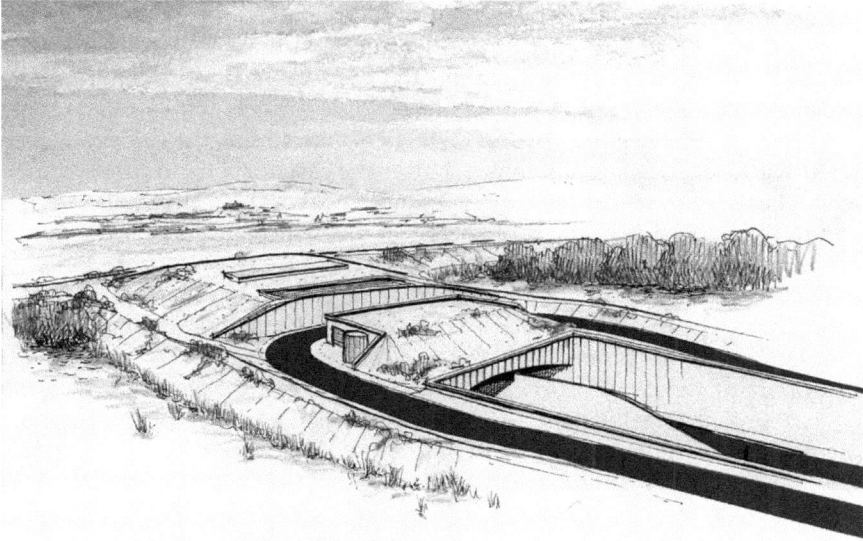

Figure 5.3 Limerick Tunnel service building within flood protection earthworks. (Courtesy of DirectRoute.)

the tunnel passes. Whatever shape or features are adopted the portals and approaches have to remain functional. There must be no safety hazards and the colors should be subdued to assist the driver's eyes in accommodating to the reduced light levels inside the tunnel as they drive toward it.

NOISE

Tunnels are inherently noisy places. In a road tunnel, the noise from the car or truck engines plus the noise from the tires on the road surfacing reverberates from the solid reflective surfaces of the walls and roof. The ambient noise level can approach 100 dBA. Similarly, in a rail tunnel, the noise from the engine traction system and from the steel-on-steel contact between wheels and rail again combines to create a very noisy environment. Although these levels are high, this is not generally an issue because the operation and maintenance regimes should be such that people do not work in the tunnel under live traffic conditions.

As far as the tunnel users are concerned, no special measures are required to reduce the sound in a tunnel. The sound is sufficiently reduced to tolerable levels by the fabric of the train or car they are traveling in. To provide a sound-absorbing treatment to the walls would form an additional maintenance liability. It would require a roughened surface that would attract the dirt and dust particles present in the tunnel. This would darken the wall compromising the lighting and also be more difficult to clean because of its roughness. As it happens, some tunnels with spray-applied fire protection have good natural acoustic properties as the fire protection is left with just such a rough finish. This is the case for the Øresund Tunnel, which has spray-applied fire protection above the level of the wall cladding and to the underside of the roof. This was a side benefit, however, and not the result of a project requirement.

There are two areas where noise has to be considered and, if necessary, action has to be taken to reduce it. These are at the portals and around any ventilation exhaust points; assessment of the noise levels at the portals will form part of the overall EIA. At the portals, the traffic emerges into the open approaches, which usually have full height retaining walls to reflect the noise. The geometry of the carriageway and the walls tends to reflect the noise upwards. In an urban situation, this can cause excessive noise levels for surrounding buildings.

When planning the tunnel, base level measurements of the ambient noise levels in the portal areas should be taken. The increase in noise levels due to the tunnel can then be established. The increase at a typical major road tunnel portal can be in the order of 10–15 dBA. The environmental authorities will set the limits for noise levels around the portal, and if the levels are too high, there are only a limited number of mitigating measures that can be taken. If the surrounding buildings are relatively low, then noise barriers can be erected on top of the approach walls. These will create a noise shadow reducing the noise levels in the surrounding area. If the buildings are higher, then the noise barriers will be less effective. Improving the sound insulation of the building might be possible, although in a modern building, scope for improvement will be limited. Conversely, the noise insulation in a modern building may be adequate to provide acceptable noise levels in the interior despite the increase due to the tunnel. The spread of noise from a portal and its approaches can also be limited by the additional ground absorption provided by the soft landscaping often installed around the approaches.

If these measures do not work, then it might be necessary to extend the covered length of the tunnel so that the portal is located in a less sensitive position. This was necessary at the Medway Tunnel in the United Kingdom, where the cut and cover tunnel was extended 250 m to move the portal away from a development area where the increased noise and air pollution would have been unacceptable to the developer. In that instance, the developer of the site contributed to the additional costs as he decided it was in his interest to do so. An intermediate solution is to use sunscreens that form a partial barrier over the approaches and may have a degree of sound-absorbing ability.

Another area of concentrated noise emission is around the ventilation exhaust points. Immersed tunnels often make use of longitudinal ventilation in which the exhaust comes out through the traffic bores and the additional noise merges with the traffic noise. If transverse systems are used, the pollutants are usually exhausted through some form of ventilation stack. The location and height of the stack may have to be varied to find an acceptable arrangement from a noise standpoint. The noise in this instance arises from the ventilation fan motors. The designer should do all that is possible in terms of selecting fan types and sound attenuators to limit the noise generated.

Noise levels during construction also form part of the EIA. Construction of an immersed tunnel involves a considerable amount of equipment and labor. The associated noise of these operations could be unacceptable in a nonindustrial environment. Background ambient noise levels will need to be established at the construction site. Through discussion with the relevant authorities, limits can be set for any increase above the existing ambient levels, and the contractor will have to work within them. There are two types of levels to set: an overall level and an instantaneous level. For example, a discrete noise in the middle of the night in a rural estuary can be very disturbing to those living nearby, although in itself the noise level may not be high. Frequently, national legislation requires a contractor to submit construction plans to assess the likely noise levels at key sensitive receptor positions. This is compared with the measured baseline levels to identify the increase in noise for specific construction operations. Authorities can then determine any necessary constraints or mitigation and give the required consents to permit the construction to proceed.

AIR QUALITY

The tunnel design needs to consider air pollution. Overall pollution from the traffic is not increased but that pollution is concentrated. If longitudinal ventilation is being used, then the pollution is concentrated around the portals, whereas with transverse or semi-transverse ventilation the pollution is concentrated around the exhaust stack. This is less of an issue as the location and height of the stack can be designed to disperse the pollution without impacting the surrounding area.

Pollution contour diagrams may need to be developed around the portal if this is an area of concern. This is a specialized task that may require numerical modeling, taking into account wind speeds and directions as well as the local geography. The eight-hour averaged carbon monoxide contours for the high growth peak-hour traffic flow will identify the extent of pollution around the portal, and changes can be made to the layout of the approaches and positioning of the portal and ventilation exhausts until an acceptable solution is reached. A balance must be struck between the benefits and disbenefits of a scheme. For example, the area around the portal that will suffer increased air pollution may not be the one where people are likely to spend extended periods of time, and this needs to be set against the possible reduction in air pollution at other locations where the traffic is reduced because of the new tunnel crossing.

Apart from carbon monoxide, there could be concentrations of lead particulates that are produced during the combustion of leaded fuels. Although this has been an environmental concern historically, it is unlikely that this will be an issue in the future because of improved vehicle emissions and the phasing out of leaded petrol. Nevertheless, in some developing countries this may still have to be considered. It is possible to install electrostatic precipitators to remove particulates in the air being exhausted from the tunnel, though these are expensive and require significant space provisions at the portal or in the service building and therefore are not a preferred solution.

Chapter 6

Marine environment

For any immersed tunnel project, it is essential that the marine environment is fully understood. Hydraulic and hydrographic surveys and studies play an important part in the planning and detailed design of any such tunnel scheme. They are used to address such issues as

- The hydraulic consequences of both the temporary and permanent works, which would include changes to the scour and accretion patterns
- The rate at which sediment is deposited in the open trench, to enable the dredging contractor to plan his work
- The loadings that need to be applied on the tunnel elements and that need to be taken into account for the selection of the appropriate marine plant, equipment, and methods

They will enable the correct decisions to be made while planning, designing, and constructing the tunnel. The overall hydrographical, meteorological, and climatic conditions at the site must be investigated and documented, and used as a basis for the project. Once these conditions are understood, they can be used to assess their various impacts on the construction operations and on the completed tunnel itself.

The location of any particular tunnel will determine which areas of hydrography and hydraulics are most important. They will vary significantly, depending on whether the tunnel is situated in inland waters or is in an estuarial, harbor, or offshore environment. For example, coastal projects will be heavily influenced by wave and swell conditions, whereas inland river crossings may be more influenced by the current and water depths and the risk of river blockage and scour. Estuarial tunnels may be most influenced by tidal variations and currents.

In this chapter, the most important hydrographic parameters are discussed, including how they feed into the project planning, design, construction, and operational stages in the life of a tunnel. The hydraulic effects that must be considered in the design of the tunnel, which result from flow conditions, are also investigated, and the specific aspects that are crucial to the success of the immersed tube method are explained. The aspects that benefit from numerical analysis and modeling techniques are noted, along with guidance for scale model testing—something that is commonly undertaken for immersed tunnels.

HYDROGRAPHY

Hydrography is the measurement, description, and analysis of natural waters and their margins. It is essential to understand the hydrography of the body of water in which a tunnel will be constructed. The aspects that must be understood and accounted for in the design and construction of a tunnel are

- Water levels
- Wind and waves
- Water density
- Currents
- Sediment transport
- Ice formation

Each of these aspects is described below. Usually, historical records can be obtained from weather stations and marine survey stations in the region where a tunnel project is chosen to be constructed. However, it is unlikely that comprehensive data will be available for the actual site chosen for the tunnel and it is normal for the interpolation of existing records to be carried out, to establish initial hydrographic parameters, and for further surveys to be carried out in the years leading up to the tunnel construction, to get precise data for the project.

Water levels

The variation in water level and, hence, the available water depth, will have a large influence on the planning of a tunnel. It will dictate construction methods in terms of floating and maneuvering the tunnel elements. For example, the tidal range may limit the time periods available for immersion activities if there is insufficient water depth to maneuver the elements at low water levels. Water levels will also govern the earthworks that provide flood protection to the tunnel approaches.

Understanding the maximum and minimum water depth is therefore important, not only for construction planning, but also to ensure that design is undertaken to establish the best possible construction techniques for the given water depth and tidal range. It is usually possible to obtain historical data on the tide levels for a standard set of defined levels. Typically, the following water levels are used, although different definitions are in use around the world:

- Highest Astronomical Tide (HAT)
- Mean High Water Springs (MHWS)
- Mean High Water Neaps (MHWN)
- Mean High High Water (MHHW)
- Mean High Water (MHW)
- Mean Sea Level (MSL)
- Mean Low Water (MLW)
- Mean Low Low Water (MLLW)
- Mean Low Water Neaps (MLWN)
- Mean Low Water Springs (MLWS)
- Lowest Astronomical Tide (LAT)

Reference can be made to British Standard BS 6349 (2000), the code of practice for maritime structures, for the definitions of the terms listed above. These levels, coupled with a bathymetric survey to establish the bed levels, will give all the water level planning information that is needed. Of course, practice varies around the world as to what water levels are routinely recorded. However, whatever levels are recorded, it is important to understand the definitions of those levels in terms of the return period, as they will not necessarily be the same as those above. The important behavior to understand is the frequency at which certain water levels occur on a daily, monthly, yearly, and long-term basis.

It is common for the mean water datum to differ from the land datum being used for the shores on either side of the tunnel. A conversion factor usually needs to be applied and it is important to be very clear in all design data, construction, and operational documentation as to what levels are being used so that the land and water levels are consistent and correctly correlated. For long sea crossings, specific methods may be required to determine the project level system. For example, on the Øresund Crossing, a common datum was established with reference to the land data in both Denmark and Sweden. The effects of the earth's curvature may even come into play for long crossings and there may be a need to undertake surveys to establish a precise datum and project reference grid.

Some data will be published on navigational charts and tide tables. Port and harbor authorities may also keep records. In the absence of historical data being available, it is likely that project-specific surveys will need to be carried out to establish the water levels at the project site. This is not uncommon, as immersed tunnels are often located away from port areas and established monitoring stations where the information is routinely gathered. If surveys are required, this should be identified very early on in the project life, as it would be beneficial to collect data for one or two years ahead of the anticipated construction date in order to get a sufficient volume of reliable data that can be used for statistical analysis.

The influence of rising water levels over time due to climate change must also be taken into account as this will determine the height of any flood protection measures. Historical rates of sea level rise, along with predictions for the future, can be accessed from the Intergovernmental Panel on Climate Change (IPCC). Various other research bodies also exist, which means that if there is no specified requirement in national codes, data can be obtained to assess the predicted rate of sea level rise for a particular site.

Global rates of sea level rise are compounded in some areas by the effects of isostatic readjustment following the melting of continental ice sheets at the end of the last glaciation. This is the effect of land that had previously been pushed down under the weight of ice rebounding upwards. Rates of sea level rise are not uniform around the globe. Even when isostatic effects are excluded, there are differences related to such things as variable thermal expansion and large-scale oceanic currents. Local subsidence due to faulting may also be a consideration. Not all waters are tidal and, for a tunnel built beneath inland waterways, different conditions apply. In rivers, lakes, and canals, the water level will be dictated by rainfall and annual variations in flows. Relating a water level increase to rain storm intensities that fit particular return periods may be an important relationship to understand. The probability of certain water levels arising during construction or the life of the tunnel structure should be understood. Seasonal variations in water level may also occur, for example, if the watercourse receives melt-water in the spring, which results in elevated water levels for a specific period each year.

An understanding of the water levels and their frequency of occurrence defines the geometry of the project—for example, the flood protection levels or the interface between the immersed tunnel elements and the cut-and-cover section, as well as both the permanent and temporary loadings to be applied to the tunnel elements.

Flood protection

The design of flood protection works needs to consider the high water levels that may occur. Typically, the height for flood protection walls or earthworks will be set after considering a particular return period. In Europe, this is commonly taken as 1:10,000 years. Some national standards specify lower return periods—for example, the Federal Highway Administration (FHWA) standards for the United States specify a 1:500 year return period. In low-lying areas, a pragmatic approach may need to be taken if the region surrounding the tunnel will be below the flood level for a specified high return period, in which case, flood gates may need to be installed at the tunnel portals. However, this does not necessarily mean the specified return period being used is incorrect. In fact, it may still be desirable to have more onerous criteria for the tunnel than the surrounding area, as flood protection has more to do with getting the tunnel back into operation. If it takes several weeks or months to get the tunnel operational, long after the flood water has subsided, then this may be a major inconvenience for the transport network.

Often, the flood levels are determined from a cost/benefit and risk perspective. Considering the extent of flood protection works and the associated capital expenditure for a number of return periods can inform the project client of the options available. After considering the risks of flooding and the potential costs and time to re-open the tunnel, the best balance can be achieved.

Towing and immersion

Where the water depth is limited and insufficient for transporting or maneuvering the tunnel elements, a number of approaches to construction can be considered, including

- Additional dredging of the watercourse to form an approach channel to the site. This may be for access within and around the tunnel site, or from a nearby navigation channel, or involve dredging of the navigation channel to the site. This can be costly, have environmental disadvantages, and may require significant time to secure permits for the dredging and disposal of dredged arisings.
- Use of the tunnel approaches for construction of the tunnel elements. This ensures that sufficient water depth is provided, but may have repercussions on the program.
- Temporary lifting of tunnel elements with buoyancy barges/pontoons to reduce draft. This may help if a transport route has particularly shallow stretches.
- Partial construction of the element to limit its weight and, therefore, its draft. This enables the element to be towed to near the tunnel site, where construction can be completed.

These measures can have a major cost impact and, in order to arrive at the best solution, accurate data on water levels and depths must be available. Often, problems with restricted water depth can be overcome by utilizing the short-term increase in water level that occurs with the tides. Operations can be timed to coincide with the highest fortnightly tide—the spring tide—to avoid otherwise expensive construction methods. Although this only allows a short window of opportunity to undertake specific operations, it can often be enough.

Wind and waves

The wind and wave conditions that can be expected at the time of construction will have a large influence on the construction planning and will also need to be taken into account in the design of the tunnel. While the tunnel elements are afloat, the wind and wave conditions will be applying dynamic loading to the structure. Depending on the magnitude of the waves, this can be insignificant or can require major temporary works, such as temporary prestressing of the tunnel element structure to resist bending effects. If wind and wave loads have the possibility of being extreme, and are not cost-effective to design for, then some contingency planning for the towing or immersing of the elements may be necessary to ensure that such extreme loadings can be avoided in the event of sudden bad weather or sea conditions arising.

As with water levels, meteorological stations often record wave heights on a continuous basis. The data is commonly recorded as the significant wave height. This is the average height of the highest one-third of the waves. It is not the highest wave that can occur. This will be obtained from a statistical analysis, often using a Rayleigh distribution. Maximum wave height information may also be available, but this would be unusual. It is likely that some specific project surveys may be necessary for coastal projects that are not protected within the confines of rivers or estuaries. Some statistical analysis will be necessary to assess the likely combinations of wave height and wave length for the exposure periods being considered for the towing and mooring of the tunnel elements. The direction of waves is also important. Depending on the current and wind directions, the waves may come from any number of directions, and each direction will have its own characteristics of height and length.

Once the wave and wind characteristics are understood, they can be used to set the design parameters and to manage risks during construction. One example is the wave criteria assumed during towing of the tunnel elements in an offshore situation. The wave criteria might be defined for a frequent occurrence that could be applied in the design of the tunnel in a similar way to a serviceability limit state criterion. Under such wave conditions, it would be expected that the structure behaves elastically and that the cracking of concrete might not be permitted. The waves for this criterion might be those of a reasonably high probability of occurring, say 5%, or 1 in 20 years. There may be another criterion developed, similar to an ultimate limit state or extreme condition, where the probability is unlikely to occur, say 1%, or 1 in 100 years. Under this condition, which might be called a survivability event, failure would not occur, but some cracking might be permitted, provided the elastic limit is not exceeded. An analysis of the wave data for the two sets of wave loading will be necessary to determine the wave height and length that apply to each case. Once determined, the design of the tunnel elements can be carried out.

At this point, the marine operations can be planned and the "weather window" approach can be developed. Weather predictions for wind speed and direction can be correlated with the wave height and direction. This would allow a procedure to be developed to decide whether the towing of a tunnel element can take place, by assessing the likely wave height, length, and current conditions that might occur during the towing period. It is likely that a lesser return period might be set for the go-no-go criteria, compared to the design values. For example, if the extreme wave height used for design is based on a 1% probability, then the go-no-go might be based on a 2% probability to build in some factor of safety. This may be an iterative process, balancing the cost of the provisions in the tunnel to resist the wave loading adequately with the risk of the wave occurring, and what this means for the operational windows for construction.

The effect of wind on construction, other than its influence on the wave conditions, is not great. When concrete tunnel elements are afloat, they generally have a low freeboard and are therefore not greatly affected. For steel tunnels, when the shell has been launched and has to be towed to a fitting-out site just with the keel concrete placed for stability, the majority of the element will be above the water line and, hence, will be subjected to the effects of winds during towing operations.

The height and frequency of the swell generated by distant storms needs to be predicted in coastal areas as well those of more locally generated wind waves. Therefore, the longer term effects of wind, the time lag between wind and wave, and the impact of storms at a distance from the tunnel site may all need to be considered.

Extreme wave conditions

There are a number of extreme wave effects that may need to be considered as well. The first is offshore deepwater waves. These may be relevant in coastal locations close to open oceans. Large swell waves generated from ocean storms may cross the site. These would temporarily increase the water level over the tunnel, increasing the hydrostatic pressure on the elements. These should not affect the temporary conditions during construction as there will be advance warning of their approach, and the necessary precautions will have been taken to safeguard the marine plant and equipment.

Storm surges often occur in estuarial sites. The increase in water level has to be taken into account in the design of the appropriate flood protection, as well as the increase in pressure on the elements.

Then, there are tsunamis. If encountered at sea, these have long wave lengths and are unlikely to affect a floating element. They may even pass unnoticed. However, closer to shore, the wave shoals and increases in height. This possibly results in large volumes of water overtopping the sea defenses and flooding inland. It may be necessary to take this into account while designing the tunnel flood protection by increasing the height of any flood protection bunds or by installing floodgates that can be closed when there is a tsunami warning. Similar to other waves, they will temporarily increase the hydrostatic pressure on the elements.

Current conditions

The forces applied due to currents on the tunnel elements during their towing and maneuvering can be considerable. Ensuring that these loads are taken into account in the design of the permanent and temporary works is vital. Large bending moments may be applied to the tunnel elements due to a combination of waves and currents while they are restrained by the mooring lines.

Current velocities will determine the design of winching, cabling, and anchoring systems in the temporary works for maneuvering and restraining the element for immersion. They will also have a significant effect on the magnitude of the towing forces for tug boats. Indeed, they will impact the selection and design of all the marine plants required for the construction of the tunnel.

The nature and timing of dredging works may also be influenced by the currents. Depending on bed conditions, high currents may cause the migration of the dredged trench and require maintenance dredging. They will be an important consideration in the design and construction planning of backfilling works around the tunnel, to ensure that materials can be placed and scour does not occur.

Understanding the current patterns in the wider region is also important, to ensure the dispersal patterns of any sediments arising from dredging are taken into account. In coastal and exposed sites, currents will be largely linked to other meteorological conditions, that is, the direction and strength of prevailing winds, coupled with the fetch, the distance of clear water in the direction of the prevailing wind.

In sheltered water, currents will arise as a result of tidal conditions and may be more predictable. The difference between the velocities of flood and ebb tides may be important in planning marine operations and thus should be surveyed and understood. Mathematical models, as discussed later in this chapter, are often used to simulate and predict current patterns in coastal and estuarial locations. In river locations, modeling may need to be undertaken if there is significant blockage of the river channel during the maneuvering and placing of tunnel elements. In addition to loadings to the tunnel element itself, the reduced channel flow area can result in increased currents. This is a particular issue in narrow watercourses, such as the canal systems in the Netherlands.

Current forces may determine the level of temporary prestressing in a tunnel element if a segmental construction is being used. Whether the tunnel elements are monolithic or segmental, the magnitude of bending moments caused by the combined wave and current loading could govern the design of the tunnel element. In this case, it might be desirable to limit the length of the tunnel elements to reduce the bending effects that arise. Large currents may mean that marine operations have to be undertaken at slack water on the turn of the tide or within restricted weather windows to avoid the maximum current flows; if so, this can be a major constraint on construction.

There are some unusual current conditions that highlight the need for understanding the specific conditions at a project site. One particular example that illustrates this is the Bosphorus Strait in Istanbul. At this location, the current reverses through the depth of the water column. This is a particular phenomenon that relates to the flows between the Black Sea and the Marmara Sea. It is highly unusual, but required specific planning for the immersion of the tunnel elements to maintain control of the element as the current reversed part way through immersion.

Salinity and water density

The immersed tunnel construction method relies on having established certainty that the tunnel elements will float and can therefore be transported to the tunnel site by towing. They then have to achieve a minimum factor of safety against uplift during the construction process, when they are ballasted and immersed into the dredged trench, and in their completed condition. The floating behavior and the ability to meet the necessary safety factors against uplift are wholly dependent on the density of water as, following Archimedes's principle, the uplift force generated is equal to the weight of displaced water. To ensure the tunnel elements behave as expected, it is therefore important to work with accurate data on salinity, and the likely range of sediment content, and the resulting range of water density.

For coastal or offshore projects, the salinity levels are quite predictable, although there is a range that should be applied in design and for construction planning. For fresh water conditions, salinity is not an issue, although the variation in water density must be known in the event there are suspended sediments that vary throughout the depth of the water column. Estuarial conditions are the most challenging for the prediction of salinity levels. The interface between saline and fresh water will vary as the tides ebb and flow.

Care is needed in maintaining the consistency of units. Salinity may be expressed as density, specific gravity, or salinity. Ranges can therefore be expressed as

- Density: 9.81 KN/m^3 to 10.2KN/m^3
- Specific gravity: 0.9850 to 1.003 (dimensionless, based on the ratio of mass compared to distilled water at 4°C)
- Salinity: 0–35,000 parts per thousand (less commonly used now)
- 0–35 as a unitless ratio K—the ratio of the electrical conductivity of sea water to that of a potassium chloride (KCl) solution

The measurement of the salinity at a project site is necessary, particularly in the tidal stretches of estuaries and rivers, where there may be quite a variation. The salinity can be expected to increase with the incoming tide, and reduce again as the tide recedes and the saline water is flushed out by the river water. Seasonal variation can occur; therefore, it is important to obtain records or readings for a sustained period to avoid over-conservative assumptions in design and for planning. The variation of salinity with depth is particularly important to understand, as this will affect the buoyancy and ballasting procedures. The incoming tidal saline water tends to flow in underneath the fresh water, forming a saline wedge at the bottom of the river.

As discussed earlier, water density plays an important part in the design to ensure that the required safety factors are achieved at all times, in both permanent and temporary conditions. Under temporary conditions, the salinity, or range of salinities, is needed to assess ballasting quantities for the immersion process and to ensure that sufficient overweight is applied to the element once the marine equipment has been disconnected.

The density will increase with depth and there is a natural tendency for more dense water to accumulate in the dredged trench. This may mean that the ballasting of the element has to increase with depth, or that additional ballast is applied from the outset to ensure that sufficient overweight of the tunnel is maintained during the immersion operation. This may require larger-capacity pontoons, winching cables, and equipment to manage the additional weight when the element is floating in the less-dense water.

During construction, lower safety factors are used for the vertical stability of the tunnel elements than during the permanent condition. This is appropriate, as the exposure periods are significantly shorter. However, it means that accurate predictions for salinity levels are necessary. Typically, tunnel elements may only be ballasted with a 2.5% overweight when they are in the casting basin, ahead of being floated up, or when they have just been placed at the base of the tunnel trench. Should there be unexpectedly dense water in the basin or in the trench, the weight of the displaced water would be greater and if this becomes greater than the overweight, it will result in the element trying to float upward.

Anecdotal stories of this occurring do exist. There is one example, in a particular project, of a salt water spike travelling along the river, which reached the casting basin where the tunnel elements were ballasted and resting on the bottom. The unexpected increase in density reduced the safety factor against the uplift and gave rise to some floating of the elements.

Suspended sediment and turbidity

As discussed in Chapter 5, additional sediment can be released into the water column by the dredging process through the generation of unsteady current and flows. This causes haziness or cloudiness in the water, known as turbidity. Depending on the environmental sensitivity of the site location, it may become necessary to limit the turbidity through particular dredging methods. To establish the criteria for suspended sediment and turbidity, it is necessary to understand the natural levels of suspended sediment that occur, both seasonally and as a result of rainstorms and flood behavior. Only after this is known can judgements be made on any necessary restrictions on dredging activities.

In order to decide whether dredging activities will have an adverse impact, mathematical modeling should be carried out on the sediment plumes that may arise from different dredging methods and the various flow and current conditions that may exist. Suspended sediment concentrations can then be predicted along with the impact on oxygen levels. The impact on marine ecology can then be predicted. If adverse impact is predicted, dredging operations can be tailored to particular conditions or times of year whereby impact can be avoided or, at least, much reduced.

Mathematical modeling can also provide useful information for the construction phase. The natural change in sedimentation will usually result in material being deposited in the tunnel trench. This must be removed before the tunnel elements are placed. The extent of this maintenance dredging is important to plan the construction. If these rates are high, the contractor will not want to leave a long length of trench exposed for a long period of time before placing the tunnel—it will simply fill up again and will have to be re-dredged. Equally, he will not want to retain the dredger at the site longer than necessary as it is an expensive piece of equipment. Thus, a prediction of the amount of maintenance dredging required is an important factor in the contractor's planning.

Ice formation

The formation of ice can cause specific loading to the roof of the tunnel in the shallow foreshores. The nature of ice formation and buildup is such that pads of ice are continuously moved towards the shorelines and the pads are pushed over each other, causing a gradual increase in thickness. This could impart a significant load to the tunnel structure and may need to be considered in colder climates.

Records of ice formation are generally available from port authorities, but the precise nature of the buildup and resulting loading is difficult to assess, so designers should take a conservative view. More guidance is given in Chapter 9, under the section on loadings.

Summary of data needed for design and construction

Tables 6.1 and 6.2 summarize the hydrographic data needed for the permanent works design and construction planning.

Table 6.1 Hydrographic data for design

Condition	Specific information	Design element
Water levels	HAT MHWS MHWN MSL MLWN MLWS LAT Increase due to climate change and isostatic adjustment Influence of rainfall on water levels Seasonal variations Storm surge, wave, and run-up effects at shoreline	Length of immersed section of tunnel Flood protection bunds Hydrostatic force in immersion joints
Wave	Wave length Wave direction Significant wave height Maximum wave height Worse combinations of wave height and length for given return period appropriate to exposure duration	Temporary prestress RC design Steel/concrete composite design
Current	Maximum current speed Current patterns and directions Variation of current speed with tides Annual variation in current speed	Scour protection
Salinity/density	Maximum density Minimum density Variation with depth Impact of turbidity due to construction Variation suspended sediments with seasonal flow or high rainfall	Factor of safety against uplift
Ice formation	Seasonal temperature variations Likely ice thickness	Maximum load to roof of tunnel
Tsunami	Water height Run-up distance	Flood protection Transient hydrostatic loading on tunnel

In addition to the surveys prior to construction and the gathering of historical data, surveys during the construction process will be required for all aspects of hydrography. This approach minimizes and controls the risk of the element or construction equipment being overloaded during construction.

Table 6.2 Hydrographic data for construction

Condition	Specific information	Construction element
Water levels	HAT MHWS MHWN MSL MLWN MLWS LAT Increase due to climate change and isostatic adjustment Influence of rainfall on water levels Seasonal variations Storm surge, wave, and run-up effects at shoreline	Window for immersion or towing Access for marine vessels Keel clearance for towing Flood protection for temporary works
Wave	Wave length Wave direction Significant wave height Maximum wave height Worse combination of wave height and length for given return period appropriate to exposure duration	Loads on winching equipment, towing lines and bollards/ anchors
Wind	Maximum wind speed Wind direction Correlation to wave heights	Loads on towing lines, winching lines, and anchors
Current	Maximum current speed Variation of current speed with tides Annual variation in current speed	Loads on towing lines, winching lines, and anchors Design of temporary cofferdams
Salinity/ density	Maximum density Minimum density Variation with depth Impact of turbidity due to construction Variation suspended sediments with seasonal flow or high rainfall	Overweight required for immersion Design of ballasting system Initial freeboard Factor of safety against uplift during: a. Initial immersion b. Sandflow

HYDRAULIC EFFECTS

Scour and sedimentation

It is necessary to consider the changes in flow patterns caused both during and after construction. Changes to scour and accretion patterns caused by changes in the water velocities can have severe impacts on the waterway and its users. The principles of initiation of particle movement are well documented and established design methods exist for assessing this.

Sandbanks may change shape, move, or even disappear completely. This can interfere with navigation and nearby moorings may start to silt up and become unusable. All these effects must be investigated by an analytical model to develop a solution that is acceptable to all parties.

As mentioned, sediment accretion rates in the dredged trench are of great importance to the contractor, so an accurate assessment of these is essential. The rate of accretion is dependent on the volume of sediment being carried by the water, the water velocities, and the shape of the trench. The profile of the trench is generally chosen to suit the maximum slope gradient that will remain stable over the duration of construction. The pattern of accretion is affected by the orientation of the trench toward the direction of the water flow and the side slopes of the trench. Depending on the geotechnical parameters of the material being excavated, these can vary from almost vertical sides in hard material to very flat side slopes of up to 1:8 in soft alluvium. Accretion is likely to occur at the base of the trench if it has steep side slopes. If the trench side slopes are at shallow gradients, this accretion may only affect the slopes and not the central part of the trench. However, if the trench is exposed for a lengthy period of time, it can tend to migrate in the direction of the current as material is deposited upstream and eroded from the downstream side.

Another effect that has to be considered is the impact on the sediment dynamics due to spillage from the dredging operations. Increases in sediment concentrations, even temporary ones, can affect the nearby river ecosystem. Often, the marine authorities will already have specified limits for sediment spillage that the contractor will have to adhere to. Sediment transport from the construction works and how, and if, any additional sediment disperses has to be modeled. Relevant spill scenarios can be determined based on the types of soil to be dredged, the types of dredging equipment proposed, and the duration of the activities. The results from the sediment transport simulation show the extent of the bed affected by the works. These effects then form part of the environmental discussions and suitable mitigation measures will need to be introduced. This also links to the overall proposals for spill monitoring during construction, which are required to demonstrate compliance with the authorities' requirements.

As an example of the type of investigation that is required, consider the Øresund Tunnel. This is part of the fixed link between Denmark and Sweden and the waterway it crosses is one of the main channels that enable water to flow in and out of the Baltic Sea. The Baltic Sea supports a very delicate ecosystem and a restriction on the consent to build the crossing was that construction had to have zero impact on the water flow in and out of the Baltic Sea, so that this balance would not be upset. The crossing was part bridge and part tunnel and required the creation of an artificial island to form the transition between the two. Demonstrating to the satisfaction of the authorities that the shape of the island had no overall effect on the water flow into and out of the Baltic required extensive modeling and consideration of many different layouts. Compensation dredging was also required to maintain the flow and compensate for the blockage formed by the island.

This is typical of the investigations required on most tunnel schemes for any reclaimed areas adjacent to the shoreline to accommodate approach structures. The size and shape will have to be chosen to minimize the hydraulic effects on the adjacent shoreline so that any effects are acceptable to the authorities and adjacent owners. Any constrictions to a watercourse will increase current velocities and increase the likelihood of scour occurring.

It is a common approach on immersed tunnel contracts for the client to insist that the existing bed level and channel profile is reinstated to the pre-construction conditions, at least in the visible intertidal areas. This can be a difficult requirement to meet if there is a large volume of mobile material in the watercourse and there are seasonal patterns of erosion and deposition, or there are known characteristics of deposition requiring a regime of maintenance dredging. In these circumstances, it would be necessary to stipulate that the required profile of the channel be reinstated by coordinate and level, based on historical records and, if necessary, hydraulic modeling of the watercourse. If the bed conditions are established and stable, then the reinstatement to existing levels is a reasonable approach to take.

In some circumstances, it may be desirable to finish the top of the tunnel above the existing bed level. This is more likely in deeper water conditions or where there is a large variation in bed levels, where there is an advantage to maintaining a shallower alignment of the tunnel. In such circumstances, the influence on scour should be assessed and the risks of damage to the tunnel from a sinking vessel or dragging anchor must be fully addressed in the design. Examples of this approach have been successfully executed at the Prevza-Aktio Tunnel in Greece and the Busan-Geoje Tunnel in South Korea.

It is also necessary to consider the risks of scour during the placement of marine earthworks. The general backfill to the sides of the tunnel is usually granular and often sand. The planning of the methods and the timing of placement of such materials need to consider the likelihood of scour occurring during the construction period. If high currents are present during the construction period, the material size/grading may need to be selected to prevent the loss of material.

Flooding

A number of issues must be considered in relation to possible flooding due to a rise in the sea or river levels. If the tunnel approaches pass through flood plains, the available storage capacity of the flood plains may be affected. The route of the tunnel may also isolate an area of the flood plain or compartmentalize it, so it is less effective. Flood studies or analytical modeling may be needed to understand and mitigate these effects. The impact of reclamation must be considered in the event that flows are constricted and upstream flooding of the tunnel could occur. This may require hydraulic modeling to be undertaken to demonstrate that no adverse consequences arise from the construction, or to quantify those that do.

Inundation of the tunnel due to flooding of the hinterland would be serious, as it is possible to fill a tunnel completely with water. In addition to severe effects on the tunnel M&E systems, the loading on the tunnel foundations will increase due to the additional weight of water inside the tunnel. This could well result in additional settlement of the tunnel. In multi-bore immersed tunnels, it is also typical to consider the effect of the hydrostatic pressure on the internal walls and the uneven transverse loading caused by not all the bores being flooded. In time, the flooding would fill all the bores, but circumstances could be envisaged where the flooding is uneven as it builds up.

Because of the seriousness of such an event to the structure of the tunnel and its operation, inundation should be avoided. During the feasibility stage, it is typical to investigate the possible flood levels around the portal areas and design accordingly, to prevent inundation. This may be achieved by having flood protection embankments around the portals to protect the road until the carriageway has risen above potential flood threat levels. Figure 6.1 shows the western approach to the Medway Tunnel in the United Kingdom. The flood protection bunds extend around the approaches, fully protecting them, until the carriageway reaches a high enough level close to the roundabout, where it no longer needs protection.

In some circumstances, the use of the surrounding flood walls and carriageway levels may not provide an acceptable solution if they cannot be raised high enough. Other options are to provide flood protection barriers or gates at the portals. These may be full height panels that are lowered from a head house to close off the portal completely, or they can be as simple as a stockpile of material further up the carriageway, which is pushed across the carriageway to keep the water out when a flood threatens. The top of this temporary barrier would be above the threatened flood level. With either of these approaches, it has to be accepted that the tunnel will be out of use until the flood danger subsides. It is also important that the flood gates, material stockpiles, and any associated equipment are regularly maintained so that they can be used when required. An unfortunate example where this was not done was at the US Elizabeth River Tunnel in Norfolk, Virginia where, because of poor maintenance, the portal flood gates did not close fully, resulting in the tunnel being completely flooded.

Figure 6.1 Medway Tunnel flood protection. (Photo courtesy of Kent County Council/ BAM Nuttall/Carillion/Philip Lane.)

Forces on tunnel elements

The following cases may need to be considered in designing tunnel elements for the temporary hydraulic conditions experienced during construction:

- Vertical bending due to waves during towing or at temporary moorings
- Horizontal bending due to currents during towing
- Horizontal bending due to currents during immersion

In addition, the following forces imposed on the marine plant must be determined to correctly size the marine plant and equipment:

- Drag/slip force occurring during
  - Towing of the elements
  - Mooring of tunnel elements during immersion
  - Mooring of elements to pontoons during immersion
  - Positioning tunnel elements with winch cables
  - Loads on lowering cables for immersion

Another effect is the influence of passing vessels, which can cause uplift forces on the elements while they are in a temporary condition on their foundations.

Vertical bending due to waves during towing or at temporary moorings

Wave characteristics are defined by a height and wave length and a clothoidal curve. It is possible to assess the impact a passing wave may have on the tunnel element using static methods, such as is described in British Standard BS6349 part 6. These methods are useful to establish the general impact that waves may have, but are not sufficiently accurate for detailed design. Dynamic modeling of the tunnel elements is preferable and accounts for the inertia, stiffness, and natural frequency of a tunnel element. It also enables a variety of wave conditions to be examined to establish which is the most severe.

The worst condition in design is not necessarily caused by the largest wave height that may be encountered. If this has a very short or very long wave length associated with it, then the loading may be small. A large wave with a very long wave length may impart little bending moment to a tunnel element. A wave length that is twice the length of a tunnel element is likely to cause the most severe effects. However, multiple combinations of height and wave length should be considered to determine the maximum load effects for design.

Horizontal bending due to currents during towing and immersion

Vertical bending moments due to waves need to be combined with horizontal moments due to currents. While under tow, the ends of the tunnel element will, to some degree, be restrained against the current by the tow points at either end and will act as a beam with a uniform lateral load applied.

In an offshore environment, it is likely that the wave loading will induce the predominant load effects and the current may be a minor component. However, in inland waterways, the current may be the predominant effect. Both should be examined to determine the behavior and necessary levels of prestressing or structural reinforcement to cater to the loads.

As an example of the magnitude of forces that may be experienced, the Øresund Tunnel was designed for a significant wave height of 1.6 m in 15 m of water depth. The design bending moments used for the prestress design in the 175 m long tunnel elements, which were derived from the dynamic modeling, were in the order of 285 MNm vertically and 235 MNm horizontally.

Drag/slip force

The force in the towing cables between the tug boats and the tunnel element is often referred to as the slip force. It may also be called the drag force or haulage/towing capacity. The force is a function of the drag coefficient of the tunnel element, which is a function of its geometry, the speed at which the element is being pulled, and the density of water. The basic formula that can be applied to calculate the slip force is

$$F = \tfrac{1}{2}.A.r.V^2.R$$

where

F = the resulting slip force
A = the exposed area of the tunnel facing the current, beneath the water line
V = relative velocity
r = the specific volume of water
R = the friction or drag coefficient

The friction coefficient is dependent on the shape of the element, but is typically somewhere between 1.5 and 4, and can become higher under unfavorable conditions, such as constrained waterways.

Other effects also have a bearing on the slip force and a more thorough approach was suggested by van de Kaa (1978) where the return current and the reduced water level at the rear of the tunnel element are considered.

The actual drag coefficient may be significantly above the calculated value due to other effects, such as

- The horizontal trim of the element.
- Deviation from alignment, that is, if the center line of the element has an angular deviation from the direction of tow, a greater drag area will exist.
- Trim dipping during tow, where the bow dips lower in the water. This can be particularly significant if a shallow freeboard exists and there is little water depth below the element, which is often the case with concrete tunnels.
- Navigation around tight bends causing a change in the current direction, compared to the tunnel element alignment.

Factors can be applied to represent these possible increases in drag, or alternatively, the behavior can be investigated with scale modeling. A factor of two or three times the calculated value may occur, depending on site-specific circumstances.

The slip force arising during towing is used to determine the required tugboat pulling capacity and to decide the number and arrangement of tugs needed. It will also be used to design towing bollards mounted on the tunnel elements and the tow cables. The slip force needs to be calculated for other stages of the construction, for example, once the element is released from the towing tugs and is undergoing the immersion process. Calculations will be needed for maneuvering the tunnel elements into position over the tunnel trench and to size the mooring lines and anchorages when the tunnel element is held in position for the immersion process.

Loads on lowering cables for immersion

During the immersion process, the ballast tanks are filled with water to change the tunnel element from positively buoyant to negatively buoyant. In addition to ballasting the element to reduce the freeboard to zero, further ballast is added to give the tunnel element a notional overweight. This overweight is typically between 300 T and 500 T and must be sufficient to enable a positive load to be measured on the lowering cable and to ensure that it remains a positive load, taking into account the possible variations in the system. The most significant of the variations that might occur is the variation in water density. Any likely variation in density that might be experienced must be accounted for in determining the ballast tank capacity, the winch cable load capacities, and the pontoon or lay barge design.

Influence of passing vessels

Vessels passing over the tunnel when it has been placed on its temporary or permanent foundations in the dredged trench, but not backfilled, could have a destabilizing effect on the tunnel elements. Depending on the size and speed of the vessel, and the clearance between its keel and the top of the tunnel, it may be possible to generate uplift forces.

If the waterway in which a tunnel is being constructed is known to be busy with large vessels, and it cannot be closed to traffic for any significant period of time, then this phenomenon must be investigated either by mathematical or physical modeling. If it is a risk to the construction, then speed limits and navigation restrictions should be imposed until such a time when the tunnel is backfilled and stable.

In the completed condition, when the tunnel is backfilled, large vessels passing over the tunnel may cause disturbance to the layer of rock protection covering the tunnel if it is not designed to resist the localized currents that are caused by the propeller of the vessel. These can be either horizontal scour currents or uplift forces and the effects could be particularly severe if the tunnel passes beneath an area where ships have to make complex maneuvers for changing direction or for berthing.

Propeller scour can be assessed using recognized methods, such as in "Erosion of bottom and sloping banks caused by the screw race of maneuvering ships" by H. G. Blaauw and E. J. van de Kaa (1978). In order to evaluate the magnitude of the scour current that arises, some basic information is needed, such as the geometry and thrust of the propeller, or thrust coefficient, the draft of the vessel, and keel clearance. Currents can then be calculated and the bed material sized to prevent the initiation of movement. Similar guidance is available from the BHRA report RR2570, "Propeller induced scour," by M. J. Prosser (1986).

Uplift forces may be generated by large propellers close to the sea bed. This was investigated for the Øresund Tunnel. The worst condition was that of a maneuvering vessel with zero speed and maximum engine power. It was found that low pressure in front of the propeller caused uplift forces on the rock armor stone. This effect turned out to be the governing criterion for the design of the rock layer over the tunnel and was more onerous than general scour and resistance to falling anchors.

MODELING

Modeling of hydraulic behavior is often necessary to fully understand the forces imparted to the tunnel element and the behavior of the watercourse. Simple static calculations give a good indication of behavior, but not all effects can be anticipated, and 2D and 3D modeling may identify further hydraulic effects that need to be taken into account. Numerical modeling with software can offer good insights into behavior, but often, scale modeling is necessary

Whichever type of model is used, it is first necessary to collect base data to enable the model to be calibrated. This will include data on

- Currents
- Water levels
- Wave heights
- Sediment transport
- Salinity

The currents and water levels must be measured over the full tidal cycle—both the daily and monthly cycles—and during different seasons, to ensure that all conditions can be considered. Sediment transport and salinity data must also be measured throughout the water column as variation with depth is important to the dredging and placing operations. The collection of data on this scale is a large task and, if possible, use should be made of existing data. Numerical modeling is constantly being improved but, in essence, the model must be capable of simulating the tidal flows and currents, the wash load sediment transport, and the bed load sediment transport. Models are available commercially, but it might require using more than one model and integrating the results. Low resolution can be used in the wider model, with a higher resolution being used in a specified area around the tunnel alignment.

First, the models must be calibrated against the field data to confirm the accuracy of the assumptions and boundary conditions in the model. This may require several attempts with different boundary conditions before the model can be shown to be simulating correctly. Once the model is working satisfactorily, alterations can be made to the model to investigate the effects of introducing changes into the estuary, for example, dredging the trench, reclaiming land adjacent to the banks, or forming artificial islands that might be required for long tunnels.

Scale modeling

In planning the towing and placing operations, a key task is to calculate the forces on the tunnel elements and the various towing and mooring lines during these operations. These marine operations are potentially dangerous and careful planning is required to develop a safe mode of operation.

The scale modeling of tunnel elements in water tanks may be carried out at hydraulic research laboratories. A number of these exist around the world and, in Europe, the facilities owned by Delft in the Netherlands, the Danish Hydraulic Institute in Denmark, or HR Wallingford in the United Kingdom have been active in this field. Typically, these facilities enable waves to be generated, bed and channel geometry to be modeled, and scale models of tunnel elements to be floated and maneuvered to assess their behavior and determine the loadings applied to them by the hydraulic conditions used.

Scale models can be used for

- Verification of loadings on the element from current, wave, and swell
- Verification of loading on towing and mooring equipment
- Verification of the stability of the tunnel elements in the various stages of floating and immersion
- Verification of drag coefficients
- Verification of behavior in specific project conditions, channel geometry

Given the importance of these calculations to the safety of the marine operations, modeling is often carried out to calculate the forces more accurately and increase confidence in the operation. Nowadays, mathematical models can simulate marine operations and are used to calculate the forces in the towing lines. Before the development of such mathematical models, physical models were used to evaluate the forces. These are still used on projects to validate the computer analysis. The models are approximately at 1:50 scale, so even the model of the tunnel element is about 2 m long and extensive facilities are needed to build and test such a model. The models replicate the conditions in the waterway, including the dredged trench, any adjacent bends in the river, and any other special circumstances.

Physical models can be used to investigate the whole sequence of marine operations. The element can be tested in a variety of positions, both in plan across and along the river. Turning the element in the river so that it goes from parallel to perpendicular to the flow can also be simulated. It can also be tested at various depths as the element is lowered into the trench. Here, the angle of the trench to the main river flow and the relative blocking factor of the element are significant. The physical model will also model the added hydrodynamic mass, which is a function of the shape of the element, its velocity, and its orientation to the flow. The models also identify any unexpected excitations in the system due to dynamic effects.

Model tests need to be very carefully designed in order to get reliable results that can be transferred to the field. Accurate geometric scaling is needed, as is scaling of the engineering properties. A common approach for scaling is to use dimensionless parameters such as Reynolds and Froude. This enables the physical parameters of length, mass, force, moment, acceleration, and time to be modeled accurately. In addition, a model will have hydroelasticity requirements that take account of stiffness, structural damping, and mass distribution.

Tunnel element models can be fitted with measuring devices to detect forces and measure load effects. It is usual to define a series of conditions for testing and these may include a number of wave directions, different restraint conditions to the tunnel element, and behavior in open water and above the dredged trench. Pontoons may also need to be modeled in addition to the tunnel element as they contribute to the mass and volume of the overall system and are as much impacted by wave and current as the tunnel element. Different ballasted conditions may also be modeled, representing different freeboards and mass of the tunnel element.

Measuring devices such as wave probes, current meters, water level gauges, and force gauges of various types to measure single or multiple components of forces will be used to record the load affects. The results from these instruments will need interpreting and scaling to get representative information for the full-size tunnel elements.

Other physical tests can be undertaken in laboratory flumes to analyze the flow behavior around tunnel elements. This might be particularly useful in narrow waterways, where the immersion process may cause considerable blockage of the channel and the changes to current velocities and the potential for scour need to be understood.

An example of such a model was that carried out for the Conwy Tunnel. The estuary in the area of the tunnel was relatively shallow—only about 10 m deep at high tide—and characterized by many sandbanks and channels. The line of the tunnel was skew to the river at an angle of 60°. During placing, the element represented a large blockage to the flow in a complex channel. The physical model was used to determine the forces on the element and mooring lines at various stages during the placing operation. The model was at a scale of 1:64 and the area of river modeled was about 1500 m × 800 m. The boundary conditions for the model were calculated from a mathematical simulation of the estuary, which was correlated against the measured depth and velocity data. The physical model was then calibrated against the mathematical results to verify that it worked correctly before the element model was introduced.

The element was then placed in the model and included all the mooring lines and pontoons so that all the forces critical to the operation could be measured. The model was tested with a range of current velocities and water depths on both the flow and the ebb tides, with the element at various locations, orientations, and depths in the trench so that every stage of construction was investigated. The results enabled the forces on the mooring lines to be established and verified that the bollards and lines to be used had sufficient safety factor. The model also interestingly identified unacceptable oscillations in some of the mooring connections at certain stages of the operation. The mooring locations were amended to avoid these.

Model tests on this scale require large-scale facilities, which are not widely available. It is a fairly expensive exercise but yields highly valuable information for complex circumstances. For straightforward applications, numerical analysis is generally sufficiently accurate to determine the forces on the element and the mooring system.

Numerical modeling

Numerical modeling can be a valuable tool in design and construction planning. A hydrodynamic numerical model will provide information for the structural design and marine operations, including the dredging works, as well as for the evaluation of environmental impacts of scour and accretion and the impact of such things as dredging spillage. Such models should be set up at an early stage in the project as they can be used to optimize the scheme and its impacts and limit the options that proceed to more detailed study.

Hydraulic mathematical models are available commercially—for example, the MIKE suite of software from the Danish Hydraulic Institute. Some of the areas where modeling is useful for immersed tunnel projects have been discussed earlier in this chapter and are summarized as follows:

- Dynamic modeling of elements in floating conditions to determine the bending moments in the tunnel element due to dynamic effects, in order to design the tunnel steel shell or the longitudinal reinforcement or prestress in a concrete element.
- Modeling of changes in flow velocity and current patterns due to construction to assess the potential for changes in the scour and sediment deposition in the waterway. Conditions that might be modeled include:
  - Temporary cofferdams protruding into the waterway from the landfalls that may constrict the flow
  - The dredged trench or a partially dredged trench

- Numerical modeling of water levels and the potential flood risk due to changes to the flow characteristics of the waterway as a result of the temporary construction stage conditions or the permanent conditions, for example, if land reclamation is carried out that changes the fundamental geometry of the channel.
- Modeling of flows around tunnel elements causing constricted flows. This may cause increased flow velocities that could initiate scour, and will impose loading on the tunnel element.
- Numerical modeling of safe navigation can be a very useful tool. This can be carried out using desktop software, but there are also simulators that can assist in determining if safe navigation can be achieved around the construction work site in the river if the navigation channel is narrowed.
- Numerical modeling of sedimentation in trenches is helpful in high sediment-carrying waterways and the bed profile can be modeled with the dredged trench to determine the amount of accretion that could be expected.

Chapter 7

Mechanical and electrical installations

The mechanical and electrical (M&E) systems of a modern tunnel are vitally important, irrespective of whether the tunnel is carrying road or rail traffic. After all the civil engineering has been finished and effectively "hidden," it is taken for granted; it is these M&E systems that interact with the tunnel users and operators. They allow the safe operation and maintenance of the tunnel and are an indispensable part of any modern transportation tunnel. Since many of the key aspects of tunnel electrical and mechanical design are common to all forms of tunnels, in this chapter, we will only give a general overview of such systems, and point out the particular issues that are relevant to immersed tunnels.

The electrical and mechanical installations in a modern tunnel are considerable. Dealing with road tunnels first, they cover the following aspects:

- Ventilation and fire safety
- Lighting
- Communications
- Control systems
- Drainage

Current safety legislation varies from country to country, but the general principles are the same everywhere—the safety of the tunnel users and the safety of the asset. The main codes and legislation covering these are found in NFPA 502 (road tunnels) and NFPA 130 (rail tunnels) from the U.S. National Fire Protection Agency (NFPA), EU Directives, Permanent International Association of Road Congresses (PIARC), and national standards. Developing a strategy for the safe operation of the tunnel in both everyday and emergency conditions is a critical aspect of tunnel design. In the United Kingdom, this is achieved by the formation of a Tunnel Design Safety Coordination Group, which has representatives from all the organizations involved. These are the operator, highway authority, designer, and the emergency services. In the United States and some other countries, this group is known as the Fire Life Safety Committee. The remit of the group is to develop systems and procedures for the safe operation of the tunnel, and to ensure that all parties are in agreement with them and are familiar with their operation. These discussions are often fundamental to the layout of the tunnel, so it is essential that such a group is formed in the early stages of planning the tunnel. A lot of the group's work will be done during the preliminary design stage and they should have completed most, if not all, of their work by the start of construction.

ROAD TUNNEL SAFETY

The first consideration in the case of a road tunnel is what traffic is to be allowed through the tunnel. Hazardous substances, which have the potential to cause considerable damage to people and property, are routinely transported by road in large quantities in developed countries. The vehicles themselves are subject to national regulations for construction and use, but after that, they are often allowed virtually unrestricted access to the highway network. Hazardous loads fall into several categories, including

- Explosives
- Compressed or liquefied gases
- Flammable liquids, such as petroleum products
- Flammable solids such as paper, plastic, and wood
- Poisonous and infectious substances
- Chemical corrosive substances
- Radioactive substances

It is worth remembering, however, that the disastrous 1999 Mont Blanc Tunnel fire, in which 39 people died, resulting in the tunnel being closed for 3 years, started in a lorry carrying flour and margarine, which were not really considered dangerous loads.

When it comes to hazardous loads passing through a tunnel, they can be allowed unrestricted access, prohibited completely, or escorted through the tunnel in a controlled manner. If hazardous loads are to be prohibited, then the consequences of an incident on the alternative route must be considered. If the alternative route, for example, is through a densely populated area, the consequences of, for example, a tanker fire on property and people could be severe. In these circumstances, it might be preferable to allow the hazardous loads to pass through the tunnel, which can be designed so that an incident can be contained more safely and with minimal damage to the tunnel structure.

Which of these operational systems is adopted would be informed by a risk analysis of the consequences of an incident. There are various risk analysis procedures that can be used. A quantitative risk analysis model (QRAM) has been developed by the Organization for Economic

Cooperation and Development (OECD) and PIARC. In the United States, the Federal Highway Administration (FHWA) has produced a guidance document, "Prevention and control of highway tunnel fires," which gives a method of predicting casualty probabilities. In the United Kingdom, a simple risk assessment is used, based on tabulating the hazards and impacts and assigning scores for each. Multiplying the two gives an overall risk priority score that can be used to decide how the risk should be managed. The details of such a risk analysis are beyond the scope of this book, but the analysis would typically consider the following factors:

- A review of the hazardous loads using the tunnel
- The alignment and cross section of the tunnel
- Identification of likely incident scenarios occurring, either in the tunnel or on alternative routes
- Evaluation of the possible consequences of the various scenarios on people and property
- Evaluation of the frequencies of such events
- Combination of traffic flows, accident frequencies, and consequences to determine the risks to people and property for both the tunnel and the alternative routes.

In general terms, the results of such an analysis usually show that the risks associated with diverting hazardous loads and not permitting them to use the tunnel are higher for incidents involving less than 10 fatalities. It is safer to let the hazardous loads through the tunnel. This reflects the safer driving conditions in modern tunnels, where accident rates are usually lower than on the equivalent length of open road. This conclusion often reverses when considering larger events where the greater risk is to allow the hazardous loads through the tunnel. This is a consequence of the confined nature of the tunnel, so the consequences of a major incident can be more severe. This does not mean that hazardous loads should automatically be prohibited, just that the tunnel systems and operational procedures have to be designed accordingly.

If the risk analysis shows that, on balance, hazardous loads should be allowed to pass through the tunnel, which is the usual outcome, then there is a decision to make about whether they are allowed unrestricted access or whether access should be controlled in some way. The most usual way of controlling access is to escort the vehicles through the tunnel, either individually or in convoys. Either way, there has to be a marshalling facility where the loads can be parked until the escort is available, and an escort vehicle or vehicles to lead and accompany the loads through the tunnel. The operation will also interrupt normal traffic through the tunnel as a safe distance needs to be established, both in front of and behind the load, as it passes through the tunnel.

The operation of an escort system is something to be carefully considered from an operational viewpoint. From the lorry driver's perspective, stopping at the tunnel control and waiting for an escort will delay his journey, so there is a temptation not to stop and report and just drive straight through the tunnel. This, of course, is easier in an unmanned tunnel. The lorry can be picked up on the tunnel monitoring systems but by then, it will be too late and the vehicle will be in the tunnel. If the tunnel is tolled, then hazardous loads can be identified at the toll plaza and the escort procedure implemented. In an unmanned tunnel, there is no robust method of identifying and stopping hazardous loads, so there is no realistic prospect of stopping them passing through the tunnel, and the tunnel and its facilities should be designed accordingly.

These aspects are of particular importance in an immersed tunnel because of the location beneath a waterway, where failure of the structure due to a major explosion or fire incident would be catastrophic.

Incident Detection and Management

A critical part of tunnel operation is the detection and management of incidents in the tunnel, which can range from disabled vehicles to major accidents and fires. These detection systems are essential to the safety of the people using and operating the tunnel, and the early detection of an incident is essential to the welfare of tunnel users as well as minimizing the damage to the tunnel itself. A wide range of detection systems can be installed in road tunnels. Automatic incident detection systems use sophisticated programming to identify unusual occurrences. These monitor traffic flow and the environment in the tunnel, and raise alarms when an incident is detected. Traffic monitoring systems range from vehicle detector loops set in the carriageway to closed circuit television (CCTV) and radar-based systems. Detector loops can be used with sophisticated algorithms to identify vehicles that have stopped in the tunnel. Similar systems to detect stationary vehicles and alert the control room automatically can also be based on CCTV installations. These use full pan, tilt, and zoom ability and provide full coverage of the tunnel.

When an incident is detected, it is important that the plan for dealing with it is well thought-out and rehearsed, so that all parties are aware of the part they have to play. Too often, tunnel incidents have spiraled out of control because those responding to it have not acted promptly or correctly. Incident management starts at the planning stage of the tunnel, when all the parties involved in the tunnel—designer, operator, highway authority, and emergency services—must come together to identify and draft plans for dealing with incidents.

Smoke detectors and visibility detectors are other methods of warning that an incident has occurred in a tunnel. Whatever system is used, however, it is important that the detection systems are monitored and that the people monitoring them know what they are looking for and what to do in the event of something happening. There is little point in having a comprehensive CCTV system monitoring a tunnel if no one is monitoring the output from that system. Similarly, all personnel must be aware of the emergency plans and their responsibilities and actions.

Emergency Escape

The provision of an emergency escape route is an important operational consideration and one that affects the planning of the tunnel at the outset as it can affect the layout of the tunnel cross section. Tunnel users must be given some safe haven in the event of a tunnel fire, so an obvious escape route with good signage, lighting, and a public address (PA) system to give direction are essential. Many people have died as a result of staying in their vehicles during a tunnel fire in the mistaken belief that it is the safest place. In an immersed tunnel, there are two main evacuation possibilities. Access can be provided to the other road bore through emergency doors in the central wall. Alternatively, a separate emergency escape bore can be provided between the traffic bores. This can provide safe access to the surface, usually by stairs or ramps at the ends of the tunnel, or to the other traffic bore, away from the incident. Immersed tunnels tend to have more flexibility in the provision of escape facilities than bored tunnels. As escape doors can be easily provided at regular intervals, the safety refuges that are commonly provided in bored tunnels are not required.

The simplest system is to have emergency doors in the dividing walls between the tunnel bores. This layout gives the minimum width of tunnel section. The doors are generally spaced about 100 m apart and they can either be hinged or sliding. Custom varies between countries and the door type and markings are often set out in national standards for uniformity. Some countries permit the doors to be locked and opened by the operating authority automatically in the event of an incident. Other countries will not permit them to be locked at all in case the automatic opening system fails just when it is needed, and this is the recommended approach. The durability and specification of sliding doors need close attention. The pulling force to open them should be limited so that they can be opened by anyone, and they should be fitted with a self-closing mechanism. They require regular maintenance to ensure that the mechanism operates properly. Hinged doors are simpler and usually fitted with a push bar opening system.

All doors would be alarmed so that the tunnel operator is alerted to one being opened. With a road tunnel, such an arrangement has one clear disadvantage as the door gives access adjacent to the fast lane in the other carriageway. The danger is that someone fleeing an incident runs straight into oncoming traffic. Clear warning signs can be put on the doors to warn users of this, but in an emergency, people will often not realize the importance of the message they are conveying. A method of overcoming this is to install a short length of pedestrian barrier opposite the doors, which prevents pedestrians walking through the doors straight into the oncoming traffic. Some authorities believe that these barriers themselves form a hazard to motorists, and prefer not to have them. They argue that the frequency of incidents is so low that the permanent danger to motorists caused by the barrier outweighs the very infrequent danger to fleeing pedestrians. Other authorities argue that the barrier is not substantially different from those encountered in many urban situations and that the advantage is with the pedestrians, with little hazard to the motorist. This arrangement may only be preferable in low-speed urban tunnels. This is the type of debate that can be resolved by the tunnel safety committee during the development phase. Questions—such as how quickly the tunnel can be closed and whether traffic can be stopped in the adjacent bore before many people exit through the door—can be reviewed with that particular tunnel and operator in mind.

If the pedestrians are to have a dedicated route to safety, then a separate bore is needed. This is the preferred option but, while giving a safer operating environment, it increases the width of the tunnel, increasing the cost. In short tunnels, they may not be required if the portals are visible as car passengers will then tend to evacuate directly towards daylight via the portal. In a typical dual carriageway immersed tunnel, a separate bore would be provided between the two traffic bores. This would be approximately 1.5–2 m wide and would be accessed by emergency doors from each carriageway (Figure 7.1). To keep the escape passage free from smoke in the event of an incident, the bore would be kept at a slight overpressure to the traffic bores. Thus, if one of the doors is opened, smoke is not drawn in to the escape bore. The bore would continue towards the tunnel portal. Here, it would either give access to ground level via a stairway or ramp or, alternatively, give access to a safe area between the carriageways at the portal. Emergency lighting has to be provided throughout these escape bores.

As space is at such a premium in immersed tunnels, the escape bore can also be used to accommodate the cables and pipework required to service

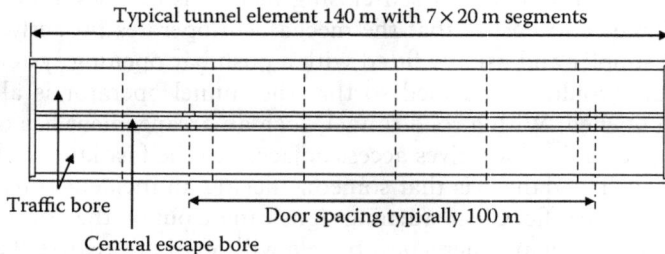

Figure 7.1 Typical plan layout of emergency escape doors.

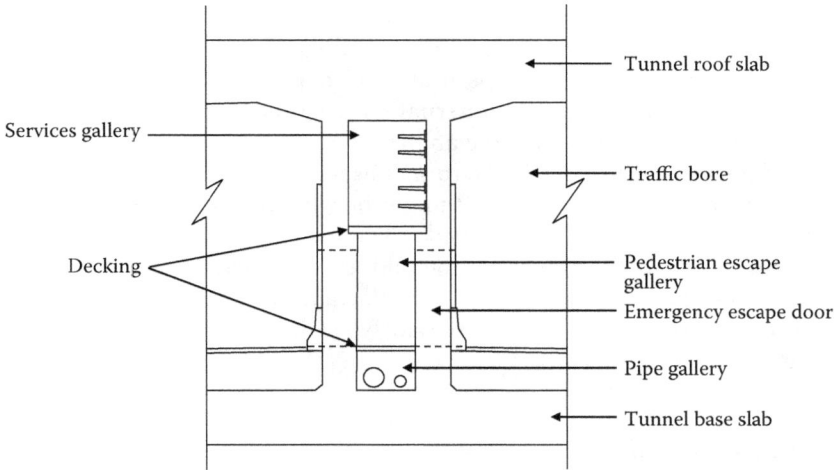

Figure 7.2 Central services/escape bore.

the tunnel systems. A common arrangement is to divide the escape bore into two or three compartments, with cable and pipework running above and below the pedestrian escape route (Figure 7.2). This central bore can also accommodate the step down transformers and switchgear needed for the tunnel systems.

Combining the escape and services facilities in this way does present some problems. The electrical systems will almost certainly require technical rooms to house transformers and similar equipment, particularly in longer tunnels. These are likely to be near the middle of the tunnel and will take up most, if not all, of the room in the central bore, blocking it as an escape route. If this is the case, a risk assessment will be required and a strategy developed, with clear signing to direct people in the right direction.

Pedestrians must have a safe route from their vehicles to these emergency exits. This should preferably be at a low level. Raised walkways, which were often used in the past as they separated pedestrians from the traffic and gave the pedestrian more reassurance, are not favored now in some countries as access to them is restricted and people with impaired mobility cannot get on to them. Even access walkways at low level should be provided with dropped kerbs adjacent to the emergency exits to permit wheelchair access. In the United States, raised walkways are still used, but this is for maintenance access and not for escape. Low-level gaps are provided in the raised walkways to give access to the escape doors.

The escape doors are usually highlighted by the use of color on the walls around them, with direction signs placed on the walls to give the direction of the nearest door. PA systems can be installed to provide information and direction in the event of an incident and emergency lighting should be provided to highlight the escape route and doorways in what could be a smoke-filled environment. The internal noise level is relevant if a PA system is contemplated. To overcome the ambient noise levels, the system will have to produce some 106 DbA if the tunnel users are to hear it. This requires some care in the planning stage. It might be worthwhile to use a PA system to gain attention if people are outside their vehicles and in danger of doing something untoward. However, the sudden interjection of such a high-volume message could cause even more alarm in the person it was intended for. PA systems are becoming more widely used, but are not yet favored by all operating authorities.

RAILWAY TUNNEL SAFETY

Circumstances are slightly different in a railway tunnel. Operationally, trains often try to clear a tunnel before stopping in the event of an emergency, but that is not always possible, so provision still must be made to evacuate the passengers. The principles are similar to road tunnels. Doors give access either to another rail bore or to a separate evacuation bore. To enable passengers to disembark from a stranded train, walkways are provided at the appropriate level so that the passengers can just step out of the train without having to climb down. A further sophistication that has been adopted is to have the emergency doors at the same spacing as the train doors. With this arrangement, an empty train can be drawn up in the empty bore opposite the doors so that passengers can simply, and safely, walk through the doors and in to an evacuation train. This type of arrangement would work well with a train that was stranded because of mechanical failure but, in a fire situation, there might not be time to arrange for a train to be stopped adjacent to the doors.

Modern trains are constructed with safety in mind and have a low fire load, so a major fire incident involving a passenger train is very unlikely. More likely is a fire in a freight train. Here, if the train is unable to exit the tunnel to enable the fire to be dealt with, then the risk is to the integrity of the structure rather than to passenger safety.

Consideration can be given to installing sprinkler or mist systems at intervals through the tunnel, the idea being that the train driver can stop under one of the systems and the fire can be extinguished. This does, however, require the train driver to know where he is, where the sprinkler is, and be able to reach it and stop. Most authorities prefer the train to continue and exit the tunnel, where the fire can be dealt with. Which system is adopted will depend on a risk analysis of the particular circumstances of the tunnel and the preferred operating procedure.

VENTILATION

Because of its importance to the final cross section, ventilation is one of the first items that should be investigated when planning an immersed tunnel. It is one of the key considerations for the early stages of design as it has such an impact on the cross section and, therefore, the cost of the tunnel.

Tunnel ventilation is a specialist subject in its own right, but the basic elements are common to all tunnels. The ventilation system provides sufficient fresh air into the tunnel to keep the exhaust emissions within limits that are safe for the tunnel users. In addition to this, and very importantly, it also has to be able to control the smoke and heat developed if there is a fire in the tunnel. The heat and smoke must be reduced to bearable levels and provide an environment suitable for the firefighters attending the fire in the tunnel. The ventilation system must also expel the smoke from the tunnel, directing it away from the users trapped in the tunnel and giving them time to escape.

For some short tunnels, the airflow produced by the traffic—the piston effect—is enough to maintain the safe levels of exhaust fumes in the tunnel during normal operation. Immersed tunnels are unlikely to be this short, so some form of forced ventilation will be required in the tunnel and, in almost every case, will be required to deal with the effects of a fire in the tunnel. If these emergency ventilation requirements are to be met, a system must be provided that allows the gases produced by the fire to be extracted, and both the speed and direction of airflow in the tunnel bores for road and rail traffic to be controlled.

A key aspect of a successful ventilation system is the effectiveness of fire detection, as this is critical for the control of a fire. Detection can be achieved either by an automatic fire alarm, by a manual alert from a user, or from the general monitoring of the tunnel with heat and smoke detectors. Manual detection systems include fire alarm panels as well as alerts via mobile phones and emergency call stations. These alerts trigger a preliminary alarm in the monitoring station, which must then initiate further action.

Automatic detection systems can detect both smoke and heat and enable the location of the fire to be established. This should then ensure that the optimum fire ventilation program is selected, based on the current status of the direction and the speed of the airflow in the tunnel. An array of fire detectors is usually deployed to detect the temperature and temperature rises, with video detection systems being used for the detection of smoke densities. Time is of the essence in successfully controlling a fire and the system should be designed to shorten the detection time to less than 15 seconds and minimize the false alarm rate.

The use of automatic firefighting systems is becoming more popular. There are various types, such as sprinkler, mist, or deluge systems—each of which operates slightly differently. The issue for immersed tunnels is to find sufficient space for the pipework and valves, which can be quite large. The water can be provided from the mains, but it is preferable to have a reservoir that is usually located at one of the portals. These systems can also influence the layout of the ventilation system and the extent of passive fire protection. They are not a replacement for passive fire protection, but they enable the development and spread of the fire to be controlled, and enable the firefighters to get closer to the seat of the fire. For example, if a high-pressure water spray system is employed in conjunction with a rapid detection system, the maximum temperature developed in the fire could be limited to less than 1000°C, or at least, the duration of the effect would be substantially reduced. This reduces the quantity of hot and noxious gases to be extracted, leading to a lower ventilation capacity. Visibility would also be better for the emergency services dealing with the incident. Their use could also potentially lead to the adoption of smaller design fire loads.

Ventilation standards

The standards required for tunnel ventilation are continually evolving and designers should consult international organizations, such as PIARC, the World Health Organization (WHO), and the International Tunnelling Association (ITA) for the latest developments. There will also be local requirements from the planning and operation authorities that have jurisdiction over the tunnel. Typical maximum values permitted in the tunnel are those below which there should be no adverse effects on people. Typical values are given in Table 7.1, but requirements are changing frequently and reference should always be made to current national legislation.

The ventilation system will be designed to prevent these levels being exceeded. In doing so, there is a limit—usually 10 m/s—to the air velocity in the tunnel. Above this, it becomes dangerous to people in the tunnel, so this, in conjunction with the amount of pollution, the cross section, and the length of the tunnel, becomes a major factor in the design of the ventilation system. The monitoring systems in the tunnel will activate the ventilation

Table 7.1 Typical limits on pollutant levels

Pollutant	Threshold value
Carbon monoxide (CO)	120 ppm for 15-minute duration
	65 ppm for 30-minute duration
	45 ppm for 45-minute duration
	35 ppm for 60-minute duration
Nitric oxide (NO)	25 ppm
Nitrogen dioxide (NO₂)	3 ppm
Obscuration (measure of visibility)	0.005 per meter

system at trigger levels below these threshold values so that the pollutants are diluted and the threshold values are not exceeded. If the monitoring systems show that the levels are likely to be exceeded, then action to reduce traffic in the tunnel or even to close it temporarily should be implemented.

Longitudinal ventilation

The simplest form of ventilation in an immersed tunnel is the longitudinal system. This uses the traffic bore itself and sufficient air is pushed through it to maintain the pollutants at safe levels. There are two main ways of providing this ventilation. The most basic is to have jet fans mounted in the roof at intervals along the tunnel. Air is drawn in at one portal, blown through the tunnel by the fans, and is blown out through the other portal. This is shown in Figure 7.3.

This system is used widely in immersed tunnels because of its simplicity and the fact that the traffic bore itself is used for ventilation, thus saving space in the cross section. In some older designs, the fans were placed at regular intervals along the tunnel above the traffic envelope. The space required for the fans is approximately one meter, so this extra height over the full immersed length was a considerable penalty. Later designs have placed the fans in niches in the tunnel roof, thereby reducing the tunnel height over the majority of the tunnel length. The fan niches may increase the height of the tunnel element locally, in which case, they preferably should not be located under the navigation channel, or some of the advantage of having shallower elements is lost. It is also sometimes possible to locate the fans in roof niches without having to raise the external tunnel roof line. This arrangement gives the lowest height of tunnel element as no extra height is needed to accommodate the fans, but it may involve a greater number of smaller-diameter fans. In a segmental tunnel, the niches

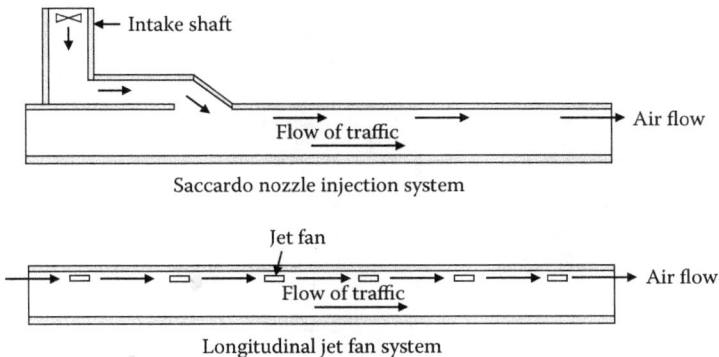

Figure 7.3 Longitudinal ventilation (developed from FHWA).

can be contained within a single segment length for ease of construction. The options for positioning fans are shown in Figure 7.4.

The pushing effect of the fans does not have to be distributed evenly along the length of the tunnel. It is, for example, possible to locate all the fans near the portals in the cut-and-cover sections, dispensing with the need to have fans in the immersed part of the tunnel, altogether minimizing the height of the tunnel and, therefore, the cost. The ventilation design has to assume that some fan power is lost in an incident—for example, the fire could be directly below a bank of fans, quickly putting them out of action.

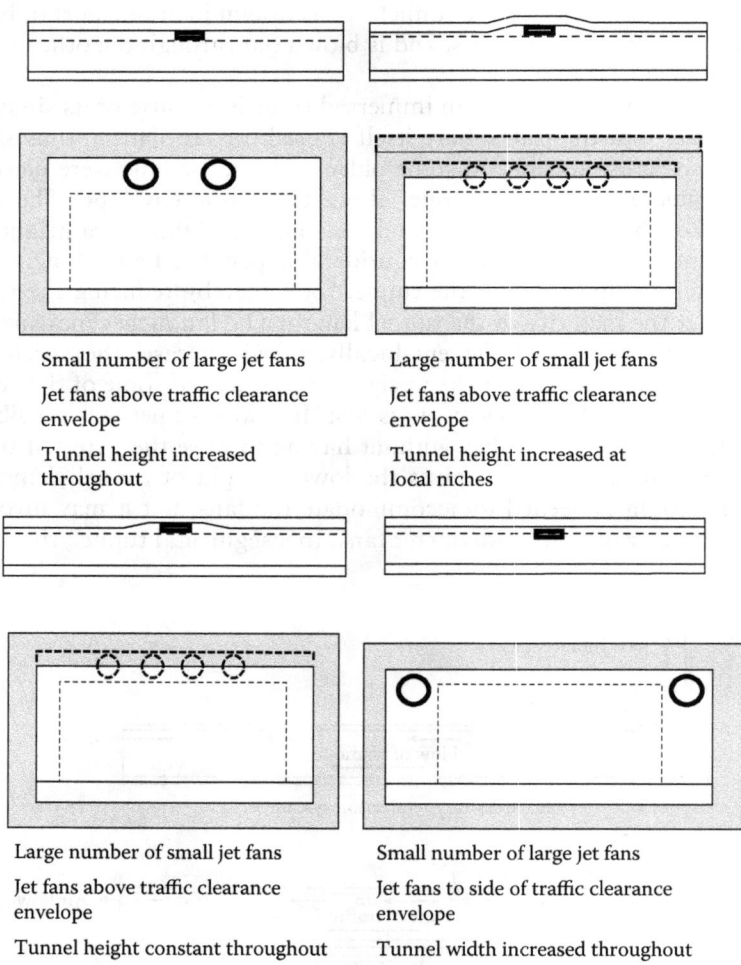

Small number of large jet fans

Jet fans above traffic clearance envelope

Tunnel height increased throughout

Large number of small jet fans

Jet fans above traffic clearance envelope

Tunnel height increased at local niches

Large number of small jet fans

Jet fans above traffic clearance envelope

Tunnel height constant throughout

Small number of large jet fans

Jet fans to side of traffic clearance envelope

Tunnel width increased throughout

Figure 7.4 Location of longitudinal ventilation jet fans—cross sections and long sections for four options.

If all the fans are clustered closely together in a cut-and-cover section, then more redundancy may have to be built into the system by having more fans.

An alternative form of longitudinal ventilation is the Saccardo system, also shown in Figure 7.3. Air is blown into the tunnel through the roof via a nozzle close to the tunnel portal. This does away with the need for fans within the tunnel altogether. The fans are located in a building above the tunnel and blow air into the tunnel at high velocity, so not only is there space saving in the tunnel, but the maintenance of the fans is simpler and safer as they are not in the tunnel.

Longitudinal ventilation is the most common form of ventilation in immersed tunnels as it does not require extra space to duct air either in or out of the tunnel. The fans do not necessarily operate all the time as the piston effect of the traffic provides a level of natural ventilation.

The length of tunnel that the longitudinal system is appropriate for is dependent on the traffic flows, the percentage of heavy goods vehicles, and the level of vehicle emissions. As vehicle engines are becoming more efficient and the level of exhaust pollutants reduces, the method becomes applicable to longer tunnels. Currently, road tunnels up to 4–6 km in length can be ventilated with a longitudinal system, with the polluted air being discharged through the portals. As immersed tunnels generally cross waterways, it is not usually possible to introduce intermediate air entry points to extend the length of the system as would be possible for a land tunnel. Intermediate ventilation islands have been considered for very long tunnels, but they are not only expensive to construct, but are also a hazard to navigation and are best avoided, if possible. At the time of writing, current studies for the Fehmarnbelt Tunnel suggest that it will be possible to ventilate this 19 km long tunnel with conventional longitudinal ventilation, with no intermediate islands.

With a longitudinal ventilation system, the air in the tunnel becomes more polluted the further it goes along the tunnel as it picks up more and more pollutants. This polluted air is generally blown out of the tunnel through the traffic exit portal. This may result in an unacceptable increase in pollution around the portal, particularly in a residential area. If this is so, then the air can be drawn out of the tunnel through the roof near the portal and exhausted through a ventilation shaft on top of the tunnel. These shafts reduce the impact of the exhaust gases at ground level by discharging the pollutants at high level, allowing them to disperse over a wider area. The shafts do have a visual impact on their surrounding area, but can be architecturally designed to be the landmark symbol for the tunnel.

An alternative is to pass the polluted air through electrostatic precipitators. These clean the polluted air by passing it between electrically charged plates. The particulates in the air stream become ionized and are electrically attracted to the collecting plate, thereby cleaning the air. The material collected on the plates is removed either mechanically or hydraulically.

There are two types of installations, depending on what the system is required to achieve. If it is to protect the environment around the portal, then the precipitators need only be installed at the portal to clean the air before it is discharged to the atmosphere. If, however, it is to clean the air within the tunnel, then precipitators will have to be installed at intervals along the tunnel and the cleaned air returned to the tunnel flow. In this way, precipitators could, in theory, be placed at 1–2 km intervals along a tunnel and used as part of the ventilation system for long immersed tunnels. These are being developed mainly in Japan and Norway and have been installed in some tunnels in both countries. These precipitators do, however, require a large amount of space, either above or alongside the tunnel to house the collecting plates. They also use large amounts of energy and the waste material has to be collected and disposed of. Consequently, their use is likely to be of overall economic or environmental benefit in relatively few specialized situations.

A common feature of longitudinal systems is an anti-recirculation wall at the portals. This is a 25–30 m length of central wall at each portal between the two carriageways. This stops the air that is being exhausted from one traffic bore being sucked straight back into the other traffic bore.

Ventilation in rail tunnels

In rail tunnels, the ventilation requirements under normal operating conditions are much less onerous than for road tunnels. Most modern railway systems use electric traction systems although, even with these, some provision has to be made for diesel engines that may be used in emergency situations and for maintenance. The critical design situation is again the fire scenario, where the system has to be designed to cope with the smoke from a train fire. The principles are the same as for road tunnels and the solutions similar, with longitudinal ventilation being the simplest and cheapest option.

A factor in rail tunnels that does not occur in road tunnels is the consideration of the pressure transients as the train passes through the tunnel. As a train passes through the tunnel—especially if it is a long tunnel or a high-speed train—the aerodynamic resistance quickly builds up. If this is not relieved in some way, then the power required to drive the train increases considerably. The usual way to relieve this pressure is to provide pressure relief ducts between the railway bores at regular intervals throughout the tunnel. Generally, however, these are not needed in immersed tunnels as they are relatively short and the blockage factor (the ratio of the cross section of the train to that of the tunnel) is not large enough. What does have to be considered in design is the overpressure caused to the tunnel structure as the train passes through. This can be determined, for example, from Eurocode 1. The designer also has to consider the stability of the tunnel

Figure 7.5 Øresund Tunnel portal with sunscreen (left) and rail pressure relief (right). (Photo courtesy of Øresundsbron.)

fixings and, in particular, any duct covers or doors within the railway tunnel, which may react to the pressure transients caused by the train.

These pressure transients have to be dissipated as the train enters the tunnel, so it is typical, where high speed trains are concerned, to provide pressure relief louvers at the portals (Figure 7.5). These consist of openings in the approach tunnel roof, wider at the tunnel entrance and gradually becoming smaller into the tunnel, which ease the pressure change as the train enters the tunnel.

Semi-transverse and transverse ventilation

There is a limit to the applicability of the longitudinal ventilation system. The longer the tunnel, the greater the amount of pollutants, and this may become too large to be diluted to safe levels simply by the air in the traffic bores. To overcome this, more fresh air has to be introduced into the system. The simplest approach is with a semi-transverse ventilation system, shown in Figure 7.6. This system requires additional bores to carry the fresh air.

Fresh air is ducted into the tunnel in separate bores and introduced to the traffic bores at low levels at regular intervals along the tunnel. These air ducts are generally larger than any required for emergency escape, so the additional space provision is considerable. They cannot be used

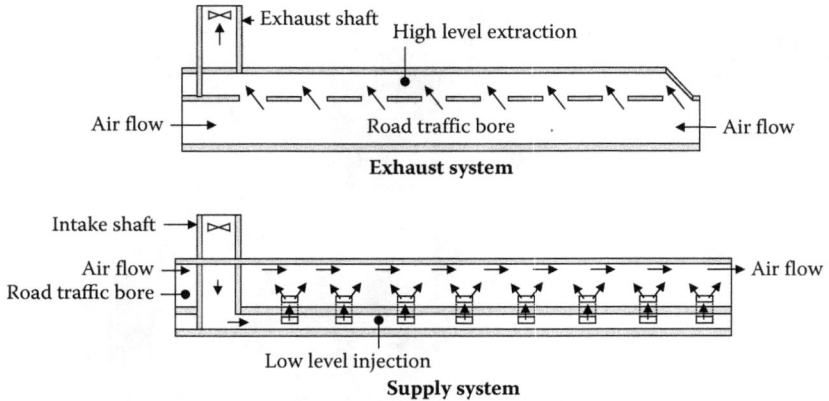

Figure 7.6 Semi-transverse ventilation (developed from FHWA).

as escape bores because the air velocities in them are too high. They can either be at the sides of the traffic bores or below them. Placing on the sides is preferable in a rectangular tunnel as additional width is cheaper to provide than additional depth. This air is then discharged through the tunnel exit portal in the same way as for longitudinal ventilation. By introducing additional fresh air into the tunnel, greater dilution of the pollutants is achieved, so the system can deal with more pollutants than a simple longitudinal system. The main drawback to such a system is clearly the additional space required for the air ducts, which adds considerably to the overall cross section and, therefore, the cost of the tunnel. Typically, these will add some 30% to the overall cross-sectional area of the tunnel. The precise value depends on the tunnel length and the amount of pollution in the tunnel.

If a semi-transverse system cannot be designed to cope with the ventilation, then a fully transverse system can be used. With this system, fresh air is ducted in to the tunnel along its length as with the semi-transverse system, but in addition, polluted air is also removed from the traffic bore along its length. The principle is shown in Figure 7.7. Thus additional fresh air is introduced along the tunnel and the polluted air is removed along the tunnel. Such a system requires yet more ducts to extract the polluted air, increasing the cross section even more. The space above and below the carriageway in a circular steel tunnel lends itself to this approach.

Semi- and fully transverse ventilation systems have different implications for firefighting scenarios in the tunnel. Fully transverse systems and semi-transverse exhaust systems enable the smoke to be extracted directly at the location of the fire. Dampers are used to control where the smoke is

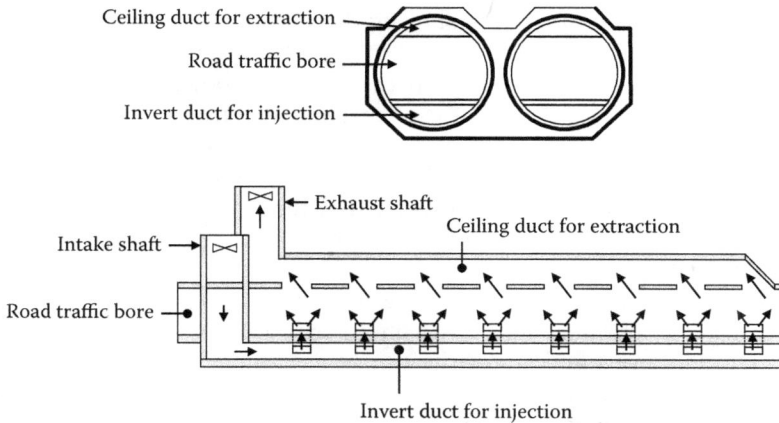

Figure 7.7 Fully transverse ventilation (developed from FHWA).

extracted. The dampers and extract ducts must be designed to cope with the high temperatures of the gases they are extracting.

LIGHTING

Good lighting in a road tunnel enables drivers to see the road ahead and to identify hazards in the road in front of them. It also provides the tunnel with a light and more user-friendly appearance, which reduces the apprehension that some drivers have about using tunnels. Practice varies around the world. Some countries, for example Norway, have long tunnels with low levels of lighting. Others, such as the Netherlands, have brightly lit tunnels. Immersed tunnels are generally in environments that require a high level of lighting.

As drivers enter a tunnel, their eyes have to adjust to the lower lighting levels inside the tunnel. This adjustment takes time, so to avoid the effect of driving into a black hole, the lighting levels are reduced progressively from a relatively bright threshold zone at the portal to the internal light level in the tunnel by a series of transition zones. Similar provisions are made at the exit from the tunnel to smooth the transition from the darker interior of the tunnel to the ambient light level outside the tunnel. As human eyes can adjust from dark to light more quickly than from light to dark, the exit threshold length is shorter than the entrance threshold.

In designing the entrance and exit lighting, the location and orientation of the tunnel must be taken into account. This is particularly so for tunnels where the portals face east or west and there are times when the sun will be low in the sky shining directly at the drivers. These circumstances will require very bright entrance threshold lighting levels. To reduce the lighting requirement in these circumstances, sunscreens are often provided over the road at the entrance portal to assist the drivers. These, however, do have maintenance considerations. In winter conditions, they are susceptible to a buildup of ice and snow, which could fall onto the traffic below. For this reason, they are not favored by some authorities, although they can reduce the amount of entrance threshold lighting required.

The lighting levels in the entrance and exit threshold zones are usually controlled automatically using photo sensors to measure the ambient external light levels and adjust the tunnel lighting accordingly. In designing the lighting system for an immersed tunnel, it must be remembered that although the tunnel will generally operate with one-way traffic in each bore, under maintenance or exceptional circumstances, it may be necessary to operate with two-way traffic in one bore. Speed restrictions may have to be introduced as drivers entering the tunnel the wrong way would be faced with the shorter exit length lighting threshold. In practice, this is unlikely to be a major disadvantage as strict speed limits are likely to be in force if the tunnel is being operated in contraflow mode.

Lighting in the tunnel is provided by luminaires in the tunnel roof above the traffic lanes. The lengths of the threshold and transition zones depend generally on the design traffic speed, and the associated safe stopping distance, as well as, possibly, the orientation of the tunnel. The alignment of the luminaires is parallel to the direction of the tunnel so that their light shines transversely on to the tunnel walls and gives a uniform level of lighting across the tunnel.

The interior finishes of the tunnel play a part and the walls should have a high reflectance value and be a relatively light color as this will reduce the number of luminaires required, with consequently lower capital cost and energy consumption. For example, in the United Kingdom, the wall reflectance is required to be at least 0.6, whereas a plain concrete finish is only 0.3. In immersed tunnels, where space is at a premium, the provision of a cladding system would require more internal space and, consequently, be more expensive, so the concrete walls are often simply painted. The paint system must provide the required reflectance value, but not give a glossy finish as this would result in problems with glare from tail lights being reflected off the wall, particularly in curved tunnels.

There is a wide variety of choice for the luminaires themselves and the technology is improving all the time. The choice should be made on a whole-life costing, taking into account capital cost, replacement cost, and the traffic management associated with this, as well as the energy consumption. LED lighting systems are now being implemented in tunnels, which is leading to further economies.

The overall efficiency of the lighting system depends on the walls retaining their reflectance. For this reason, as well as general aesthetics, the tunnel should be washed regularly to maintain the reflectance assumed in the lighting design. The interior of a tunnel is a harsh environment generally, with the vehicle exhaust pollutants, combined with the detergents used in tunnel cleaning, and the general mixture of dust and road salts brought in by the traffic. The timing of the cleaning operations is normally determined by experience as it depends on the traffic flows in the tunnel and the types of vehicles using the tunnel. The luminaires have to be cleaned as well, and for this reason, they are designed to be resistant to water penetration when they are subjected to high-pressure water jets during cleaning.

CONTROL SYSTEMS

An overall control system is essential to the safe working of the tunnel. A range of sensors is provided in the tunnel to monitor the environmental conditions, traffic flows, and to detect incidents within the tunnel. The outputs from these are normally collected by a Supervisory Control and Data Acquisition (SCADA) system. This receives and processes the information and automatically determines the appropriate operating environment for the tunnel. So, under normal conditions, the operation of the tunnel will be automatic, but always with a provision for manual intervention, if required.

The design of such SCADA systems is outside the scope of this book, as it is common to all types of tunnels. System capability and reliability are ever improving, but there are a few general principles that should be adopted. The system should be designed with a high degree of redundancy and must be fail-safe in that the failure of a particular system should result in the associated equipment automatically going into a safe mode. An uninterruptible power supply should back up the SCADA system and consideration should be given to the effects of fire or power failure. The equipment also needs to be robust enough to operate in the tunnel control room environment, which may look like an office, but is really an industrial application.

POWER SUPPLY

The power requirements of a modern tunnel are considerable, being in the order of several MW. To provide a reliable supply to the tunnel, it is common to have two independent feeds to the tunnel that are from different parts of the local network. Each supply will be sufficient to power all the tunnel systems, the idea being that if one supply goes down, the system automatically routes to the other to maintain the tunnel systems. As the two supplies are on different parts of the network, they should not go down together.

However, there is always a possibility of both systems failing simultaneously, and to cope with this, back-up diesel generators are provided. These cannot maintain all the tunnel systems, but are designed to provide a reduced level of lighting and pumping. Generally, the generators are not required to power the ventilation system. To provide continuous power, a battery back-up system (an uninterruptible power supply, or UPS) is also provided, which cuts in immediately after there is a loss of supply and maintains the minimum systems until the generator powers up.

DRAINAGE

Immersed tunnels nearly always cross waterways, so the lowest point in the tunnel will be at approximately its center. It is therefore necessary to provide drainage sumps to collect water in the tunnel and accommodate pumps that will discharge the water from the tunnel. Immersed tunnels are a very dry form of tunnel construction, but because they are placed directly in the water, it is often thought that they will leak to some degree. The opposite is true. The construction of the tunnel elements is carried out above ground and is generally of high quality with a high standard of watertightness, so it should be watertight with any subsequent leakage through the joints being minimal. With very few exceptions, immersed tunnels are virtually free from leakage through the fabric of the tunnel. Drainage systems are therefore designed with no allowance for leakage water inflow. Of far greater significance is the water from cleaning operations, accidental spillages, or broken fire mains, and fire mains running during testing or firefighting.

A drainage sump is provided at the low point of the tunnel. On longer tunnels, there may be more than one low point, requiring more than one drainage sump, or a number of sumps to enable staged collection and discharge through the tunnel. Because space is limited, the mid-tunnel sump is made as small as reasonably practical; a capacity of 50 m³ has come to be adopted as the norm for road tunnels. However, the specific conditions of each tunnel should be addressed to see if this is adequate. This size is a compromise that works effectively as a sump, but is small enough to be accommodated within the tunnel structure without requiring any blisters or compartments on the outside of the tunnel. Sufficient space for the sump can be found within the ballast concrete and by locally thinning the base slab.

Water is fed to the sump through gulleys and pipework in the tunnel, which collects any washing water or spillage. The pipework is set in the ballast concrete. The depth of the gullies can set a minimum limit for the thickness of internal ballast concrete. Other systems are provided to channel any other water that seeps into the tunnel, either through joints, in service ducts, or through the ballast concrete, to the sump. In the sump are a series of duty and standby submersible pumps that then pump the accumulated water to a larger sump, generally situated adjacent to the portal. From there, it is discharged to the local drainage system via oil separators or a treatment plant.

The sump is provided with level sensors that automatically control the discharge pumps. With the possibility of accidental spillages, there is a risk that hazardous liquids, such as petrol and oil, could be spilled in the tunnel and be collected in the mid-tunnel sump. These could result in an explosive mixture building up in the sump. To prevent the risk of such an explosion, the sumps are provided with sensors that will identify a concentration of hydrocarbons. These sensors are connected to automatic safety systems that will flood the sump with inert gas or foam to prevent an explosion occurring. The sump pumps would also usually be automatically disabled so that the liquid can be contained in the mid-tunnel sump until arrangements are made for its discharge, for example by bringing in a tanker and pumping directly into that rather than using the tunnel discharge system.

A common issue with the mid-tunnel sump in road tunnels relates to the covers of the sump. Access has to be provided to the sump for cleaning purposes and to maintain the pumps. It might be possible in some tunnels to do this from the side verges or from a central gallery, but often, this requires some covers in the carriageway, and the covers are larger than normal drainage manholes. Generally, covers in the carriageway should be located away from wheel tracks but, due to the size of the sump covers, this is not always possible. Considerable care is required with the seating and bedding of these covers or else they are liable to come loose and start to rock under traffic loading. If this does happen, then the covers would have to be reset, which means closing the tunnel for a period to carry this out. This is inconvenient and results in a loss of tunnel reputation, and possibly, loss of revenue for a toll tunnel.

Allied to this cover problem is the question of whether to provide a cutoff drain in the carriageway at the portal. These transverse drains across the full width of the carriageway at the portals are designed to intercept any storm water running into the tunnel and discharge it into a sump adjacent to the portal. This reduces the pressure on the mid-tunnel sump. The designer should give careful consideration to whether this is necessary—current thinking is to omit them. Transverse drains are a maintenance liability, especially if there are problems with their seating, and they are liable to become clogged with silt, requiring frequent cleaning. The normal carriageway kerb drainage can be designed to cope with storm flows in the approaches, with an increased number of gullies collecting the flow and discharging it to the portal sump. Only a small quantity of the approach flow gets into the tunnel directly via the road surface and this is within the capacity of the mid-tunnel sump.

Chapter 8

Tunnel approaches and service buildings

At each end of an immersed tunnel, it is necessary to construct some in situ works to form the tunnel approaches. These tunnel approaches generally comprise a length of cut and cover tunnel construction containing the tunnel portal and a length of open ramp structure that retains the ground on either side of the alignment as the road or railway lowers toward the tunnel portal. Some of the most complex engineering issues that arise on immersed tunnel projects occur in the tunnel approaches. This is due to the varying geotechnical conditions that are often encountered and the desire to create dewatered excavations in which to build the approach structures. Because of the proximity of the waterway, the approach structures are always built below, or partly below, the groundwater table. The method of excavation therefore requires the groundwater flow to be cut off by some means of temporary works, so that water cannot enter the excavation. In addition, the tunnel approach areas are often used as a temporary construction yard in which to build the immersed tunnel elements, and the need to optimize the arrangement to suit both temporary and permanent requirements may generate some unusual design conditions to consider.

The choice of permanent works and temporary works solutions for the approach structures can have a major influence on the cost and program of any particular project. It is therefore very important to make the right choices early on. Making good use of the temporary works by incorporating them into the permanent works may yield cost savings, although this can result in some complex construction sequences and design details.

As well as the structures containing the road or railway, the tunnel approaches are usually the home of the operation and control center for the tunnel. This can vary from a set of simple plant and equipment rooms to a fully manned operation center with maintenance crews and equipment, breakdown recovery vehicles, workshop, and storage facilities. These can be above ground, below ground, architecturally splendid, or unobtrusive and discreet. Depending on the need and the requirement of a tunnel owner and operator, all these solutions can be accommodated with some careful planning.

This chapter describes some of the complex issues associated with the approach structures and sets out the wide variety of choices commonly available to assist the designer or construction planner in determining the solution that is appropriate for their particular project. The options and possibilities for service buildings are also described.

FORMS OF APPROACHES

The form of approach structures may be dictated by the constraints of space or the method of achieving groundwater cutoff for construction. Equally, the presence of the water table and the method to resist uplift may be factors in deciding the appropriate form of structure. However, the forms of structure for the cut and cover section and the open ramps are chosen from one of two basic types: (1) in situ construction within open cut or (2) construction using embedded walls and in situ slabs. There are many texts and references that describe these forms of construction in detail. This chapter aims to provide an overview of the possible forms of construction and to identify any particular features to be aware of when these forms are used in immersed tunnel crossings. One example where the approach ramp structures were floated into position and immersed in the same manner as the tunnel elements is also briefly described.

To construct the tunnel approaches within a dewatered excavation, either suitable impermeable strata must exist, or the form of construction needs to provide a watertight barrier. The presence of impermeable strata enables the most cost-effective solution to be developed. Typical options for achieving groundwater cutoff are shown in Figure 8.1. These are shown for a cut and cover tunnel but are equally applicable to open approach ramps. A further constraint with respect to groundwater is the potential to connect the watercourse with a ground aquifer. If water is extracted from the aquifer for drinking water supply, then saline intrusion must be prevented. This could lead to the choice of structural form being constrained.

The arrangement and form of the tunnel approaches are often dictated or heavily influenced by the available space that exists, in which to construct the tunnel. Immersed tunnels are increasingly being built in urban situations. This can mean the site is heavily constrained by property, industry, or existing infrastructure alongside the tunnel, which must be preserved. Width limitations may dictate that approach structures are formed with embedded walls as opposed to open earthworks. Piled walls or diaphragm walls can be installed close to properties or structures, provided care is taken regarding the impact of vibration and ground settlement caused by wall deflections.

Adjacent properties must be protected from settlement because of dewatering of the tunnel approach excavations. Surveys of the properties to establish existing conditions are necessary if these risks arise, and modeling of

Type of solution	Ground conditions	
Open earthworks excavation In situ construction	Fully impermeable ground or above water table	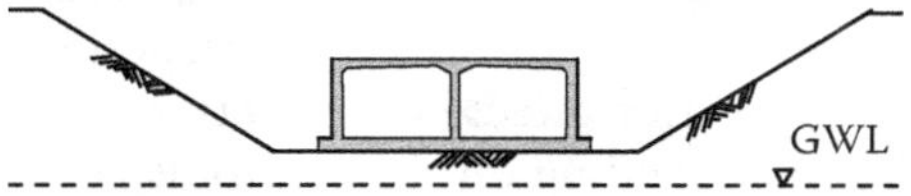
Open excavations with dewatering–sump pumping or deep wells In situ construction	Low-permeability ground	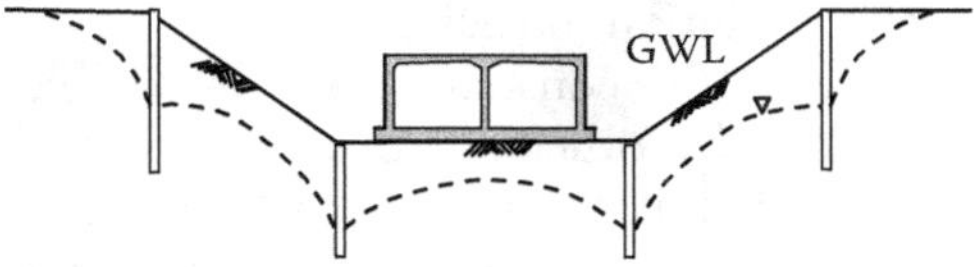
Embedded wall with in situ slabs Embedded wall with natural ground at base of excavation In situ construction Temporary cut-off wall with open earthworks within In situ construction	Impermeable strata at depth	
Embedded wall with tremie concrete base slab Embedded wall with grout plug at base of excavation	Permeable ground	

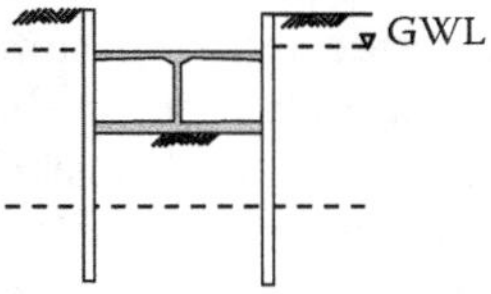

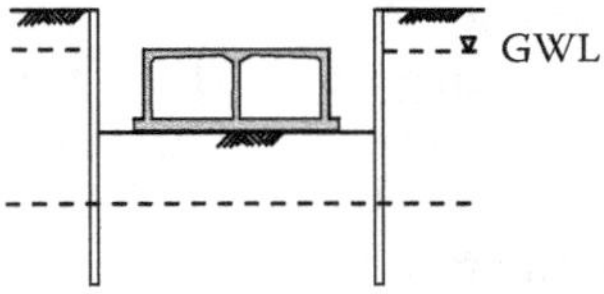

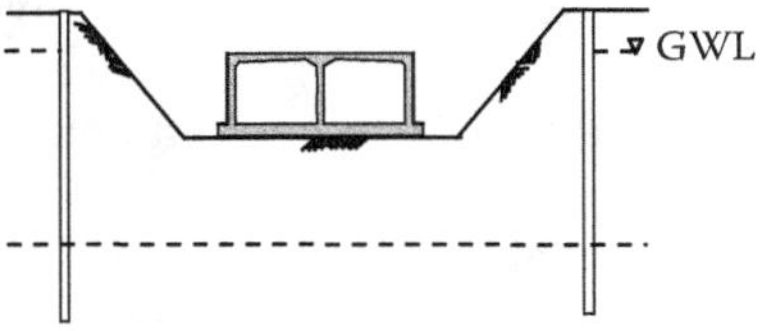

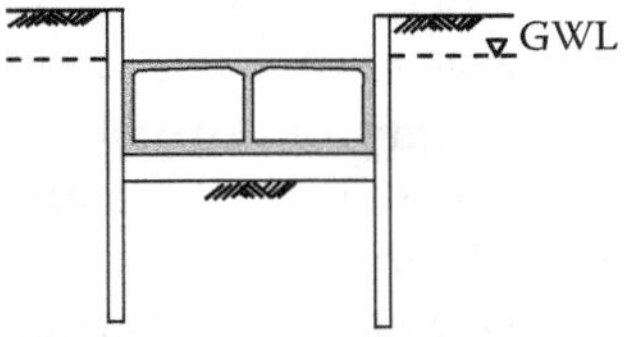

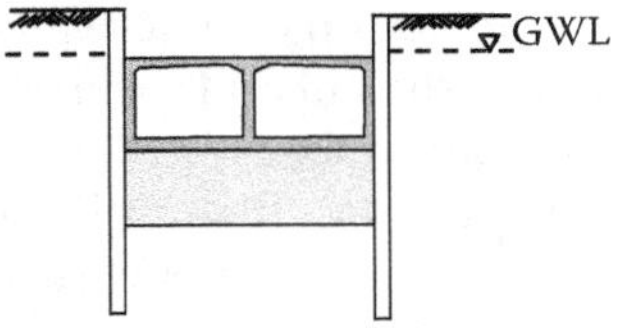

Figure 8.1 Forms of approach tunnel.

settlements to predict behavior during construction is needed. Continuing surveys throughout construction and settlement monitoring of properties and the surrounding ground are required. Acceptability criteria need to be developed, and trigger values for an action/response plan need to be agreed on by all parties affected. Protective works such as pregrouting could be employed depending on the suitability of the ground, but these techniques can be expensive and time-consuming. Forms of construction that do not require large-scale dewatering may be more cost-effective and present a lower risk solution in such situations.

Easements for construction may be needed if the works extend into land owned by others. Use of land outside the project boundaries can be arranged, but the necessary legal agreements need to be in place. It is unusual for third parties to accept permanent features such as ground anchors passing below their properties, so forms of structure need to take such restrictions into account.

Building alongside existing tunnels has its own specific set of issues regarding maintaining the stability of the existing tunnel during construction. Two recently constructed tunnels illustrate well the type of engineering problems that arise and that need to be overcome. The Tyne Tunnel in Newcastle, United Kingdom is a 1960s cast iron–lined bored tunnel that carries two lanes of traffic. Because of rising traffic congestion, a new crossing of the Tyne was proposed to double the capacity of the crossing, this time using immersed tunnel techniques. Because of land constraints and the alignment of the road network, the northern approach of the new immersed tunnel crosses above the old bored tunnel. The separation between the two structures is only a matter of meters, so in order to protect the existing tunnel from concentrated loading or from the effects of ground settlement, a reinforced concrete raft formed by a grid of deep beams was formed between the two tunnels. This also served to allow the tunnel structure above to be designed for relatively uniform settlements and avoid the risk of differential settlement close to the existing tunnel.

The second Coen Tunnel in Amsterdam was conceived out of similar desires to improve traffic flow and upgrade the highway network coming into Amsterdam from the west. A new five-lane tunnel has been constructed alongside an existing immersed tunnel. In general, the separation between the structures is between 10 and 20 m. To protect the existing tunnel from lateral instability during the dredging of the trench for the new tunnel, a stiff temporary underwater retaining structure had to be constructed between the old and the new tunnels. Detailed geotechnical analysis was carried out to assess ground movements and the impact to the old tunnel during dredging. It is a difficult exercise to model all construction activities accurately and agree with what would be acceptable movements for the existing structure, because this requires detailed ground and structure information and a sophisticated modeling approach. It is also difficult because an assessment has to be made of a 50-year-old structure to determine what amount of movement it could tolerate without its serviceability or durability being compromised. Constraints of the tunnel approaches prevented a different alignment from being taken, so the contractor was forced into dealing with this circumstance.

CUT AND COVER TUNNELS

As seen in Figure 8.1, the form of cut and cover tunnels generally falls into one of two categories: the in situ box constructed in an open earthworks excavation or an embedded wall structure.

In situ box construction

When temporary retaining works have been constructed such that the area of approach can be excavated below the groundwater level, a reinforced concrete box structure can be constructed using traditional in situ techniques (Figure 8.2).

Conventional formwork and concreting methods can be used, although there are some specific areas that need consideration. The foundation for the length of a cut and cover tunnel must take a number of factors into account. It is often the case that in the permanent condition, the net forces are directed downward, owing to fill over the box. However, as the alignment rises, the fill depth reduces, so if there is a high groundwater table,

Figure 8.2 Cut and cover tunnel construction.

the buoyancy of the tunnel may become an issue that must be addressed in the design. As the structure emerges from the ground, there may then be a gradual reduction in the buoyancy again. However, this pattern is not always the same, and careful analysis of stability against uplift should be undertaken over the full length of the cut and cover structure.

In the temporary condition, the weight of the box in a dewatered excavation may require that the settlements have to be controlled, as the bearing pressures are considerably higher than those of the adjacent immersed tube. Because of the low bearing pressures, the immersed tube generally does not require any significant foundation works, despite often being in very poor-quality soils. The cut and cover tunnel, on the other hand, may require substantial foundation works to control settlements in similar poor-quality soils, and it is common for piling to be required beneath the approach structures.

If piling is provided beneath the cut and cover tunnel, there will be a sharp change in the support to the tunnel at the transition to the immersed tube. This needs to be carefully analyzed for the permanent condition and for the short-term settlements that the immersed tube may experience during installation and backfill. Backfill around the cut and cover tunnel can be quite conventional. Free-draining granular materials that can be well compacted alongside the tunnel are preferable. Any requirements for groundwater cutoff using the backfill materials can be incorporated. If construction is in a seismic zone, the choice of materials to avoid liquefaction applies in the same way as for the immersed tunnel.

The watertightness of the structure should be commensurate with that provided for the immersed tunnel. Although it is arguable that the cut and cover tunnel can be accessed in the future, should its waterproofing cease to be effective, the cost of excavation and repair would be considerable, if not prohibitive. It is therefore desirable to adopt a similar philosophy to the immersed tube section. This may mean application of waterproofing membranes around the external tunnel perimeter or the use of cooling within the concrete to control temperatures during curing, in order to eliminate early age cracking.

Waterproofing of the joints between the sections of the cut and cover tunnel should be treated in the same way as the immersed tube. The in situ construction may lend itself to different types of waterbar, but the same philosophy of sealing should be applied. This is to ensure that a double seal against water is provided, often by an external waterbar in combination with an internal waterbar. The internal waterbar is often of a groutable form, which gives added security against leakage.

Joint fillers are needed to accommodate the predicted movement of the cut and cover tunnel. Cut and cover tunnels restrain the longitudinal movement of the immersed tube by their self-weight and resistance from ground friction. There must be sufficient resistance to the immersed tunnel forces so that the thermal contraction and expansion does not cause a gradual lateral shift of the structure. It is therefore normally the case that movements within any individual joint in the cut and cover length arising from temperature variations are relatively small.

Embedded wall structures

Embedded wall structures are often used to form approach structures as they have two distinct advantages:

1. Space for construction is minimized.
2. The embedded walls may be utilized for both temporary and permanent situations.

The forms of embedded wall are numerous. Some examples include

- Diaphragm walls
- Contiguous bored pile
- Secant pile
- Permanent combi-pile

All solutions will require a base slab and a roof slab spanning between the embedded walls to form the cut and cover box. Depending on geometry, ground conditions, water table, and the contractor's preferred way of working, the slabs can be installed using either top-down or bottom-up techniques. The principles of both these methods are shown in Figure 8.3.

The top-down form of construction is attractive, in that it minimizes the temporary propping required between the embedded walls as excavation takes place. The walls are installed first and then the top slab at ground level. The structure then utilizes the top slab as the prop in the temporary condition, while excavation is carried out beneath it. Temporary props may need to be installed beneath it as excavation proceeds, but this is not always necessary. Then the bottom slab of the structure is installed, using the ground to form and support the underside.

Another great advantage of the top-down construction method is that construction can continue above the top slab at the same time excavation is being carried out below, and either temporary or permanent road/site access can be established over the tunnel quickly. This may offer schedule benefits, particularly if the service buildings and plant and equipment rooms are located above the cut and cover tunnel or if there are significant earthworks. A disadvantage of this method is that access for plant and equipment for excavation and base slab construction is restricted, but with adequate planning this can be managed effectively.

Bottom-up construction   **Top-down construction**

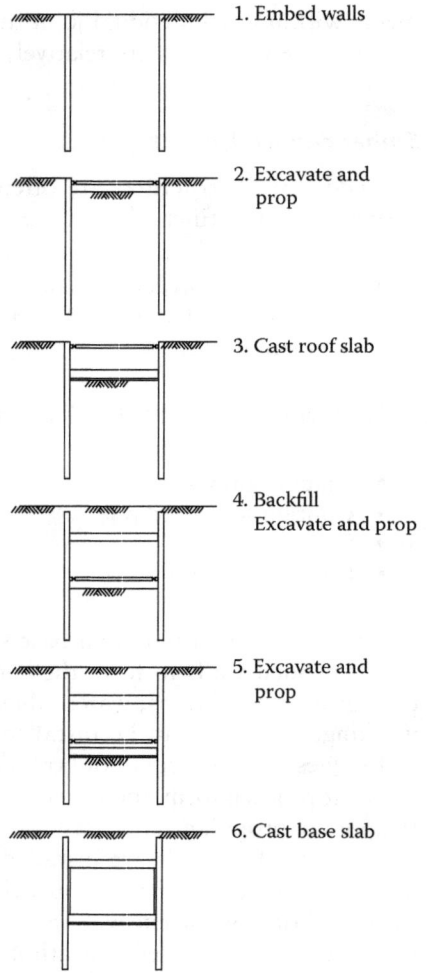

1. Embed walls   1. Embed walls

2. Excavate and prop   2. Excavate and prop

3. Excavate and prop   3. Cast roof slab

4. Excavate and prop Cast base slab   4. Backfill Excavate and prop

5. Cast roof slab   5. Excavate and prop

6. Backfill   6. Cast base slab

Figure 8.3 Top-down and bottom-up construction sequences.

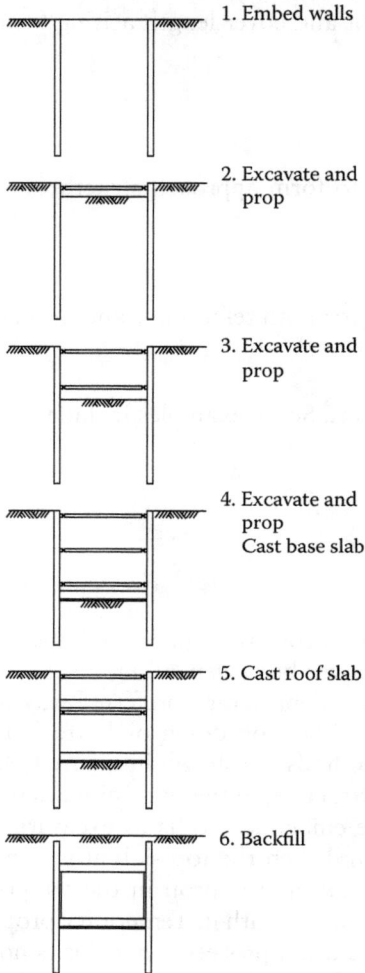

The bottom-up form of construction is the more conventional approach, whereby the walls are constructed and then excavation proceeds directly. Temporary props are installed at a high level to brace the walls as excavation proceeds, and once the base slab is constructed, the top slab falsework and formwork are erected, and the top slab cast and then the temporary props can be dismantled. Temporary props usually take the form of steel sections, either deep I-sections or tubular sections. This method is easier for excavation and hence cheaper in that respect. However, with respect to temporary construction materials, it is a little more expensive because of the need for temporary propping, falsework, and formwork to cast the top slab.

There are a number of design considerations that apply to embedded wall structures irrespective of their construction method. The sequence of working must always maintain the base stability of the excavation. The base of the excavation can be at risk if the water pressure applied to the underside of the base exceeds the resistance offered by the ground. If the embedded wall extends down into impermeable material, the thickness of the impermeable layer in the base of the excavation will determine if this is stable. If excavating in permeable materials, it may be necessary to pregrout the base of the excavation to ensure it is watertight and a sufficient depth of plug exists to counter the water pressure. Such grouting can be applied whether bottom-up or top-down construction techniques are employed.

Some forms of embedded wall are not fully watertight. This can be the case with diaphragm walls and secant pile walls. Although overall durability of the structure is not necessarily a problem, a degree of leakage or seepage might be expected at joints, and provision is often made for a secondary/lining wall in front of the piled wall. This serves two purposes: to provide an improved esthetic finish and to provide a gap to allow drainage of any leakage water.

There are a number of important design issues to be addressed on embedded wall structures with respect to watertightness and drainage, including the following:

- Waterstop detail between D-wall panels: Contractors tend to have their own waterstop arrangement that suits their equipment. Designers need to accept this but specify a performance level for the joint or any principles such as single or double waterstops.
- Watertight connections between slabs and walls: Use of membranes and hydrophilic sealants are generally necessary.
- A degree of acceptable leakage: This may need to be specified, depending on the form of construction, such that drainage can be provided accordingly and there is a quality level set for construction.
- Positive drainage to allow for potential leakage: Facing/lining walls, drainage channels, and weep pipes need to be detailed. Care is needed in the design to ensure the principles of drainage are correctly reflected. For example, if the design assumes a wall to be fully drained, the provisions for drainage need to be consistent with this. In detailing any drainage channels, adequate provision must be made for maintenance. If the drainage blocks, then water can bridge across from the structural wall to the facing walls. This ultimately will find its way to the internal surface of the tunnel and is likely to damage any internal coating or lining.

If steel combi-piles—discrete large-diameter tubular steel piles with vertical sheet piles between them—are used, the connections between slabs and the steel sections need careful detailing to ensure that watertight connections can be made. The durability of the steel sections needs to be addressed. Coatings will provide only a certain life, and sacrificial thicknesses or protection with cathodic protection might be needed. This type of solution is somewhat unusual, and it is more common for a permanent structure to be formed within the piled wall, using the combi-pile wall as a means to support the sacrificial external formwork.

Clear decisions need to be made on how the structural connections are intended to behave between the structural walls and slabs. It is common in the design process to consider connections to be pinned and to design for the maximum bending moments that result from this approach. However, joints may still have a degree of stiffness, and this should also be considered to ensure the load effects arising in the walls are correctly designed for. It is therefore usual to design for an envelope of stiffness for joints.

APPROACH RAMPS

Approach ramps may be constructed in the same manner as the cut and cover tunnels, either by in situ construction within a dewatered excavation or by utilizing embedded walls and a propping slab. However, there can be a distinct difference in behavior between the cut and cover and ramp sections. Cut and cover sections, if covered by earthworks or combined with service buildings, may have sufficient self-weight to resist uplift owing to groundwater. The structural elements of the ramp sections have relatively little self-weight and are usually naturally buoyant. The buoyancy can be overcome by adding ballast concrete beneath the road or railway or by making the main structure oversized to increase its weight. Adding ballast is by far the most cost-effective of these two methods. Although the structure needs to be deepened to accommodate the ballast, it is much cheaper to make relatively small alterations to the reinforced concrete structure and add in mass concrete than to make the reinforced concrete structure much larger overall. This is because of the relative costs of the reinforced and mass concrete. However, a third alternative, which can be cheaper still, is to introduce a foundation with tension piles or anchors that provides the necessary resistance to uplift. This approach minimizes the materials used and results in the lightest, most efficient reinforced concrete structure, which usually offsets the cost of the additional materials associated with the discrete piles or anchors penetrating into the ground below.

The ability to utilize a tension structure for the tunnel approaches is clearly dependent on the ground conditions present and being able to generate friction with the chosen tension element. A wide variety of solutions exists. Steel piles can be used, driven into the underlying ground, and cast into the base slab of the ramp structure with a well-detailed connection to ensure the longevity of the connection. Driven precast concrete piles can be used, or bored cast in place piles, with starter bars giving a good connection to the base slab.

The quality of the connection between the tension piles and the base slab is important for durability. All possible failure mechanisms should be considered when designing the piles to ensure that this connection does not become overstressed and suffer distress or fail owing to unpredicted behavior. Horizontal movement of the ramp structure due to thermal expansion and contraction must be considered with uplift forces. There have been some examples where tension piles have failed in shear at their connections to the slabs because of horizontal forces, with the result that they could no longer fully resist uplift forces from the groundwater. This type of behavior was experienced at the original Coen Tunnel in Amsterdam and required some extensive repair and reballasting during the tunnel's refurbishment some 50 years after it was originally constructed.

In competent soils, there is the possibility of using ground anchors for uplift resistance of the ramp structures. One of the biggest examples of their use was on the Øresund Tunnel, where the ramps were restrained using a grid of some 2300 large-diameter high-tensile bar anchors, with lengths of up to 17 m, anchored into the Copenhagen limestone underlying the tunnel approaches (Figure 8.4). The durability of the system was important. The anchors were buried beneath the structure, with no access for inspection or remediation over their life. Corrosion protection for the ground anchors was provided by a double protection system to ensure that a minimum 100-year design life could be achieved. A percentage of the anchor heads were left accessible so that they could be tested over the life of the tunnel to ensure there was no unexpected reduction in the bar tension. This approach should be taken whenever bar or strand ground anchors are used. In addition, a degree of redundancy should also be allowed for to ensure the structure remains stable if a small percentage of failures occur over the design life.

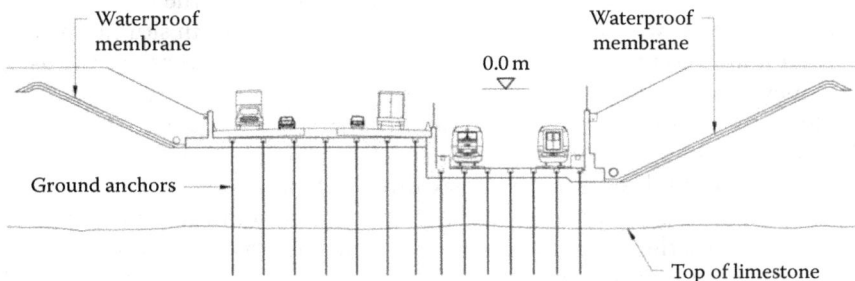

Figure 8.4 Anchored ramp structure for the Øresund Tunnel.

Whichever system is used, the approach ramps may need to be designed for different conditions through the construction period. When the approach ramps are dewatered to enable construction of the permanent structure, there will be no uplift from the groundwater, so the structures will be inclined to settle. If the underlying ground is soft material, the foundation will need to be designed to prevent unacceptable levels of settlement occurring before the dewatering is switched off and the uplift pressures become active. Therefore, it is common for piled foundations to the ramp structures to be designed for a short-term compression load regime and a long-term tension load regime.

An innovative solution was used for the Jack Lynch Tunnel in Cork, Ireland, in the mid-1990s, where the contractor chose to float in one of the approach structures in the same manner as an immersed tunnel element. There was no cut and cover tunnel, just a floated U-shaped ramp element, which transitioned directly to the immersed tunnel. This had particular advantages for this site as it avoided the need for extensive temporary earthworks and the need to install a potentially difficult temporary groundwater cutoff solution that was high risk and not guaranteed to be successful because of the underlying voided limestone. This has not been done elsewhere, and it is really a contingency solution where risks are otherwise high. It is unlikely to be adopted as a routine construction method for immersed tunnels.

There were a great many design and construction challenges that the design and build contractor had to consider. Because of the function the ramp serves, it is a naturally tapering structure and therefore would not float in the same way as a conventional tunnel element. In addition, it had no roof to mount equipment on, so it was a floating trough structure, referred to as a boat unit. Although unusual, this was overcome and the contractor devised a method to float, transport, and immerse the boat unit. This included some temporary wall extensions, to ensure there was a freeboard achieved, and additional ballast boxes externally to the main U-shaped structure, to enable sufficient weight to be achieved after immersion such that the minimum factors of safety against uplift were met. As the boat element was to be immersed to the highest point on the tunnel alignment, the immersion operation was carefully timed to utilize high tides and enable the element to be placed. The boat was immersed against a small section of in situ ramp. The normal immersion philosophy was followed, and a rubber gasket was used to achieve a watertight seal between the boat unit and the slab. Because of the low water pressures, a softer natural rubber gasket was used in place of the usual Gina SBR rubber gasket. Once placed and ballasted, the first tunnel element could then be immersed against the boat unit in the normal way.

The design of the boat structure required some specific measures. In particular, because the U-box was quite wide and subject to uplift water pressures, the base slab would tend to hog, causing the structure to flex and open out. To prevent this, the base slab required a high level of stiffness and required transverse stiffening beams. This made the structure quite expensive compared to a conventional approach ramp structure but was necessary because of the circumstances of the project site.

When the road or railway alignment in the tunnel approaches has risen from the tunnel to a certain elevation, it is possible to switch from a structural solution to an earthworks and ground membrane solution. This requires a watertight, tough flexible membrane that is buried in the earthworks around the approach ramp that prevents groundwater intrusion into the tunnel approaches. It can pass beneath the ramp structure, but this is unusual as it would require additional excavation and dewatering to install. More commonly, it attaches to the sides of the ramp structure and extends into the earthworks to the side. This allows the height of the structural retaining walls to be kept lower than the level of groundwater table, as they are no longer the means of providing watertightness (Figure 8.4). A ground membrane will still be subject to uplift pressures and has to be checked against prescribed factors of safety to prevent it displacing.

This solution was used on the Øresund Tunnel, which had a very wide approach ramp because of the combined road and railway. A structural solution for such a width would have been very expensive, so a membrane solution was adopted. The membrane needs to be sufficiently low beneath the road to allow a full depth of pavement construction and to enable road drainage to be installed. Similarly for a railway, it must allow enough depth for the railway foundation and ballasting as well as track drainage and installation of service ducts and draw pits. Some leakage of the membrane was experienced on the Øresund Tunnel project after construction that led to some ongoing maintenance costs through increased pumping from the tunnel drainage sumps. Overall, it was still probably the most cost-effective solution. Nevertheless, it illustrates the difficulty with this approach, in that it is highly workmanship dependent. The membrane must be tough and yet workable in order to lay. On-site welding is required, and this is potentially a source of leakages. Care must then be taken in trafficking the membrane by foot and by machine once a sufficient layer of fill is placed over it. Sand layers are needed above and below the membrane to minimize the possibility of puncture.

INTERFACE WITH IMMERSED TUNNEL

One of the first issues to decide at the preliminary design stage is the location of the junction between the immersed part of the tunnel and the in situ approach structures. The immersed tunnel method requires that the tunnel element is floated into position, but this in itself would not rule out dredging a channel into the riverbank and extending the immersed part of the tunnel under the bank. The main governing criterion though is the water level.

Immersed tunnel elements are large but have a relatively low bearing pressure on their foundations because they are underwater and only apply their submerged weight. Over large parts of the tunnel, this may be less than the weight of the soil that has been removed and the immersed tunnel results in a net unloading on the underlying soil. If part of the tunnel element rises above the water level, then the element becomes heavier and exerts greater pressure on the foundation. For this reason, during the early development of immersed tunnels, the top of the immersed tunnel roof was kept below the lowest possible water level, the lowest astronomical tide, plus an allowance for meteorological effects so that the load on the foundation could never increase owing to part of the element being uncovered at low water. This effectively defines the maximum length of the immersed tunnel in any given waterway.

Designing this way kept the bearing pressures even and reduced the amount of differential settlement and shear transferred between the elements. With many tunnel alignments, this meant that the junction between the immersed and in situ tunnel was some distance into the river. Such an arrangement requires extensive temporary works in the river to construct the in situ section. Large steel cofferdams or earthwork bunds are needed to enclose the construction area. The area inside them has to be kept free of water trying to flow both through the walls and up through the floor. In sandy areas, this would require a considerable dewatering effort. The temporary works extending into the river might also be unacceptable to the river authorities because of their impact on current flows, navigation, or the sediment transport regime. For all these reasons, there is often pressure to extend the immersed part of the tunnel so that the connection can be made under the land or at least as close to the bank as possible. One way of achieving this would be to lower the vertical alignment, but that would increase the cost of the whole crossing unnecessarily. A better method is to allow part of the end tunnel element to be above low water.

This has ramifications on several aspects of the design. As previously noted, it affects the settlement and shear transfer between the elements, and it also reduces the force in the Gina gasket. None of these issues are insurmountable, and over time designs have been developed that have pushed the boundary higher. Initially, the top was kept below mean tide level, but elements have now been placed where the top of the tunnel element is at high tide level so that it is always above water level at some stage of the tidal cycle. Placing an element with the roof slab at one end above water level is also possible. The element can be maneuvered into position at high tide with one end above water and then allowed to settle to the correct level as the tide falls.

The increasing sophistication of the structural and geotechnical analyses required and improvements in marine plant have enabled this progression in design. Another factor in determining this boundary is the extent of the works required around and over the tunnel at the riverbank. The riverbank will have to be reinstated to prevent the hinterland being flooded. Often this will require a flood protection embankment to be reinstated over the tunnel. Such an embankment places a high vertical load on the tunnel. Again it is not generally desirable to put such a heavy load on the immersed section as it increases the bearing pressure and its corresponding settlement. Such high loads are better carried by an in situ structure which, if necessary, can be of diaphragm wall or piled construction. A balance has to be struck between all these factors to arrive at the optimum location for the interface between the immersed and in situ sections of the tunnel.

CONSTRUCTION SEQUENCE

The first immersed tunnel element is normally placed against the in situ section on one side of the waterway, which means that construction of that in situ section has to be sufficiently far advanced to receive it. A typical construction sequence for this is as follows:

1. Construct a temporary earthwork bund or steel pile cofferdam in the river to reclaim the land required for the end of the in situ construction.
2. Install cutoff walls to prevent water coming through the earthworks. Steel pile cofferdams are largely watertight and would not require this.
3. Excavate within the reclaimed area, installing a dewatering system as excavation progresses to keep the excavation dry.
4. Construct the seaward end of the in situ section and seal it with a watertight bulkhead.
5. Construct an earthwork bund or piled cofferdam over the completed tunnel to reinstate the riverbank. This will include a cutoff wall to prevent water flowing through the bund into the excavation.
6. Remove the outer temporary earthwork bund, or cofferdams, so that the end of the in situ section protrudes out underwater and is ready to receive the first immersed element.
7. Place the first element against the exposed end of the in situ section.

This sequence is illustrated in Figure 13.9 in Chapter 13, which describes the earthworks. At the other end of the tunnel, the approach will be constructed similarly to enable the immersed tunnel to connect with it. If the tunnel elements are placed first, before the cut and cover tunnel is constructed, then a bund is constructed over the last tunnel element. This bund will also connect to the general earthworks around the approach. It then forms an enclosed area, which can be dewatered to allow construction of the cut and cover tunnel adjacent to the last immersed tunnel element.

CONNECTION TO BORED TUNNELS

Some immersed tunnels have been constructed with a transition directly from the immersed tunnel to a bored tunnel, without any intermediate traditional cut and over construction. This has been necessary where there is hard rock in the foreshore areas that does not lend itself to cut and cover techniques or where the tunnel continues at depth inland. This latter reason is why the technique has been used for metro tunnels in Hong Kong and for the Bosphorus Tunnel in Istanbul.

The technique to achieve this type of connection requires a specially modified immersed tunnel element or caisson-type structure that allows the tunnel boring machines (TBM) to drive directly into it. The Sha Tin to Central line crossing of Victoria Harbor in Hong Kong is being planned using a caisson structure, whereas the recently constructed tunnel beneath the Bosphorus utilized a modified tunnel element. There must be a chamber at the end of the immersed tunnel element to receive the TBM, which has a low-strength material through which the TBM can drive, similar to a soft-eye construction. Behind this there needs to be a bulkhead to maintain the watertightness of the tunnel element beyond. There must also be facilities to create a seal between the tunnel lining ring constructed by the TBM and the concrete structure of the immersed tunnel element.

It is possible to backfill the end of the immersed tunnel with selected granular backfill material to enable the TBM to drive through. Alternatively, a weak tremie concrete plug can be formed that the TBM could also cut through. The use of tremie concrete is favored as it gives security against instability and locks the end of the immersed tunnel in place while the TBM is driving into it. The concrete mix requires careful design to give a mix that is stable and sufficiently viscous to reduce segregation, but has the high levels of workability and pour rate needed for these underwater conditions. It is common to leave the shield from the TBM in place and weld this around its perimeter to a steel profile cast into the immersed tunnel reception chamber. However, the precise detail of this will depend on the TBM used and the details of any shield.

A typical construction sequence for forming such a connection between an immersed tunnel and a bored tunnel is as follows:

1. Dredge/blast socket into rock.
2. Immerse tunnel element with the reception chamber placed in rock socket.
3. Underfill the tunnel element with firm foundation, probably grouted.
4. Backfill around the tunnel element with granular material or tremie concrete.
5. Drive TBM into the reception chamber.
6. Complete segment lining as far as possible with precast ring segments.

7. Inflate initial seal for watertightness against the shield.
8. Complete any external grouting around the shield.
9. Remove outer bulkhead.
10. Weld the shield to cast-in steel profile for permanent watertight seal.
11. Dismantle TBM within the tunnel and remove.
12. Complete in situ concrete lining in end of the TBM tunnel extending into the reception chamber.
13. Remove inner bulkhead.
14. Continue with internal finishing works in the tunnel.

The details can be seen in Figure 10.10 in Chapter 10, which describes tunnel joints. Construction tolerances are an important consideration in determining the geometry of the reception chamber at the end of the receiving tunnel element. The most critical of these is the directional accuracy of the TBM. This will govern the clearance to be provided around the perimeter of the shield. There must be good confidence in achieving this construction tolerance as the chamber will be set tight to this tolerance with no additional clearance. It is important to minimize this space as a more reliable watertight seal can be formed.

If more than one bored tunnel is entering the immersed tube, the tunnel element reception chamber may need to be flared to accommodate converging alignments of the TBMs. The TBM tunnels will only be able to get within a few meters of each other as there is a requirement for a pillar of material to remain between the tunnels. The width of this pillar is greater than the normal width of the dividing wall in an immersed tube, so the remaining convergence of alignments will need to happen over the length of the immersed tunnel element. As the tunnel element width varies over its length and the reception chamber has an enlarged cross section to provide a collar around the TBMs, the element geometry is not uniform and the floating behavior needs to be examined to ensure the marine plant can manage the element successfully. The increased air volume at the reception chamber also increases buoyancy, so factors of safety against uplift need to be considered locally so that one end of the element is not at risk of floating up again after it has been placed.

The soft plug in the end of the reception chamber needs to be engineered to suit the TBM but would typically be a low-density aerated concrete or mortar that provides a secure watertight plug, but which may be removed by the cutting teeth or discs of the TBM. The sealing arrangement often features an inflatable seal that can close against the shield. This type of seal is chosen as it can close a large gap and can be activated remotely when access to the void is not possible.

If this type of connection is required, it is likely there will be a sharp transition from rock to soft ground. This will need careful consideration in design for any differential settlement effects. The details described above effectively form a continuous connection between the immersed tunnel and the bored tunnel. There is a risk that the bending effects in the first tunnel element may be severe and may require either of the following:

- A foundation that gives a gradual transition from hard to soft founding materials
- Adjustment of joint positions to suit the settlement profile and to prevent the differential settlement from causing severe bending moments in tunnel elements

An alternative would be to have a very short reception chamber and a tunnel immersion joint as close to it as possible.

If a caisson structure is used to connect the two types of tunnel, the techniques are similar. The caisson is placed first and has a soft eye on one side to receive the TBM and a bulkhead and Gina sealing plate on the other side to receive the immersed tunnel element. Such caissons can be extended above water to serve as ventilation towers in the completed tunnel.

USE OF TUNNEL APPROACHES AS THE CASTING BASIN

Often there is no suitable location for a large casting basin adjacent or close to the tunnel site. One option to overcome this, rather than fabricate the elements some distance away and tow them to the site, is to use the tunnel approaches as the place to build the elements. The approach structures have to be built in situ so that all the labor and facilities would be there anyway, so this method can be efficient and cost-effective with respect to site establishment and management of the construction works.

The approach structures are only the width of the tunnel, so if the tunnel elements are to be built there, they will be built in a long line. This can place a physical requirement on the length of approach needed for construction, and normally the length of in situ tunnel would not be sufficient to accommodate many elements. In such a case, the seaward end could be fitted with a removable gate or cofferdam and the elements constructed in batches. The gate is removed after each batch is constructed, the elements floated out, the gate replaced, the approach dewatered, and the next batch constructed. This is a balance between geometry and time to determine if it is feasible to build the elements within a suitable timescale for the overall tunnel construction program. Generally, the technique is used for shorter tunnels with relatively few elements.

For example, the Guldborgsund Tunnel in Denmark had only two elements constructed, one at a time in the Lolland approach. On the Limerick Tunnel in Ireland, it was possible to extend the excavation for the northern approach landwards such that all the five elements could be built at the same time in one line. Building the elements this way presents some other challenges to construction. The main one is to the program. Construction of the in situ approach structure cannot be started until the elements have been built and floated out and the approach resealed and dewatered. Thus, that particular approach is going to finish later and will probably extend the whole construction program. Balanced against this delay to the program is the cost saving of not having to build a separate casting facility.

Building the elements in the approaches means some changes to the layout of the approaches. First is the depth of the excavation. The approach structures are, by their very location, on the higher part of the tunnel alignment. The elements, however, have to be built at a depth where the water level would be close to the top of the element. To build the elements, therefore, the approach area will have to de deepened. This can require extensive temporary walls to contain the excavation within a reasonable area. It also increases the extent of any dewatering operations, although these would have been necessary in a separate casting facility anyway.

Construction of the elements themselves is similar to building them in a regular casting basin. Once they have been completed and floated out, construction of the approaches can begin. First, the basin must be made watertight and pumped dry. This will require sealing around the last tunnel element. The end of the last element has to extend into the dry section of the approach so that the in situ approach structure can be built against it.

There are basically two methods of doing this: (1) structural and (2) using an earthworks solution. The structural method is normally used when the exit from the approaches to the river is contained between structural walls, forming a short canal. The gap between the immersed tunnel element and the structural walls is sealed. This can be achieved using a combination of steel piles and sealing gaskets. The design of the canal will generally be such that the gap to be sealed at the sides of the element is about 2 m wide, requiring a couple of steel tubes welded together. When the approach side is dewatered, the hydrostatic head on these closure panels will be large, possibly up to a 15 m head of water at high tide. The side closure panels will need to be braced, often by diagonal bracing fixed to the floor of the approach.

The underside and top of the element also need to be sealed. To seal the underneath, some preparatory work is needed while the elements are being constructed. A cill beam can be constructed across the neck of the channel close to where the end of the element will be. In the vertical plane, the top of the beam will be approximately 150 mm below the underside of the element. A rubber bunny-type seal, so called because it resembles rabbits' ears, is cast into the top of the cill beam, so that when the element is placed, it compresses the bunny seal and forms a watertight seal under the element.

The top of the element is sealed by constructing a wall on top of the element. This can either be a purely temporary steel wall or could be a reinforced concrete wall that becomes part of the permanent reinstatement of the river wall. In this way, a seal is provided all around the element, enabling the approaches to be pumped dry.

As the approaches are pumped out, the end of the last element is exposed and loses the hydrostatic force that is being applied to it. In such circumstances, if there is not enough frictional restraint to prevent movement of the tunnel element, the overall compression on the gaskets between the elements is lost as the gaskets expand with the moving of the last element. Frictional restraint around the elements will limit this, but this type of movement must not be allowed to occur at all. Thus, it is normal to provide the end of the last element with some sort of bracing or temporary prop to prevent any longitudinal movement and retain the compression in the gaskets. This can be as simple as concrete wedges inserted between the end of the element and the floor of the approach. The Medway Tunnel in the United Kingdom was constructed like this and used the base slab of the approaches to brace against (Figure 8.5).

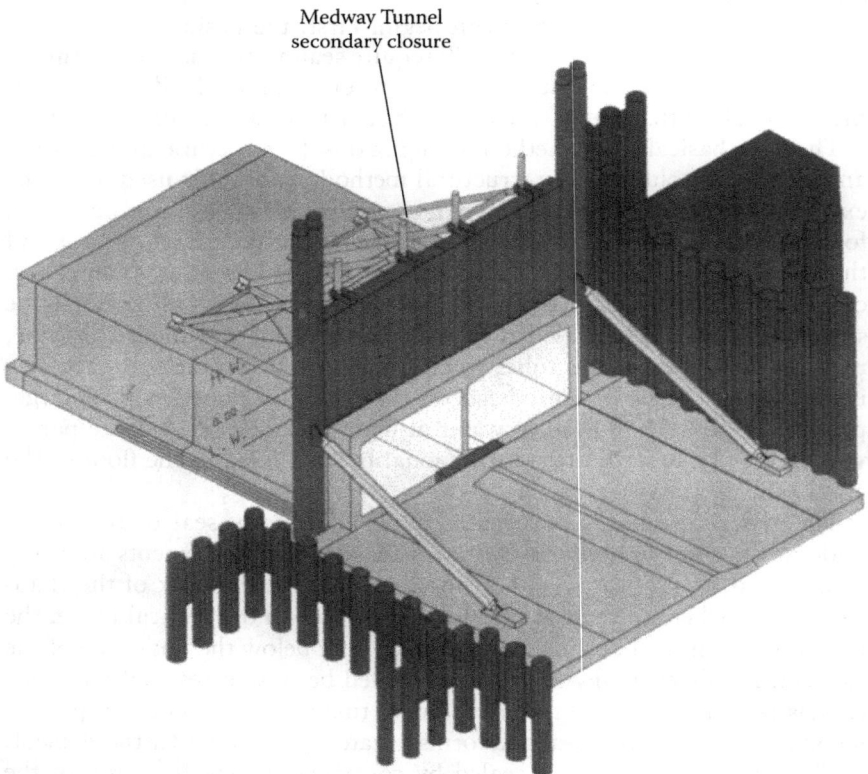

Medway Tunnel
secondary closure

Figure 8.5 Medway Tunnel end closure detail. (Courtesy of BAM Nuttall.)

The earthworks method can be used if there is more room around the end of the last element. It can also be used with a conventional casting basin. When the last element has been placed together with its foundation and backfill, an additional amount of backfill is placed over the element, generally along the line of the permanent riverbank to be reinstated. The weight of this backfill provides sufficient frictional restraint to prevent the gaskets decompressing as the approach area is dewatered.

Attention is required to the detail of the backfill to avoid possible weak points in the watertightness of the earthwork seal around the element. A suitable cohesive material should be sufficient at the sides and above the element to prevent water penetrating. Under the element, the sand or gravel foundation is a potential water path into the approaches. This can be sealed using a cill beam type of approach similar to that used in the earlier structural method. Alternatively, the permeability of the foundation material can be reduced over a length of 4 or 5 m by introducing a band of more cohesive material under the end of the element. The introduction of this fill material around, and particularly above, the last element requires a detailed analysis of the forces in the element and the associated settlements.

Once the approach area has been dewatered, construction of the approaches can start. First, the levels will have to be built up as the bottom level will be lower than required for the approaches. The area will be filled to the underside of the approach structures; then they can be constructed in situ using conventional methods.

SERVICE BUILDING FACILITIES

All transportation tunnels require a degree of operational control, which may be carried out locally to the tunnel itself or remotely. This will be dictated by the local highway or rail authority or tunnel operator and often the police and emergency services. Requirements vary greatly from country to country and from tunnel to tunnel. The service building may therefore, just be a simple collection of technical rooms or be a fully functioning operational and control center with permanent staff for operation and maintenance of the tunnel.

The first thing to consider regarding the building is the operational philosophy and whether the building will be manned or unmanned. If unmanned, the room schedule will typically comprise rooms for the following equipment:

- Uninterruptible power supply (UPS)
- Generator
- Transformer
- HV and LV switchgear

- Equipment stores
- Radio and telephone
- Communications and control
- HVAC air conditioning plant

Rail tunnels may have rooms containing specific rail control systems such as rail safety systems, interlocking systems, remote control systems, and transmission systems. Depending on the ventilation system adopted for the tunnel, there may also be fan equipment rooms and air scrubber rooms, and if manned, there will also be

- Control room
- Rest rooms
- Locker rooms
- Kitchen facilities

There may also be workshops, offices associated with tolling or vehicle recovery, and space for maintenance and operational teams over and above that needed for tunnel control. Floor areas vary considerably from tunnel to tunnel depending on the operating philosophy, the ventilation solution, and the size of the tunnel. Buildings are often of a significant size in relation to the proportions of the tunnel approach and tunnel portal. Their appearance and location, therefore, needs careful thought.

Service building location

Service buildings have to be located close to the tunnel; several options are available. They can be placed on top of the tunnel, giving direct access to the tunnel for cabling and maintenance, or close to the portal. In the latter case, they can be alongside the tunnel approaches at low level, and therefore barely visible from outside the approaches. If placed on top of the tunnel, their visual impact can be minimized by using low-rise buildings that are effectively screened by the flood protection embankments round the portal. If the tunnel alignment permits, they can still be on top of the tunnel but also be completely underground. This arrangement was adopted for the Øresund Tunnel as shown in Figure 8.6. This very sizeable service building is completely out of sight yet gives direct access to the tunnel for the cables and pipework and also allows direct access for maintenance personnel.

The relationship to the tunnel portal is the primary consideration in locating the service building. Many tunnel operators would like to have a view of the road and tunnel portal from the building. This is not essential with the typical CCTV installations that are now being used, but it could influence the siting of the building.

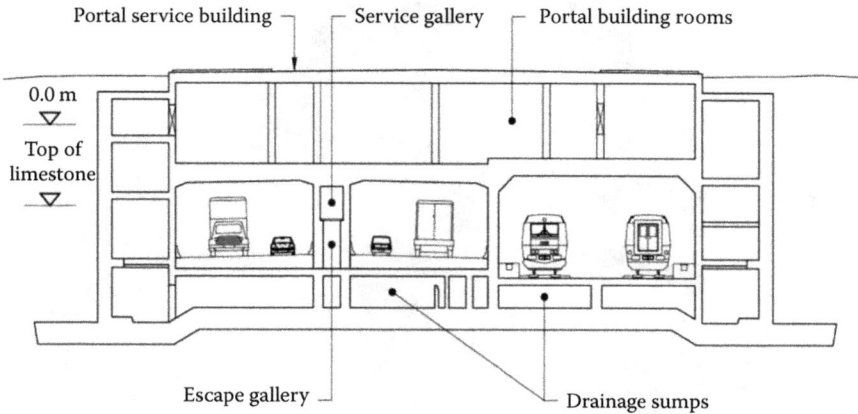

Figure 8.6 Øresund Tunnel below ground service building.

Service buildings built on the line of the tunnel have the advantage of requiring less land take and using the tunnel structure as their foundations. Although a building that is integral with the tunnel might seem a complex solution at first glance, it is in fact very efficient and can be quite simple. It can allow direct access into the tunnel from the building and service galleries for maintenance staff, and it also enables the shortest possible routes for cabling back to the plant and equipment rooms. The disadvantages are small, in that a building on top of the tunnel can be more intrusive visually, but this depends on the elevation of the building since, if land permits, the building can be spread along the cut and cover tunnel and incorporated into the earthworks and landscaping areas around the tunnel. This was done very successfully for the Limerick Tunnel (Figure 8.7). Siting the building on top of the tunnel will give rise to increased or more variable loading to the tunnel, which in turn could give rise to differential settlement of the structure, so specific foundation solutions may need to be implemented.

Offline service buildings have the advantage of permitting greater flexibility in their design as they are not constrained by the footprint of the tunnel. This can give more architectural opportunity and perhaps more opportunity to mitigate visual impact if the site is a sensitive one. The disadvantages are that a greater land take will be needed, which is not always available, and separate accesses may be necessary. From a functional and cost perspective, the longer cable routes between building and tunnel are less desirable, and if ventilation towers and plant rooms are needed at the tunnel portal, then there may be a number of split function buildings required. This can be made to work and has been the approach followed for some tunnels, with a series of separate cheaper simple buildings rather than one combined building. If the tunnel is tolled, toll booths can be at some

Figure 8.7 Limerick Tunnel service building. (Photo courtesy of NRA/DirectRoute.)

distance from the tunnel portal and it is common to have separate buildings to deal with the tolling operations.

Aesthetics

The architecture of the service building is likely to be important to the local community. The location of the tunnel in terms of its urban or rural environment will play a part in choosing the right approach. An example is the Limerick Tunnel, which is seen in Figure 8.7, where because of the low-lying area, the project owner desired that the tunnel building should not protrude above the level of the flood protection bunds alongside the River Shannon. This minimized the visual impact greatly and kept the rural nature of the setting intact.

A number of tunnel projects have used naval architecture to influence the form and appearance of the service buildings quite effectively. Where ventilation towers are required, it is difficult to minimize the visual impact, but some tunnels have made a dramatic statement with the architecture rather than trying to disguise them.

LAYOUT OF TUNNEL PORTAL AREAS

Determining the layout of the portal areas requires a number of functional issues to be considered. These differ significantly between rail and highway tunnels.

Railway tunnels

Railways tunnels generally have quite functional portal structures. The tunnel approach is not highly visible to rail passengers, so aesthetic considerations do not need to be given high priority. This means solutions will be primarily driven by cost, and this usually results in simple U-shaped reinforced concrete ramp structures and cut and cover structures that simply provide the minimum clearance requirements around the railway.

Consideration needs to be given to emergency access into the tunnel for emergency services and for evacuation and maintenance access. However, these requirements can usually be met with simple walkways alongside the rail tracks and no additional space is generally provided. Where possible, it is more cost-effective to locate plant and equipment at ground level out of the tunnel approach and to locate switches and crossings at ground level some distance back from the tunnel portal.

Ease of maintenance and cleaning is an important consideration, as well as safety, so material selection and surface finishes should be designed with this in mind.

Highway tunnels

For highway tunnels, the layout of the portal area needs to consider the following:

- What access is needed to service buildings?
- Where are service buildings located?
- Are lay-bys or hardstandings required in tunnel approaches for emergency vehicles?
- Are crossovers required in the central reserve for traffic management?

In addition, the appearance of the tunnel portal to tunnel users is more important for a road tunnel, and the locality will affect the layout and may drive the aesthetics of the tunnel portals. Considerations include

- Are there land constraints that restrict the width of the tunnel approach?
- Is the tunnel urban or rural?
- Are there driver comfort issues to be considered (e.g., avoiding black hole effects)?
- Is there a desire for a particular architectural design for the tunnel portal?

Individually, these are perhaps not all peculiar to immersed tunnels, but the combination of the functional requirements and the influence of locality tend to combine to lead to common solutions.

Access

Access will be needed to the service buildings, and if these are located above the tunnel at the portal, a service road is needed. Typically, a service road will loop over the top of the tunnel portal, so that access into the service area comes from one direction and access out is returned via the opposite carriageway. This avoids the need to return to the same carriageway and have to pass through the tunnel following a visit to the service building. However, the optimum arrangement needs to be developed with respect to the local road network and desired methods of operation.

Parking areas for operational and maintenance staff may be needed, along with access to the service building for loading/unloading and for removal or maintenance of large items of plant and machinery. Provision of holding areas for service vehicles might be necessary, so there is an area for vehicles to wait ahead of carrying out maintenance work, and an area to tow broken down vehicles to, clear of the main carriageway. It is difficult to locate this at the same place as the service building as, for a service building close to the tunnel portal, it may be difficult to get direct access to the tunnel carriageway. These facilities are more often located further up the tunnel approach road separate from the service building area. If a tolling station is provided, it is likely to be located there.

Lay-bys, hardstandings, and hard shoulders

It is important to keep the space within the approach ramp structure to a minimum in order not to increase costs, so it is common practice to minimize the presence of lay-bys, hardstandings, and hard shoulders within the ramp. However, requirements might be dictated by national design standards and there may be other reasons for including some of these features, such as emergency services wishing to have an emergency lane or hard shoulder of sufficient width for them to get emergency vehicles to the tunnel portal quickly in the event of stationary traffic in the tunnel approach.

The need for these features should be established through discussions with emergency services and the tunnel operator to ensure that the necessary safety provisions are in place and the emergency services and maintenance staff can respond to an incident in the tunnel as they desire.

Flood protection

The layout needs to consider the flood protection aspects covered in Chapter 6. Primarily, this requires a flood protection bund or wall around the perimeter of the ramp up to the point where the road alignment rises up to the flood protection level.

Tunnel portal aesthetics

We have already discussed the aesthetics of service buildings. The actual tunnel portal will also need to be considered. Again the requirement is different for every tunnel and will be dictated by the surrounding environment, owner's preferences, and sensitivity of the local community and tunnel users. Some tunnel approaches are highly constrained and are limited in what can be done other than selecting appropriate materials and finishes. Other tunnels constructed in green field sites or in open environments provide greater opportunity to apply imagination and to respond to the environment, owner, and community. One aspect of aesthetics that is often considered is eliminating or reducing the impression that the driver is plunging into a black hole from an open environment. This can be achieved with a number of simple measures, such as

- Gradual transitions of retained height approaching the portal. This might involve a mix of retaining walls and earthworks.
- Lighting within the tunnel approach and in the transition zone within the tunnel.
- Use of daylight screens.
- Use of painted surfaces at the tunnel entrance and within the tunnel.
- Avoiding steep gradients in the approach. This can only be applied in a limited fashion as it will have a big impact on the project cost if gradients are too flat.
- Keeping the transition between open cut and cut and cover tunnel as high on the alignment as possible. This will also be driven predominantly by cost.
- Use of planting and horizontal alignment curvature in the approaches to the portals.

A high-level or architectural design was carried out for the Øresund Tunnel. This resulted in a high-quality finished construction that was not dramatic but functional. White finishes were required with other selected materials at the tunnel entrance. The portal structure roof was set at ground level with plant, equipment, and operational rooms arranged within the tapering space beneath. Some tunnel owners have made more bold use of architecture at the tunnel portals. This has been done on a number of tunnels in the Netherlands with striking results. The use of planting and landscaping can greatly improve the appearance of tunnel approaches.

Chapter 9

Design principles

Immersed tunnel structures are designed to limit state design principles in the same way as any other modern piece of transportation infrastructure. In many respects, the structural design is not unusual, albeit some of the applied loadings may be. This chapter describes the process for designing a tunnel with the necessary characteristics to be floated and immersed. It also describes the loads that should be considered in the design of an immersed tunnel—in particular, the unusual loads that are specific to the circumstances of an immersed tunnel—and the methods for the analysis and design of the structure for its temporary and permanent condition. Some noncodified aspects of design are also described.

STRUCTURE SIZING

In Chapter 4, we described the early planning process for deciding the arrangement of the tunnel and the basis for performing the preliminary sizing of the tunnel. This takes us to the point where the functionality of the tunnel is determined; the basic layout of traffic tubes is chosen; and some key parameters, such as the alignment and ventilation system and other fire life safety equipment, have been decided on. Before the design moves to a more detailed phase, it is necessary to firm up these preliminary decisions. In essence, you need to know what your tunnel is carrying and the space it requires before you can do much else. Once this is decided, we can move on to sizing and designing the structural members and looking at the stability and buoyancy characteristics of the tunnel.

The first consideration in sizing is to fix the internal traffic envelope for the vehicles that will be using the tunnel, or the envelope required for railway rolling stock. In a road tunnel, the traffic envelopes are usually dictated by national design standards. Most highway design standards will set out minimum headroom and width requirements based on the speed of traffic, lane width, and minimum headroom. When selecting the traffic clearance envelope, the designer should be clear whether the height allows for flapping tarpaulins on tall trucks (typically about 250 mm allowance) as this is sometimes excluded from the basic clearance. Once the traffic clearance envelope is defined, other items can be added. These include any horizontal and vertical clearance above the footway, walkway, or hard strip that sits alongside the running lanes. The type of footway and its width will, again, normally be set out in national highway design standards. The requirements vary considerably from country to country. The United Kingdom and Ireland have used curbs and low-level footways in all their immersed tunnels, whereas in continental Europe, a simple hard strip and New Jersey barrier is much more common. In the United States, a raised walkway is typical and is preferred, to allow safe maintenance access into the tunnel by foot. This provision needs to be agreed on early and added to the traffic clearance envelope.

For a rail tunnel, the kinematic envelope must be obtained for the railway rolling stock that is planned. For a new railway, this may need to be generated by the rolling stock supplier, but for established railways, it may already be in existence. Kinematic profiles will be required for both the straight line condition and the canted condition if there is any horizontal curvature on the rail alignment through the tunnel. Spatial allowance must also be made for any overhead electrification that can require substantial space for the catenary support. Low-height support systems are available for tunnels, but are not commonly used and would need to be agreed with the train operator in advance. The depth of the trackform also needs to be assessed. Usually, the trackform within tunnels is a nonballasted system, with either concrete sleepers set into an in situ slab, or a booted block system that is similarly set into concrete. This will have a minimum depth below the lowest rail level.

There will be a minimum space requirement for the mechanical and electrical plant and equipment. This is described in Chapter 7, but the key items to consider are whether a separate service tube is going to be provided, or whether space needs to be found under the roadway for services (this is not typically an issue for rail tunnels). In addition, if the tunnel is to be longitudinally ventilated, with ventilation fans spread throughout the length of the tunnel, some space above the road or railway may be needed to accommodate them or niches provided in the roof slab so that the fans can be set into these recesses, thereby taking up less space. Otherwise, a minimum clearance of about 300 mm is usually proved above the traffic clearance envelope to accommodate signs, lighting, and cabling.

The allowance for the alignment needs to be included. This might be a super elevation if the highway is curved horizontally, or if not, a simple cross fall of typically 2–2.5% to ensure the carriageway will drain satisfactorily should any liquid be spilled onto it. In addition, some extra horizontal width may need to be added if the curvature is tight and space needs to be provided to achieve the required sight lines. Depending on how the tunnel structure is to be built, there may also be a need for some increased width to accommodate a curved alignment within a straight tunnel element or straight segments making up a tunnel element. Although this may be only a few millimeters, it should not be omitted as all the requirements added together could come to a significant additional width that is needed in the tunnel cross section. This would apply equally to road or railway tunnels.

The final item that needs to be allowed for is construction tolerances. This is made up of three parts:

1. Element placing tolerance: Steps in the alignment of the tunnel elements or angular deviations from the intended alignment may arise due to the construction process, as shown in Figure 9.1. Typically, elements can be placed with a horizontal accuracy that is within ±50 mm. The amount to allow for will depend on the particular contractor and the equipment he wishes to use to immerse the element. This will also depend on the length of element—a longer element needs to have a greater allowance as the risk of misalignment may be higher. In the absence of a contractor available at the time of design, the ±50 mm is a reasonably safe starting point.

2. Settlement tolerance: Settlement tolerance may need to be allowed to cater for unexpected differential settlements between elements when first placed onto a gravel bed or when elements are released onto a sand foundation, before the shear key connection is formed between the elements (Figure 9.2). A typical allowance in the internal clearances might be 25 mm. Global settlements are less critical to alignment, but the designer may choose to add an allowance for that too. Typically, a global settlement would be no more than 100–200 mm, and the inclusion of additional space in the section would only be applied if the settlement would give rise to a risk that the alignment criteria for the roads or railway might not be achieved.

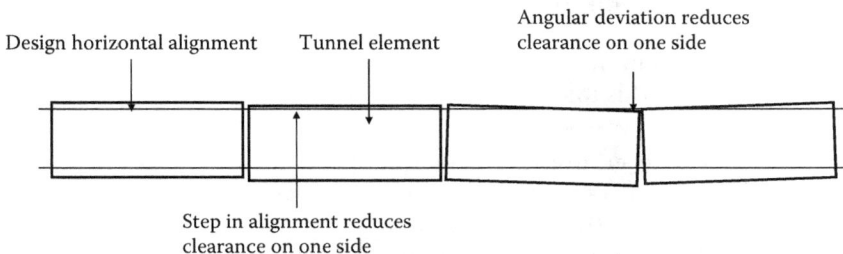

Figure 9.1 Effect of horizontal placing tolerance (plan view on tunnel elements).

Tunnel element

Design vertical alignment

Foundation

Settlement occurred

Extra space allowance needed in headroom or there
will be insufficient clearance for vehicles

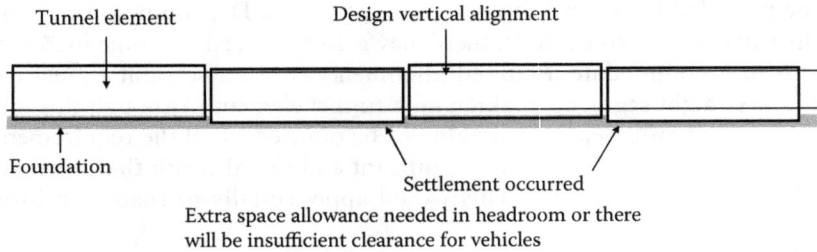

Figure 9.2 **Effect of settlement on tolerances (side elevation on tunnel elements).**

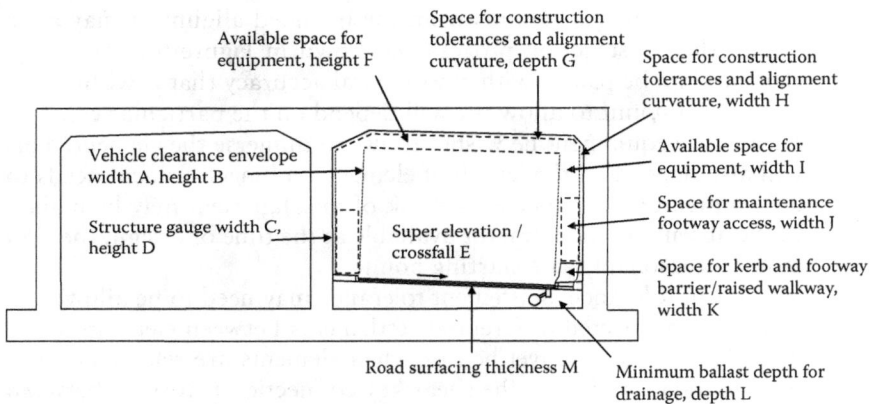

Available space for
equipment, height F

Space for construction
tolerances and alignment
curvature, depth G

Space for construction
tolerances and alignment
curvature, width H

Vehicle clearance envelope
width A, height B

Available space for
equipment, width I

Structure gauge width C,
height D

Space for maintenance
footway access, width J

Super elevation /
crossfall E

Space for kerb and footway
barrier/raised walkway,
width K

Road surfacing thickness M

Minimum ballast depth for
drainage, depth L

Figure 9.3 **Internal clearances for a typical road tunnel.**

3. Surface tolerance: Finally, an allowance should be made for deviations in the finished surface of the structure. In the roof of the tunnel, it may be possible to accommodate this in the space allowed for
mechanical and electrical equipment, but at the lower level, the traffic
clearance envelope and the footway horizontal clearances are usually
tight up against the structure gauge and some provision should be
made horizontally to ensure that the final construction complies with
standards. Variations in the top surface of the structural base slab can
be taken out in the ballast concrete, so no additional vertical height is
required for this tolerance.

The best way to ensure that sufficient space has been provided in the
tunnel cross section is to prepare preliminary drawings showing all of the
space provisions and the buildup of clearances and tolerance allowances
so that the sizing can be verified. Figure 9.3 shows the various elements of

Table 9.1 Typical space provision and tolerances to account for in internal clearances

Dimension	Typical value	Clearance for
A	n × 3.65 m + hardstrip width	Traffic width: n = no. of lanes, 3.65 m is typical lane width
B	5.25 m	5.0 m traffic height plus 0.25 m allowance for loose flapping truck tarpaulins
C	A + H + K	Total width required
D	B + E + F + G + L + M	Total height required
E	2.5%	Minimum carriageway crossfall
F	300 mm	Space for M&E equipment (excluding vent fans which may require additional space unless housed in niches)
G	75 mm	10 mm surface tolerance 25 mm settlement allowance 15 mm alignment curvature 25 mm element placing tolerance
H	80 mm	10 mm concrete surface tolerance 20 mm alignment curvature 50 mm element placing tolerance
I	As required	Space for equipment above footway—not usually critical unless ventilation fans are placed in this space
J	Width as footway Height 2.3 m	Space for maintenance personnel access and car passenger safety in the event of breakdown
K	1.0 m	Minimum footway width
L	600 mm	Minimum depth of ballast to accommodate drainage
M	75–100 mm	Two layers of wearing course with allowance for overlay, unless it is assumed renewal will include planning out of the old surface

spatial provision that have been described and that need to be allowed for a road tunnel. A typical set of tolerance allowances is given in Table 9.1. Note that some of these values will vary, depending on the national design code requirements.

Similar provisions must be made in the cross sections for rail tunnels, although the envelope may look quite different. This can be seen in Figure 9.4, which shows the rail kinematic envelope used for the development of the Bosphorus Tunnel for the Marmaray project in Istanbul.

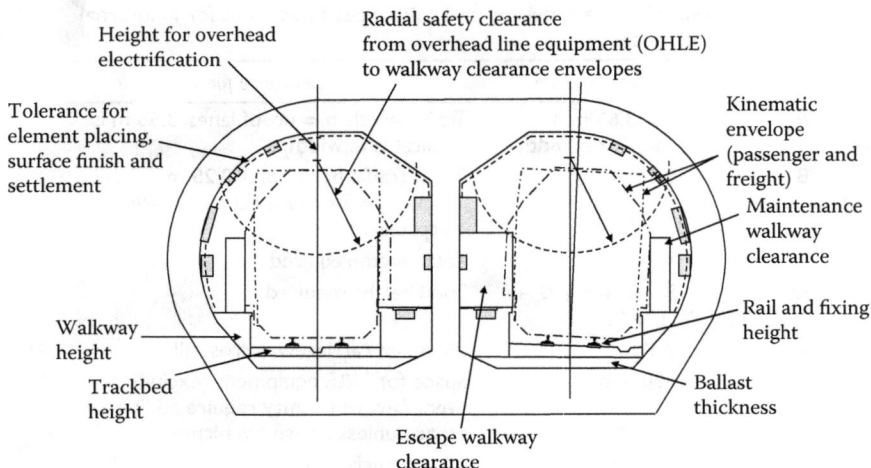

Figure 9.4 Internal clearances for a typical railway tunnel.

STABILITY AND BUOYANCY

Immersed tunnel elements must be designed to be able to float and be transported safely and be immersed using ballasting techniques that involve manageable quantities of water ballast. It must also be possible to maintain a net downward force using permanent ballast in the final condition. To ensure that the tunnel remains stable when constructed and that there is no risk of upward movement due to the uplift forces, it is necessary to assess the net downward force compared to the net upward force. The ratio of this downward force divided by the upward force is known as the factor of safety against uplift.

The most important early part of the design process is the sizing of the structure to satisfy the requirements of each stage of construction. At each stage, the buoyant behavior needs to be checked, considering the loads that are applied at that particular time and the floating (or sinking) characteristics that are desired. The study of the buoyant behavior will provide two conclusions. It will verify that the sizes of the structural members, combined with the internal air volume, will produce a structure that can be constructed as an immersed tunnel, and it will confirm the ballasting arrangements, both the sizing of the temporary ballast tanks based on the weight of ballast water that must be applied to immerse the tunnel elements and the space required for the permanent concrete ballast.

The process for assessing the freeboard that will be achieved when floating, or the factor of safety (FoS) against uplift when immersed, is a simple application of Archimedes's principle, whereby if the weight of displaced water is greater than the weight of the structure, the structure will float. If,

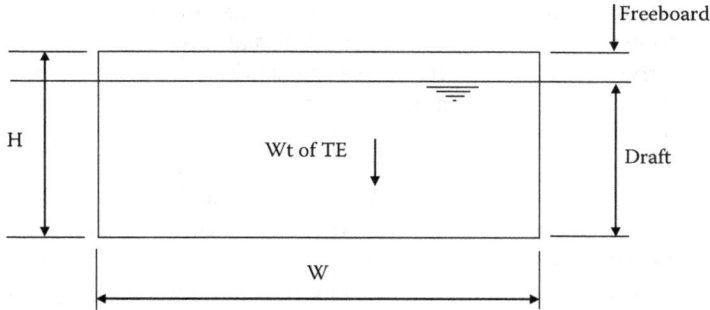

Figure 9.5 Calculation of freeboard.

however, the weight of the structure is greater than the weight of displaced water, the structure will sink. In simple terms:

$$\text{FoS against uplift} = \frac{\text{Weight of structure}}{\text{Vol of displaced water} \cdot \rho w}$$

For a simple rectangular tunnel element (TE) of length L, such as in Figure 9.5, the freeboard while floating is calculated as

$$\text{Freeboard} = H - \frac{\text{Wt of TE}}{W \cdot L \cdot \rho w}$$

For irregular geometry, the formula will need to be developed accordingly. The weight of the tunnel element will vary according to the stage of construction and should include the structure, ballast, and all pieces of temporary equipment.

To achieve the weight balance, the initial thicknesses of walls and slabs must be selected to calculate the weight of the tunnel structure. Simple span/depth ratios can be used to make a first assessment. A ratio in the order of 10:1 is appropriate for this purpose, noting that base slabs are generally slightly thicker than roof slabs because of the higher hydrostatic loading applied to the base of the tunnel. Equally, the member thicknesses used on previous immersed tunnels can be used as a guide, provided they are of similar water depth to the one now being considered.

In evaluating the buoyant behavior, the nominal weights of the structure and water should be considered, with no load factors applied. Instead, the variation in material densities should be considered and a light condition and heavy condition assessed. The light condition considers the maximum water density, to give the maximum weight of displaced water, and the minimum structure material densities. The tunnel element will sit at its highest in the water in this condition, with the maximum possible freeboard. The heavy condition is the reverse and considers the minimum water density that might be encountered and the maximum structure material densities, to get the heaviest possible condition and to determine the lowest position of the structure in water, with the minimum possible freeboard.

There are a number of techniques that can be used to optimize the structural sizing. It is clearly most cost-effective to minimize the internal air volume of the tunnel as this, in turn, minimizes the size of the structure and, hence, its cost. It is undesirable to have excess space in the tunnel or additional air volume purely to achieve the necessary floating characteristics. Put simply, the more air space in the tunnel, the more concrete and steel is needed to hold it down. The first exercise is always to fit the structure gauge as closely as possible to the required clearance envelopes for traffic and equipment in the tunnel. Then, the depth of ballast concrete that is required should be assessed. On double skin steel shell tunnels, the ballast weight is not an issue because it is simple to adjust the external volume between the steel shell and the form plates. In concrete tunnels, it is more difficult because the ballast is contained within the tunnel. There is a balance between adding space into the tunnel and adding weight from the concrete. There will be a certain minimum depth of ballast concrete required, in any case, to accommodate drainage pipework and gulleys that will collect surface water from the roadway or railway trackbed. Beneath a road, the typical minimum thickness will therefore be about 500 mm. The actual depth of the ballast concrete will depend on the weight required to achieve the factors of safety against uplift in the permanent condition.

Weight can be added into the structure by increasing the wall and slab section thicknesses. This is only cost-effective to a degree as it would not be economic to have very thick slabs that require little reinforcement to give them the desired structural capacity. If the tunnel is particularly light, rather than add significant depth of ballast concrete at the base of the tunnel, it can be more cost-effective to extend the base slab beyond the external walls. Often, these extensions are referred to as toes. This enables the weight of backfill that is placed over the toes to be considered weight in the buoyancy evaluation. Even though this is the submerged weight of the backfill, it can be a cost-effective way to keep the size of the tunnel structure to a minimum and can be particularly useful for tunnels with relatively thin walls and slab thicknesses. This is shown in Figure 9.6. Note that the friction effects and the weight of fill over the tunnel should not be taken into account.

A number of tunnel projects have utilized heavy aggregates in the ballast concrete to increase the density and avoid the need to deepen the tunnel structure. This is a viable solution if such aggregate materials can be obtained.

When designing the buoyancy characteristics of a tunnel element, it is necessary to consider the element as a whole rather than just looking at a typical cross section. This is important as there are discrete loads along the tunnel element, such as bulkheads, ballast tanks, ballast water within the tanks, and equipment mounted on the roof of the element for the immersion operations.

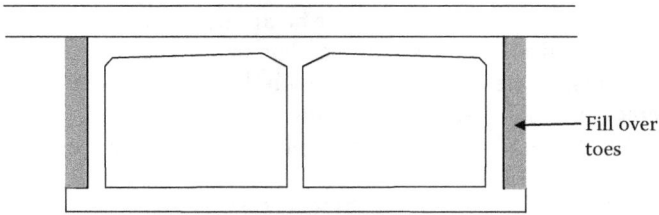

Figure 9.6 Fill over toes can improve FoS against uplift.

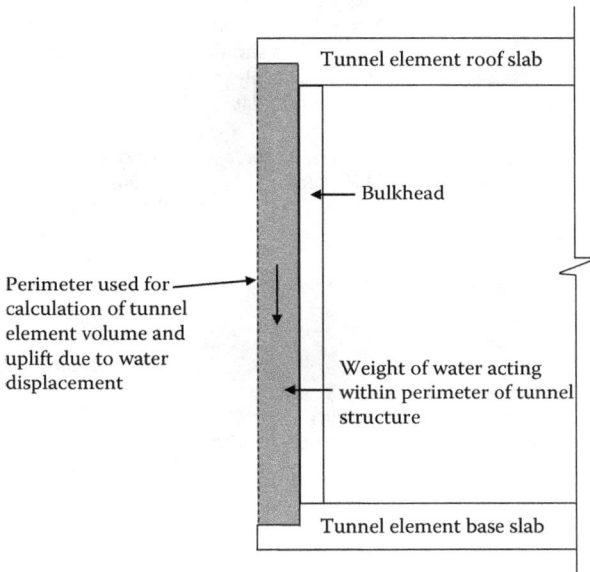

Figure 9.7 Water in bulkhead space.

These loads are quite significant and must be included. They could be reduced to a weight per linear meter of tunnel element, but it is probably easier to deal with real loads and the whole element. In addition, the tunnel cross section may not be uniform along the length of an element, with features such as recesses, openings, or voids that could affect the overall weight. The behavior of a tunnel element can be quite sensitive and it would be easy to miscalculate the factors of safety or the freeboard while floating if all the loads and variations are not taken into account. Another important load to remember is the weight of water that is within the perimeter of the tunnel at the bulkheads. The bulkheads are often set a short distance in from the end of the tunnel element and it is important not to assume there is air right up to the end of the elements; instead, the appropriate weight of water within the tunnel perimeter should be applied at the ends of the element (Figure 9.7).

Factors of safety against uplift must be applied in the design process for a number of conditions. These are set out in the Table 9.2. During construction of the immersed tunnel, they include the condition when elements are ballasted in the dry dock or casting basin ahead of float-up, the initial immersion when elements are first placed onto temporary supports or

Table 9.2 Stability and buoyancy conditions to be checked in design

Stage of construction	Criteria to be achieved	Loading considerations
Initial ballasting in the casting basin or dry dock	Minimum factor of safety against uplift prior to float-up: FoS = 1.015 (or min earth pressure of 1 kN/m²). This should be based on an assessment of likely variation in salinity and risk of uplift occurring.	Light condition: Minimum structure weight Maximum water density Initial ballast water volume
Tunnel element float-up	Minimum freeboard for calm water conditions (typically 150 mm)	Heavy condition: Maximum structure weight Minimum water density Actual water ballast assuming adjustment is possible
Transporting elements out of casting facility	Minimum keel clearance for exiting casting basin/dry dock	Heavy condition: Maximum structure weight Minimum water density Actual water ballast
Fit-out for towing	Target freeboard for stability Minimum keel clearance for tow route	Additional loads added: Survey towers Control tower Immersion pontoons Heavy condition: Maximum structure weight Minimum water density Actual water ballast
Towing	Minimum freeboard for wave conditions	Loads as previous stage
Fit-out for immersion	Minimum freeboard	Additional loads added: Access shafts Heavy condition: Maximum structure weight Minimum water density Actual water ballast
Immersion	Target overweight (typically 300–500 tons)	Additional ballast to achieve overweight Consider both light and heavy condition to ensure immersion operation can manage either circumstance
Release	Minimum FoS against uplift of 1.025 (or min earth pressure of 1–2 kN/m²)	Light condition: Minimum structure weight Maximum water density Increase in ballast water volume

Table 9.2 Stability and buoyancy conditions to be checked in design (*Continued*)

Stage of construction	Criteria to be achieved	Loading considerations
Sand flow	Minimum load on temporary support to be maintained Maximum jack capacity	Light condition: Minimum structure weight Maximum water density Fixed ballast water volume Sandflow uplift pressure Heavy condition: Maximum structure weight Min water density Max possible ballast volume during operation
Ballast exchange	Minimum FoS against uplift of 1.025 (or min earth pressure of 1–2 kN/m²)	Light condition: Minimum structure weight Maximum water density Reducing ballast water volume with increasing ballast concrete—each step in exchange process to be checked
Permanent condition	Minimum FoS of 1.06 (or min earth pressure of 1–2 kN/m²) Minimum FoS: 1.06 (minimum components considered) 1.1 (all fill and permanent parts considered)	Light condition: Minimum structure weight Maximum water density Final ballast concrete, installed in stages

onto a gravel bed, after the immersion of the elements for the longer-term temporary condition and during ballast exchange, when the water ballast is replaced by concrete ballast. Finally, the permanent condition with all loadings applied should be considered.

The factor of safety should be calculated for an individual tunnel element to take account of any specific features within that element, for example, drainage sumps, which may result in different concrete and air volumes in the tunnel. For a concrete segmental immersed tunnel in the temporary condition, it is normal to calculate the factor of safety for each individual tunnel segment and ensure that the minimum factor achieved for a segment is 1.04, provided that the overall factor for the tunnel element is 1.06. In calculating the factors of safety, only the weights of the permanent materials should be used. This will be primarily the self-weight of the tunnel structure, the weight of the ballast concrete (or ballast water in the temporary condition), the weight of the backfill, and the weight of the displaced water. No live loads are considered. Partial load factors of 1.0 should be used. The following should be excluded from the factor of safety calculation:

- The effect of friction between the walls of the structure and the backfill
- Cladding systems

- Fire protection materials
- Road pavement, curbing, barriers
- Internal fittings such as doors, M&E systems
- Internal partition walls, unless forming part of the main structure
- Any removable items

It would be normal to provide for a partial removal of backfill over the tunnel in the event that the roof of the tunnel needs to be exposed by the controlled removal of cover material for inspection or repair, or in case over-dredging or an extreme scour event results in material removal. Each tunnel should be considered individually to assess the likelihood of this occurrence, taking into account the environment, the waterway characteristics, and the specifics of the tunnel, and the criteria for design set accordingly. It is typical to allow the factor of safety to drop down to 1.06 in these conditions.

It would be unusual to remove the backfill from the sides of the immersed tunnel; so, if the tunnel has toe slabs protruding at the base slab level, the weight of the backfill above the toes—up to the roof level of the structure—can be included. For the condition where a tunnel element is placed onto temporary supports, the factor of safety against uplift should be calculated to ensure the required value is met, but the stability of the element will be controlled by monitoring the load in each of the temporary supports and adjusting the ballast weight, as necessary, to maintain a minimum down-ward force. This is done during the early operations of immersing the tunnel and under-filling. Under-filling by sand flow may generate some uplift forces, and this must be counter-balanced with additional ballast to maintain the minimum downward load on the supports. A common factor of safety used for this temporary condition is 1.025, but a lower value could be used, based on a risk evaluation, for infrequent extreme events.

The buoyancy assessment made in advance of construction must consider the likely variations in material densities. This is particularly important for the water density and the concrete density. In the absence of test data, typical values used are given in Table 9.3.

If the designer is working with a contractor on a design and build contract, it is usually possible to refine the values for concrete, based on preconstruction mix testing. The mass concrete density can be determined and adjusted to suit the anticipated reinforcement density. However, a degree of caution

Table 9.3 Typical material densities

Material	Maximum (kN/m³)	Minimum (kN/m³)
Water	10.1	9.9
Structural concrete	24.5	23.0
Ballast concrete	23.5	22.5
Backfill (submerged)	10.0	6.5

must be exercised as the factor of safety is determined using nominal values with no factors of safety applied to individual densities, so one should not be over-optimistic as to the concrete densities that can be achieved. If monitoring data is available for salinity, it may be possible to refine values if, for example, it is known that the watercourse never runs with fully fresh or fully saline water. Note, however, that the presence of suspended sediments in the water will increase its density and this must be taken into account also. The assumptions made by the designer would normally be verified during construction by taking density test cores from the concrete and making as-built surveys to confirm the geometry of the structure. This information is used to verify the actual weight of the structure and the displacement of the structure as construction progresses and hence, the factor of safety actually achieved.

Tunnel approaches

This bouyancy design process also applies to cut-and-cover approach tunnels and approach ramp structures, where the ground water level will be high in relation to the structure and similar upward forces will apply. The approach structures are rather more straightforward although the various stages of construction must still be considered. Often, the critical condition to be considered is when temporary ground water lowering, which may be required to construct the structure in a dewatered excavation, is turned off and the ground water level is first allowed to rise around the cut-and-cover tunnel. The full weight of the tunnel may not apply if the construction is not fully complete, so the factor of safety against uplift will likely be less.

Where approach structures are designed as gravity structures and use only their self-weight to resist the uplift, the following factors of safety against uplift can be used:

- Approach structures temporary condition 1.10
- Approach structures permanent condition 1.15

Factors of safety are slightly higher than for the immersed tunnel as there may be greater variation in structural thickness (particularly if constructed using embedded walls, such as diaphragm walls) and the variability of ground water levels needs to be taken into account. Design codes vary the factors of safety required and the values may be higher than stated above in some cases. Again, partial load factors should be equal to 1.0 and no allowance for friction between the structure and the backfill should be taken for gravity structures. Additional conditions may be set if there is a high tidal water level variation or ground water variation at the site, and it is not uncommon to define a set of criteria for an extreme high water event, with a factor of safety slightly lower (1.05) for what could be termed a "survivability" event. Similarly, during construction, a lower factor of safety of this order of magnitude can be considered.

The assessment of whether the backfill may be removed may be different to that of the immersed tunnel, depending on the form of the structure and whether there are constraints, such as an adjacent infrastructure, that may prevent this. A site-specific assessment is needed to determine the appropriate conditions to be considered in the stability design.

Approach structures may well have tension-resisting elements, for example, if constructed using embedded walls, or if tension piles are used beneath the structure. Clearly, here friction is appropriate to be considered. There are well-developed design methods for embedded wall structures and the design of pile groups acting in tension, that ensure a suitable overall factor of safety is achieved for the ULS condition. In simple terms, the condition that needs to be satisfied is given as

$$\text{Self-weight } G \times \gamma_G + \text{friction } F/\gamma_F > \text{uplift } P \times \gamma_P$$

Taking Eurocodes as the example, γ_G is set at 0.9 for favorable permanent load effects and 1.0 for unfavorable permanent load effects. Favorable variable loads are not considered and the partial load factor for variable unfavorable effects is set at 1.5. The partial factor to apply to friction, γ_F, is typically between 1.25 and 1.4, according to the type of tension element being considered.

Occasionally, a global factor of safety of 1.1 is specified based on a simple ratio of stabilizing loads to disturbing loads, based on nominal values. This corresponds to the aforementioned approach, considering gravity structures under stable conditions. It should be noted that design codes may have further safety factor requirements for the tension capacity of individual anchors or tension piles that should also be adhered to.

When reliance is placed on ground anchors or tension piles, then a sensible assessment of the service life of these elements, along with corrosion protection specifications, needs to be made. Any predicted reduction in performance over the design life will need to be taken into account in the uplift calculations and a degree of redundancy may need to be provided in the design to ensure the required factor of safety can be achieved in the event there are localized failures. A decision on whether the ground characteristics are suitable for this method must also be made, taking into account any client-specific or nation-specific requirements. For example, different views on the appropriateness of the use of ground anchors in cohesive materials exist because of the risk of long-term creep and achieving an effective bond between the anchor and the ground. While this can be mitigated by measures such as grouting, the risks are perceived as high and such a solution may not be acceptable to all.

A varying ground water level in the tunnel approaches where structures are only partially submerged will vary the uplift forces acting on the structures, causing a cyclical upward and downward load. Provided this is accounted for in the design for stability and for the foundations, it will have no deleterious effect on the tunnel structure, but should always be considered in the design.

If the approaches of the tunnel feature a buried ground membrane to prevent ground water seeping up through the earthworks and into the tunnel approach, then the stability of the membrane must be considered to ensure the water pressure buildup beneath it is not greater than the applied load from the fill above. A suitable factor of safety to achieve in this circumstance is 2.0.

Floating stability

So far, we have only discussed vertical stability and the need to keep the tunnel element either afloat or weighed down on the sea bed. There is another aspect to stability, which is stability against overturning of the element while floating. The element may be subjected to horizontal wind and current forces or may not float evenly if the geometry of the element is unusual. It is necessary to check that the element remains stable under such conditions and is not at risk of overturning or rolling. There are two checks to be carried out in this respect:

1. Calculating the metacentric height and ensuring it is greater than one meter above the center of gravity of the element.
2. Calculating the righting moment for a specified maximum angle of heel, typically around 30 degrees, to ensure there is always a stabilizing moment for the given freeboard.

Metacentric height

The metacentric height is defined as the distance between the metacenter (M) and the center of gravity (CG). The CG is that of the structure and does not change its position, whatever the inclination of the tunnel element. The M is defined as the point on the vertical axis of symmetry passing through the CG that, when inclined due to heeling of the element, intersects with a vertical line taken from the center of buoyancy (CB). The center of buoyancy is the CG of the body of displaced water providing the uplift, and hence generates a restoring moment to the heeling effect. As the tunnel element is inclined, the CB displaces sideward as one side of the structure displaces more water. Figure 9.8 shows the positions of the CG and the CB for a rectangular element without any heeling angle. Figure 9.9 shows how this relationship changes when heeling is introduced.

The metacentric height is the distance along the inclined line from CG to M. Provided this distance is greater than one meter, there will be a restoring moment resulting from the offset position of the CB and the element is

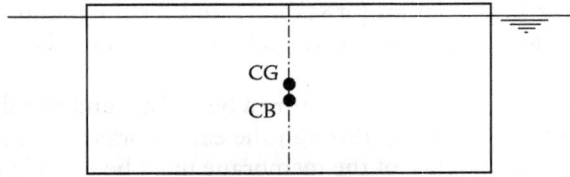

Figure 9.8 Position of CG and CB of floating rectangular element with even trim.

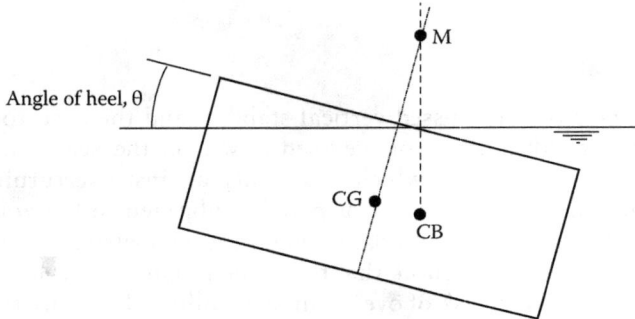

Figure 9.9 Position of CG, CB, and M of floating rectangular element with heeling.

considered to be stable. The metacentric height requirement varies considerably for different types of vessel and floating structures, but the figure of one meter is a common minimum figure used in the design of floating dry dock structures and pontoons (as per BS6349 Part 6), and has become an industry norm, even though it is not defined by any code or standard. If the metacentric height reduces below this figure, then the risk of instability increases and if the M is below the CG, then the element is unstable and will roll over. For a circular element, the CB will not move sideways as the element rotates, and thus CG and CB will always be on the same vertical axis and the metacentric height will always be zero. Hence, there is no restoring moment and the circular element is unstable and likely to roll. It will need other stabilizing measures, such as fixity to the immersion pontoons, for the time it is afloat.

Righting moment

The righting moment caused by the offset of the CB shown in Figure 9.9 will vary with the angle of heel, θ. It should be checked that there is always a righting moment up to an angle of heel in the order of 30 degrees, or the maximum angle of heel anticipated as the result of trim or wind and current effects, whichever is greater. To achieve this will require a certain minimum freeboard. If the freeboard is too low, then the element will dip below the water line and the offset between the CB and the CG will not increase significantly further and the righting moment will tend to decrease.

Usually, the righting moment is plotted graphically against the angle of heel for various levels of freeboard so that decisions can be made as to the minimum freeboard to select. These charts are sometimes known as Carene diagrams.

These two checks are very straightforward to undertake and are simple spreadsheet calculations using areas and geometry. However, they are very important—particularly if the tunnel is a narrow tunnel—as there may be specific measures needed to ensure that the elements are stable, either to be taken in design, or to be addressed by the marine contractor in the planning phase.

More developed formulae can be used to take account of the water ballast within the element. The principles are described in BS6349 Part 6. The free surface effect will have a minor influence on the calculation because, as the element heels, the CG of the body of water ballast will shift.

DESIGN LOADINGS

Permanent loadings

The permanent loadings that must be considered in the design of an immersed tunnel structure fall into five categories:

1. Self-weight of the tunnel structure
2. Hydrostatic forces
3. Permanent ballast
4. Finishings
5. Backfill

The self-weight of the structure is the most critical because it is the biggest influence on the floatability of the element. Too heavy and the element will not float; too light and it may not be possible to accommodate sufficient ballast to provide the necessary factor of safety against uplift. The sizing of the structure to obtain good buoyant behavior has been described previously. Once the basic sizing is known, the self-weight can be calculated for use in structural design and settlement analysis.

The density of the materials that are likely to be used in construction should be assessed at an early stage, in particular, which aggregates are likely to be used and what would be the likely density of the resulting concrete. During the buoyancy analysis of the section, a range of material densities must be considered to test the sensitivity of the structure to possible variations in material density. For a preliminary estimate, the weight of reinforcement in the section can be taken as 110 kg/m^3 for a segmental concrete element or 150 kg/m^3 for a monolithic concrete element. This will have to be confirmed later as the detailed design progresses, but will be sufficiently accurate for the preliminary design.

The ballast will usually be either mass concrete placed within the tunnel section or rocks placed in a ballast box on the tunnel roof. Again, a range of possible densities must be considered to test what flexibility exists in the design. The ballast does offer some possibility of rectifying the mistakes made earlier in construction. For example, if the element is built too light, then there is the possibility of increasing the density of the ballast to achieve the required factor of safety against uplift. Heavyweight aggregate could be used in the ballast concrete. Conversely, if the element is too heavy, lightweight aggregate can be used in the ballast concrete.

For buoyancy analysis, the weight of the finishings that are considered permanent loadings must be assessed on a case-by-case basis. Only finishings that would not be removed during maintenance should be included. Therefore, a typical list of permanent loadings would be as follows:

- Structural concrete and structural steelwork
- External waterproofing membrane and any associated protective concrete
- Reinforcement and other permanently cast-in items
- Ballast concrete
- Any regulating course where used, but not the carriageway wearing course

The weight of ventilation equipment and other M&E plants is not generally included in the permanent loadings as it has to be removed and replaced as part of the maintenance regime. Anything that was to be included would have to be clearly identified in the maintenance manual to ensure that it was not removed without taking some compensatory action.

Load from the backfill should be included for structural design and settlement analysis. For the buoyancy analysis, the amount of loading due to the backfill that should be included in the permanent loadings is often open to debate. The safest approach, and one taken by many authorities, is that the frictional resistance of the backfill on the tunnel walls is not included when considering the buoyancy of the element. However, when considering the settlement of the elements, it is common to include the frictional downdrag of the backfill on the tunnel walls. Any fill above the tunnel can be considered a permanent loading, provided it is protected by a rock protective layer so that it cannot be scoured away. Similarly, it is often admissible to include some, say 50%, of the rock protection as a permanent load. Whether the weight of any fill above the rock protection can be considered depends on the particular circumstances. If the tunnel is particularly deep, then some of this backfill, say up to 3 m below the finished level of the backfill, could be included. This would have to be accompanied by a risk analysis investigating the probability of the backfill that is being considered as permanent remaining in place. For the structural and settlement analyses, all loads should, of course, be included.

Variable loads due to vehicles

Variable loads that need to be considered in the design of the tunnel structure include live loading due to road vehicles and railway rolling stock travelling through the tunnel. The loadings to be applied include:

- **Transient vertical loads:** The vertical loading from road traffic vehicles and railway car axle loads will not generally be significant to the design of the tunnel structure unless the road or railway is supported on a suspended slab. This may be the case in circular steel shell tunnels or where there are ventilation ducts beneath the road or railway or where there are features such as drainage sumps creating significant voids beneath the slabs. The loads to be applied should be in accordance with the relevant national standards for the design of highway structures or railway infrastructure. Project-specific loading may need to be considered if the tunnel is carrying a railway or metro with bespoke rolling stock designed for that project. For a road tunnel, the possibility of the tunnel being part of a heavy load route needs to be checked.
- **Wheel loads and patch loads:** Similar to transient vertical loads, the localized loads due to vehicles will need to be considered at any suspended slabs or duct covers. The magnitude of the loads to apply will be as per the applicable national design standards.
- **Braking loads:** Braking loads due to vehicles in the tunnel should be considered, but are not likely to be significant in a conventional rectangular tunnel. Unlike a bridge structure where the braking loads need to be considered for the design of composite structures (beam and slab decks) and for bearings supporting deck slabs, there is no such requirement for tunnels, as the loads are transmitted through the structure to the ground via friction. The only area this could affect is in the detailing of the carriageway surfacing and the layering of ballast concrete, where care must be taken to avoid the delamination of thin layers. In a circular tunnel with a suspended slab, there may be a need to consider braking loads, if the slab is supported on bearings, and how load is transferred to the main perimeter of the tunnel structure.
- **Impact loads:** Impact loads arising from vehicle collisions may be important for internal walls that separate the road and rail tubes and that provide emergency egress corridors. Such walls must be impact resistant, and national design standards will specify the loading that is to be applied in the design. External walls are unlikely to be affected by impact loads due to the ground support behind them. Most tunnels have curbs, barriers and footways that are sometimes raised. Impact loading will nevertheless apply to the structure walls, as there will not be sufficient set-back distance behind the barriers to consider the wall protected.

- **Suction/pressure loads:** Suction and pressure loads that are caused by air movement associated with road vehicles or trains, particularly high-speed trains, passing through the tunnel need to be considered for the internal finishings, such as cladding and service duct covers, to ensure these remain stable under the suction load and do not suffer any fatigue effects due to vibration. The main structure will not be affected. Loads can be derived from design standards such as the Eurocodes, but it is not uncommon for sophisticated modeling of air pressure variations to be carried out to determine the load effects more precisely.

Temperature loads

Although the tunnel is buried, seasonal temperature variations will cause it to move due to expansion and contraction of the structure. These movements have a direct influence on the design of immersion joints and segment joints. In addition, temperature differences between the inside and the outside of the tunnel will impose distortions to the tunnel structure that induce bending moments and shear forces in the structural members, which will need to be considered in the design of the slabs and walls of the tunnel. Although temperature loads are not often the governing load condition, they cannot be ignored. There are also temperature loadings that have to be considered in the construction phase, including

- Overall temperature range (global behavior)
- Temperature gradients between the inside and the outside of the tunnel (global and localized behavior)
- Thermal movement during curing

Overall temperature range

The temperature range that must be considered in design has an important influence on the design of the tunnel joints. When the final tunnel element is immersed, the geometry of the tunnel structure is effectively locked. The subsequent variation in temperature that the structure experiences will determine the amount of expansion and contraction that occur, and hence, the movement capacity that is required to be provided at the tunnel joints. Therefore, a judgment must be made on the likely range of temperatures that will be experienced during the immersion of the tunnel elements as this forms the starting point for subsequent thermal movement. This is not necessarily the same as the range in water temperatures that can be experienced, as the temperature inside the tunnel will have an influence. However, the water temperature will be the dominant influence.

If design is being carried out independently from construction, then a large temperature range may need to be considered if the immersion activities may be carried out at any time of year. If the designer can take account of the actual construction planning, a reduced range of temperatures could be used in design and the design of the joints would be a little more cost-effective. In either situation, the important task is to assess the range of temperatures the tunnel structure may experience over its life compared to the range it is most likely to experience during construction, and to design the joints with capacity to handle the maximum differences. For example, the Øresund Tunnel was designed for a temperature variation of ±15° Celsius for the longitudinal analysis.

Temperature gradient

The temperature gradient between the inside and the outside of the tunnel must be assessed as differences in temperatures across the wall and slab sections will cause bending moments and shear forces in the structural elements. This is the simple result of one surface of a slab or wall expanding or contracting a greater amount than its opposite surface. The differential expansion causes bending in the member, which will be restrained by the stiff elements of the structure, such as the wall–slab junctions, and it must therefore be designed for the associated load effects.

The magnitude of the gradient that should be applied is not the difference between the maximum and minimum temperatures that might be experienced. It is the difference between the internal and external temperatures. These will relate to each other in that the maximum internal temperature may occur with the maximum external temperature or water temperature. Some analysis is needed of the likely air temperature variation on a seasonal and daily basis, along with the variation of water and ground temperatures. Typically, the gradient to be considered across the structural member is in the order of ±10°C or ±15°C, which can be applied either as warm external temperature combined with cold internal temperature, or vice versa, as a cold external temperature and warm internal temperature.

There is a specific temporary condition to consider when the tunnel is floating and the top slab may be exposed to sunlight and warming, whereas the rest of the structure is below the water line and the temperature of the inside of the tunnel is influenced by the water temperature.

Temperature difference during curing

For wider tunnels, there may be further temperature considerations during curing. Chapter 11 deals with temperature rise and the impact on cracking in detail, but there are some other peripheral effects that may need to be considered, particularly on segmental concrete construction.

Because tunnel segments are match-cast against each other, there will be relative movement of the newly poured segment against the previously poured segment as the temperature of the concrete rises during curing. There are shear keys around the perimeter of the tunnel and they will force the first tunnel segment to move with the new concrete, which may cause cracking. Equally, the first tunnel segment may cause restraint to the movement of the new concrete. Either way, the larger the tunnel segment, the greater the potential problems. This was experienced on the Øresund Tunnel, which was over 40 m in width, and measures had to be incorporated to ensure that prestressing ducts were not damaged by the movement, and to protect the concrete shear keys from being over-stressed in the temporary condition. The measures are relatively simple and comprise carefully placed compressible sheet materials typically used in construction joints, to allow some relative movement of the tunnel segments and avoid imposing loads on the shear keys in directions that were not intended. Similarly, a compressible layer around the prestress ducts to one side of the segment joint over about a 1 m length is sufficient to prevent the ducts from being damaged.

This lateral expansion and contraction has not been seen to affect the cast in groutable waterbars that feature in segment joints. The waterbars have a degree of flexibility due to their central bulb, which is designed to allow for some movement, and the grouting facility should compensate for any defects that arise.

Water levels, wave, and current loading

Wave and current loading is more important for the tunnel in the temporary condition than in the permanent condition, when tunnel elements are afloat and subject to direct loading from waves and currents, both vertically and horizontally. This is described further in Chapter 6. Dynamic modeling and/or scale modeling may be required to obtain accurate load effects for design. These will be used to design the tunnel element in the temporary condition. For a monolithic concrete tunnel, loading will be used to determine if the longitudinal reinforcement is sufficient to resist the bending effects. For a steel tunnel, there may be several stages of construction through towing to the final immersion, as concrete ballast is gradually placed, to check the structure has adequate capacity to resist bending moments generated by waves and currents. For segmental concrete tunnels, this loading will be used to design the temporary prestressing of the segments. This is described further later in this chapter.

Deep water swell waves may have an impact on the stability of the tunnel and the backfill around the tunnel if the wave height is sufficient for the tunnel to be affected at depth. This was investigated for the Busan-Geoje link in South Korea, where the long-term significant wave height was 9.2 m, with a corresponding period of 15 s. It was feared that the waves would give rise to the instability of the tunnel and cause uplift. It was concluded after undertaking analytical modeling that these were not significant to the tunnel, but the rock protection over the tunnel where the tunnel was above the existing sea bed level needed to be of a more stable material than conventional stone, and concrete armor units were used over the three tunnel elements that were affected. The susceptibility to uplift was a function of the porosity of the backfill and protection over the tunnel. This serves to remind the designer that the local sea conditions must be fully understood and assessed to ensure all possible loads on the tunnel structure can be accommodated.

Ice loads

Ice loading is a phenomenon that not many tunnels will experience. The need to consider this will come from researching historical weather records. If the watercourse beneath which the tunnel passes is prone to icing over, then further research may be necessary to understand if there is a risk of ice accumulation at the shoreline.

In marine conditions where the sea may freeze, the effect of the swell means ice forms in pads on the surface of the water. These pads move independently and may slide over one another with the motion of the swell and in the same way as waves break to the shore, the ice gradually accumulates at the shoreline. This can result in some considerable depth of ice, perhaps several meters in extreme conditions, and the loading to the tunnel at the shoreline can be significant. The shoreline is usually the point of highest overburden to the tunnel, so this is an important maximum load condition to consider, both in terms of structural capacity to carry the additional load and for the short-term settlement to the tunnel.

The Øresund Tunnel considered the likelihood of ice buildup at the foreshores as freezing of the Øresund Strait has occurred many times and ice at the shorelines is common. A height of 12 m of ice was considered the additional loading on the flood protection bunds above the cut-and-cover approach tunnels. A method of analysis was suggested that could be used in lieu of the 12 m requirement that was proposed by F. T. Christensen in the *Journal of Coastal Research*, Volume 10, Number 3, Summer 1994. The ice pile-up was considered to apply a distance of 30 m seaward off the shoreline and 24 m landward off the shoreline. The density of the ice pile was taken as the volume multiplied by the specific weight of ice (9 kN/m^3) and factored by 0.7. This shows that in regions where this risk of ice build-up exists, the loadings can be quite significant.

Accidental loads

Flooding

The tunnel should be assessed for the extreme event of it being flooded. This is not so uncommon and many tunnels have suffered a degree of flooding. Whether it is caused by rainfall, a burst water main, or inundation due to sea or river water level rise, the result is the same. The loading should be applied and considered as a short-term effect, and it is normal to consider a situation where pumping has failed and the tunnel is full of water.

The principal effect of the flooded condition is to cause settlement of the tunnel. The internal hydrostatic loading to the walls of the tunnel should be considered, particularly if it is possible for a single tunnel tube to flood, but this condition does not often govern the design.

Loss of support

This is not so much a loading as a particular support condition to the tunnel structure. It is an unintended situation, and so is included here under accidental load. This condition is sometimes considered for a tunnel if there is any risk of the foundation layer being incomplete or suffering a loss of material over time. This could occur for one of two reasons:

1. The bedding layer migrates into fissures in the underlying strata. This has been seen as a risk in tunnels where the trench was dredged in fissured limestone and a fine sand bedding layer could migrate into the fissure. It is possible to place geotextile or colloidal concrete to provide a layer over the fissured surface, but this is an expensive operation. Generally, it would likely be more cost-effective to select material that would be coarse enough to prevent this happening. Nevertheless, it is a risk that must be assessed and, if necessary, an appropriate support condition derived.
2. There could be a degree of incomplete under-filling of the tunnel. In the early days of sand flow foundations, there were some instances found where complete under-filling had not occurred. If the design of the tunnel is under the control of the contractor, he has the opportunity to match the risk of this occurring against his actual construction methods. It can lead to quite a severe load condition, so it is worth developing methods such that the risk is really minimized and the loss of support condition can be ignored. If the design is prepared in advance of the contractor's input, then some reasonable assumptions may have to be made as to what is achievable in the particular site conditions, taking into account the marine conditions.

Typically, the tunnel would be designed for a loss of support over a 20 m length. For a gravel bed foundation, this condition should not occur except if the bed is discontinuous by design. Clearly then, any gaps in the bed should be taken into account in the design to assess any impact on settlement or structural loading.

Ship impact

There are a number of load effects that might be considered due to a vessel sinking on to a tunnel. This may be an impact load if the tunnel is above the river or sea bed level, or a uniform load applied to an area above the roof of the tunnel representing the condition of a vessel coming to rest on the bed (Figure 9.10). The magnitude of any impact loading resulting from a ship collision should be calculated using dynamics, based on the size of likely vessels using the waterway and an assessment of the impact area and how load translates through the tunnel backfill and protection layer to the structure.

This was a load consideration for the Preveza Tunnel in Greece, which had a length of the tunnel above the sea bed. The tunnel was protected by earthworks effectively forming an embankment around and over the tunnel. Equivalent static loads were calculated on the basis of research carried out into the estimation of bow collision forces for vessel collisions with offshore structures. More recently, the Busan Tunnel in South Korea has had to be designed for similar circumstances as the tunnel is above the sea bed level at one end. In this instance, the capacity of the tunnel to withstand vertical and horizontal impact forces was determined on the basis of a 5 m by 5 m impact area. To keep the applied loading to within the capacity of the tunnel, the width of the earthworks embankment surrounding the tunnel was extended to dissipate the loading. Without this loading, the earthworks would have only extended to a distance of 3 m to either side of the tunnel at the roof level. To absorb the impact loading satisfactorily, the earthworks embankment was extended to 45 m beyond the tunnel. In addition, the number of shear keys between the tunnel segments was increased in order to transfer the load satisfactorily through the structure.

Whereas impact loading may or may not be applicable to any particular tunnel, the uniform surcharge load applied over an area of the tunnel roof representing a sunken vessel should always be applied. A review of the vessel size and type is necessary to assess the likely load that may be transferred

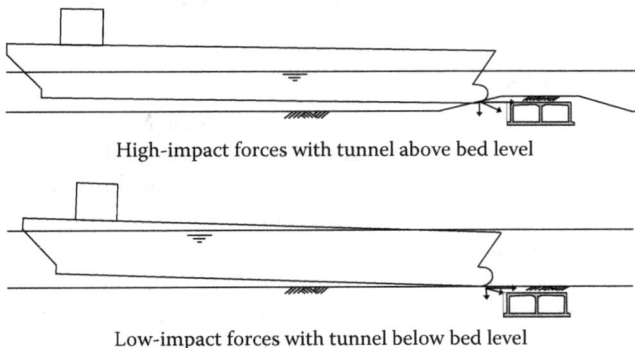

High-impact forces with tunnel above bed level

Low-impact forces with tunnel below bed level

Figure 9.10 Impact forces from a sinking ship.

Sunken ship uniform bearing pressure over sea bed and tunnel

Sunken ship increased bearing pressure over tunnel where
tunnel is above bed level

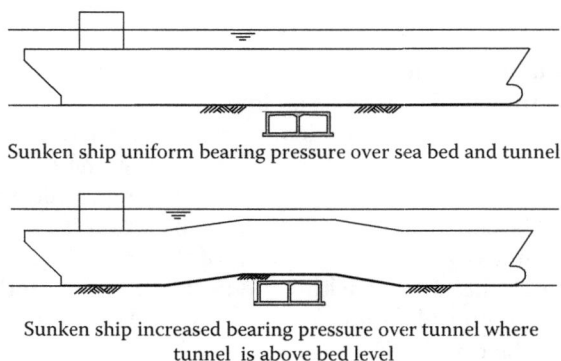

Figure 9.11 Increased loading due to sunken ship over a high tunnel.

to the roof of the tunnel structure as a surcharge load. Different types of vessels may impart different loads, that is, cargo or bulk carriers may be different to container vessels. The International Tunneling Association (ITA) State of the Art Report gives some guidance on how this may be calculated. It should be considered whether the overall bed profile is such that the sunken vessel will breach over the roof of the tunnel if its weight is not supported on either side, shown diagrammatically in Figure 9.11. This will significantly increase the loading.

For major shipping routes, typical loads to be applied are in the order of 100 kN/m² for the full width of the tunnel and over a length of tunnel equivalent to the width of vessel. The worst-case scenario must be considered for the particular structure; for example, on a segmental tunnel, the worst loading condition would be if the sunken ship load is applied to one segment only, such that the load is transferred through the adjacent tunnel joint shear keys.

Care should be taken, when applying this loading, not to consider multiple extreme events that would be very low probability, for example, sunken ship load combined with a loss of support condition at the same point. This needs to be dealt with when deciding the appropriate load combinations to apply.

Falling and dragging anchor

Ship anchors represent a risk to the tunnel structure. Many types of anchors are designed to bury deep into soft sediments. When they drag, which is not uncommon, there is a risk that they will snag on the tunnel structure, causing damage or, possibly, displacement to the tunnel. While it is possible to determine a load that could be applied to the tunnel by considering the breaking loads of anchor ropes or chains, it should be the designer's objective to make it unnecessary to apply this load, through the provision of adequate protective measures. The rock protection layer over the tunnel should be designed to dislodge the dragging anchors and prevent them impacting on the tunnel, so it is normal practice not to apply any horizontal loading resulting from this to the tunnel. Methods for designing the rock protection layer are described in Chapter 13. However, when an anchor falls to the river or sea bed, there is a potential vertical impact force to be applied to the roof of the tunnel.

To determine the magnitude of vertical load, the penetration of the falling anchor into the cover material must be calculated. Some research on the types of vessels using the waterway is necessary to determine the maximum anchor size that may fall. Then, the recognized methodology for determining the depth of penetration of a falling anchor can be applied, as given in Euro-International Concrete Committee (CEB) Bulletin 187 "Concrete Structures under Impact and Impulsive Loading," 1988. A simple conservation of energy approach can also be adopted, but experience shows this is less conservative and the CEB bulletin is therefore recommended to determine the depth of penetration into the stone protection over the tunnel roof.

The load imparted from the anchor to the tunnel roof can be calculated using a static equivalent load method based on an average impact velocity and a minimum penetration depth of 0.5 times the penetration, calculated using the CEB method above. The impact velocity is calculated by a simple drag coefficient method for a falling body.

A dynamic load factor (DLF) is obtained based on the ratio of the time for the anchor to come to a halt over the natural oscillating period of the tunnel roof structure. With this ratio known, the DLF can be taken from the graph in Figure 9.12.

The dynamic impulse force is then simply calculated as the mass of the anchor multiplied by its velocity, and this multiplied by the DLF gives the

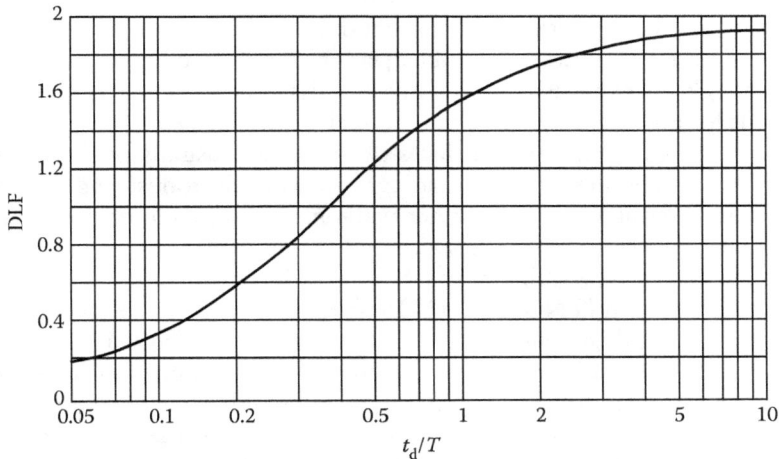

Figure 9.12 Dynamic load factor for calculation of equivalent static load.

static equivalent load to be applied. This should be applied on the basis of a 45° load distribution through the cover layer to the roof of the tunnel.

The load from a falling anchor does not so much govern any aspect of the tunnel structure design as the design of the rock armor layer. The roof of the tunnel is usually a robust thick slab with a high content of reinforcement. However, the falling anchor calculation is used to determine the thickness of the covering layer of rock. It is undesirable for the anchor to pass through the rock layer and impact directly onto the structure as this could simply cause local damage, cracking, or penetration of a membrane or steel shell. Therefore, the rock layer is designed to stop the penetration of the anchor and the resulting load transferred to the tunnel roof is used to check for adequate resistance to punching shear failure. The minimum thickness is generally set at the maximum penetration depth of a possible anchor, plus 10%.

Fire

Fire loading is used in the design of the ventilation system and the design of passive fire protection material applied to the inside surfaces of the tunnel. This is described in more detail in Chapter 15. There is no specific load case applied for a fire situation as part of the structural design.

Explosion

Explosion within a tunnel is a reasonable accidental load condition to design for. This may account for a conflagration caused through ignition of a spillage of flammable materials or failure of a pressurized container carrying LPG or CO_2, creating an explosion known as a boiling liquid expanding vapor explosion (BLEVE). It would be unusual to design for the effects of a bomb blast due to terrorist activities, but this may be the case for specific circumstances and its necessity should always be considered by a tunnel owner.

Although an explosion is very much a dynamic effect, it is normal to account for an explosion within the tunnel due to accidents by applying an overpressure to the internal surfaces of the tunnel, usually in the order of 100 KN/m^2. The main tunnel structure is relatively robust and can cope with this without a problem. Inner walls may require closer attention as they may be thinner and more susceptible to damage at this level of overpressure. Recent research by The Netherlands Organization for Applied Scientific Research (TNO) and TU Delft (Delft University of Technology) in the Netherlands has suggested this overpressure is insufficient to represent a BLEVE and an overpressure of 500 KN/m^2 or more should be used. However, to design for such a pressure would require the thickening of the structural members that could easily result in the tunnel sections being too heavy to float. Probabilistic risk assessment may be useful in determining the appropriate overpressure to consider for a particular project.

Propeller scour

Propeller scour is a loading that should be applied when designing the rock protection layer that is placed above the tunnel. Concentrated high-velocity currents can be induced by the propellers of ships that are maneuvering above the tunnel, particularly if there is a low keel clearance between the vessel and the bed. The size of the stone material in the rock protection layer will need to be determined such that there is no risk of movement of the rocks and the scour of the backfill around the tunnel is therefore prevented. Methods have been published to determine the current velocities induced at bed level based on the size and speed of the propeller and the clearance from the bed.

Construction stage settlement

Often, casting basins are purpose-built for the construction of the tunnel elements. These may be on the line of the tunnel in the approach ramp, adjacent to the tunnel on available land, or at some distance away from the site, depending on where land may be available. An earthworks basin may have a compressible foundation and the effects of settlement during the construction of the tunnel must be studied.

If a tunnel element is constructed sequentially from one end to the other, settlement may be occurring as the weight of the tunnel builds up. There may be a situation where the whole tunnel box is complete at one end, base and walls are complete at the center, and just the base slab is complete at the other end. Because of the time taken for construction, short-term settlements may happen in this period and cause more significant loading to the tunnel than the permanent case, where the reduced weight of the tunnel acts on the final foundation.

The effect of the settlement may be to overstress the structure such that the temporary condition becomes the governing design case, or to lock in stresses that reduce the capacity of the structure to carry load in the permanent condition. The concrete will be at a very young age and the available strength of concrete may need to be considered in relation to the speed of the settlement to assess this condition properly.

A contractor may be able to compensate for these effects by modifying the construction sequence, for example, building the full extent of the tunnel base slab first to achieve a degree of consolidation in the ground beneath them before starting the construction of the walls and roof slabs. Constructing tunnel segments in the hit-and-miss bays to even out settlement has been used as an effective method of limiting locked-in stresses due to differential settlement. The idea with this is to build every other 20 m to 25 m long segment. The bearing pressure bulb from each segment constructed will extend beneath the adjacent missing segments and cause a degree of pre-consolidation in the ground. When the infill segment is cast, the amount of settlement experienced should be less. For monolithic tunnels, particular care needs to be taken because the infill segment will have continuous construction joints at each side, which will cause tension in the structure due to early age shrinkage.

Improving the foundation of the casting basin may be the best way to control settlement, but there is clearly a cost/benefit assessment to be made with this approach and, generally, a combination of partial improvement and carefully sequenced construction is used. Some simple measures can be considered initially, such as using geotextiles, granular materials, and no-fines concrete in the top surface of the casting basin. More significant works such as piling or ground improvement may be necessary, but introduce large costs.

TEMPORARY LOAD CONDITIONS

There are a number of temporary load conditions that will need to be considered. When floating, the net weight of a tunnel element will be controlled by ballasting to ensure the element floats with a predetermined freeboard, or sinks with a pre-set overweight during the immersion process. Although the net load will be either upwards or downwards in either situation, there will be a distribution of loads along the tunnel element that will cause an overall bending effect.

The tunnel element needs to be considered a beam in suspension with the following loads applied to it:

- Overall buoyancy applied as a uniformly distributed load (UDL)
- Ballast tanks
- Ballast water within the tanks
- Ballast concrete (may be partly installed ahead of floating)
- Bulkheads
- Water in the ends of the elements outside of the bulkhead (if the bulkhead is set in)
- Survey towers
- Ballast pipework and other internal fittings
- Weight of pontoons (if supported at any time on the floating element)

Typical arrangements showing many of these loads can be seen in Figure 9.13 and Figure 9.14.

Some of these loads are considerable and, for long tunnel elements, can induce large bending effects in the tunnel. They need to be considered in all temporary stages as follows:

- Float-up and maneuvering out of the basin
- Fitting out for towing
- Immersion
- On temporary supports after immersion

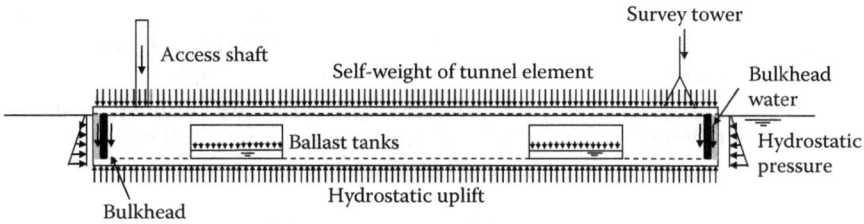

Figure 9.13 Loads on tunnel element while afloat.

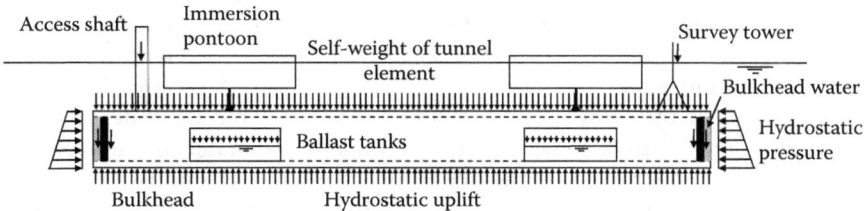

Figure 9.14 Loads on tunnel element suspended on immersion rig.

In the case of a segmental concrete tunnel element, there will be temporary prestress clamping the segments together during these operations. Often, the wave conditions encountered provide the governing loads for the design of this prestress, but in sheltered waters, it may be the distribution of the temporary loads that governs design. Part of the early design process for construction is to optimize the positions of these temporary items, particularly the ballast tank arrangements and the temporary support positions such that an economic prestress design can be achieved.

For steel tunnels a similar series of construction loads must be applied. These will include the launching load in circumstances where the steel shell is constructed on a slipway and slid into the water and staged concreting while the shell is afloat.

DESIGN CODES

There are many national and international design standards that can be used for the structural design of immersed tunnels, but few that deal explicitly with them. National codes that can be applied include Eurocodes, along with their national application documents, which are now implemented throughout western Europe. British standards are still used in a number of Asian countries even though they are now replaced by Eurocodes in the United Kingdom. Japan has its own guidelines and the Netherlands also has some specific national guidance. The United States' American Association of State Highway and Transportation Officials (AASHTO) standards are also widely used. The Federal Highway Administration (FHWA) has published a road tunnel design manual that is regularly updated and provides guidelines on how to apply the AASHTO load and resistance factor design (LRFD) codes to an immersed tunnel structure.

There are, however, a great many aspects of design for immersed tunnels that are not covered by published codes, or where design methods exist, but are little known. Some of these are listed as follows:

- Dynamic loading due to waves: BS 6349 offers some guidelines for carrying out a simple static analysis of a floating structure subject to wave impact. This is a quite conservative approach and could be used to gain a broad understanding of the effects that wave loading will have on the structure, but a dynamic analysis is recommended to obtain more precise loadings.
- Design of the joint gaskets: These are not covered by codes and the design needs to be looked at from a method and materials perspective, which is understanding the materials' characteristics and determining the sealing pressures, taking into account the stages of construction, temperature variations, and tolerances that can be achieved in construction. This is described in Chapter 10.
- Buoyancy: This needs to be approached from first principles, as set out earlier in this chapter.
- Passive fire protection: Suppliers testing data and furnace tests specific to the particular project are used to demonstrate that the surface temperatures and temperatures of the reinforcement within the concrete are controlled to the required levels by the proposed fire protection materials. Complex finite element (FE) analytical software exists that can model the transfer of heat through the fire protection material and through the concrete structure, but this is not widely used. Sufficient experience exists with manufacturers to determine the required thickness of protection material for a given fire curve.
- Rock protection against falling anchors: The CEB formula can be applied.
- Rock protection propeller scour resistance: Guidance can be found in the British Ports Association publication RR2570 "Propeller Induced Scour" by M.J. Prosser, February 1986, and in the Delft Hydraulics Laboratory publication no. 202 "Erosion of bottom and sloping banks caused by the screw-race of maneuvering ships" by H.G. Blaaw and E.J. van de Kaa, July 1978.
- Sunken ship loading: If this needs to be calculated from first principles, then the ITA guidance in their State of the Art Report is useful.
- Durability models: The most common approach currently taken is to apply a model such as DuraCrete or Life365 to determine the basic criteria for the structure to meet a required serviceability level at the

end of its design life. This approach enables the selection of the key durability parameters, such as chloride diffusivity and cover depth.

- Temporary prestressing for segmental concrete tunnels: Codes can be applied, but the criteria to be achieved in design are not well codified. Guidance is given later in this chapter.
- Seismic design of joints: This is not covered by design codes. Guidance is given later in this chapter.

LOAD COMBINATIONS

Load combinations and partial load factors defined in the design codes tend to cater to conventional structures such as bridges. An immersed tunnel has a different set of loadings and risks, however, and an assessment of the probability and criticality of a set of loads combining is needed for the various limit states that should be considered. For example, the importance of hydrostatic loads and earth pressure loading is much greater for an immersed tunnel than for a bridge. Also, live loading is far less critical for an immersed tunnel than it would be for a bridge structure. Therefore, it is appropriate to reconsider the load factors. There are also some very specific loads that apply to an immersed tunnel—particularly, the accidental and extreme loads—and the method of combining these with the normal loads such as self-weight and earth pressures needs some particular thought. The ITA issued some guidelines on the load factors for specific loadings that are particular to immersed tunnels in their State of the Art Report in 1993. However, the methodology for applying load factors according to the limit state design principles has developed since that time and hence the guidelines are now out of date.

Eurocodes and AASHTO are probably the most widely used design codes around the world, and so, the load combinations and factors applied for both of these are summarized below.

U.S. load combinations for AASHTO for LRFD design

The FHWA have published recommendations in their design manual for road tunnels, which contains tables for load combinations. It offers a good start for designers who can carry out this exercise and determine an appropriate set of combinations and factors. They are prepared for use with AASHTO LRFD design codes. Only the limit state load combinations that are relevant to an immersed tunnel structure are provided. These are

- Strength I: Basic load combination for strength and stability.
- Service I: Load combination used to check for stresses, deflection, and flexural crack control.

- Extreme Event I: Load combination used to design for the most severe earthquakes to ensure life safety and survivability of the structure, taking progressive collapse into account.
- Extreme Event II: Load combination used to design for earthquake, ship sinking, or anchor impact, individually. Note that these are not combined.
- Extreme Event III: Load combination used to design for a rare event with the simultaneous combination of loads.

For the design of other elements of a project, reference should be made back to the AASHTO LRFD Specifications (2010), where other limit state load combinations may need to be considered. The FHWA recommendations for combinations and load factors for immersed tunnel structures are given in Table 9.4. In addition, recommendations are given for the load combinations and load factors to be used when evaluating immersed tunnel sections for construction loads as given in Table 9.5.

Reference should be made to the FHWA Technical Manual for Design and Construction of Road Tunnels—Civil Elements and the AASHTO LRFD specifications for further understanding of the loadings and the limit state principles used. There are also other aspects to design, using the LRFD approach, that must be applied, such as further factors relating to ductility, redundancy, importance, and resistance. The basic equation to be satisfied in the LRFD approach is given as

$$\Sigma \eta_i \gamma_i Q_i \leq f R_n$$

where $\eta = \eta_D (\text{ductility}) \times \eta_R (\text{redundancy}) \times \eta_I (\text{importance})$, $\gamma = \text{load factor}$, $Q = \text{load}$, $f = \text{resistance factor}$, and $R_n = \text{nominal resistance of member}$.

The normal AASHTO design methods apply to determine the nominal resistance values for the structure. The FHWA gives important advice in relation to the resistance factors, noting that immersed tunnels cannot be considered as precast construction. Despite utilizing dry docks and casting basins, the construction methodology is still effectively an in situ technique.

Eurocode design

In Europe, the load combinations and applicable load factors have developed more on a project-by-project basis. When designing to Eurocodes, the following load combinations are recommended.

Serviceability limit state combinations

For the serviceability limit state, the following combinations of actions are considered.

Table 9.4 Load factors and combinations for AASHTO LRFD

Loading			Strength I	Service I	Extreme event I[b]	Extreme event II[c]	Extreme event III[d]
					Limit state		
Dead load	DC	Max	1.30	1.00	1.00	1.05	1.20
Permanent water load	WA_P	Min	0.90				
Downdrag	DD						
Dead load—surfacing	DW	Max	1.50	1.00	1.00	1.05	1.20
and utilities		Min	0.65				
Horizontal earth pressure	EH[a]	Max	1.50	1.00	1.00	1.05	1.40
Vertical earth pressure	EV	Min	0.90				0.90
Surcharge	ES						
Locked-in effects from construction	EL		1.00	1.00	0.00	1.05	1.00
Vehicular live load	LL	Note f	1.50	1.00	0.00	0.00	1.25
Vehicle dynamic load	IM						
Vehicular centrifugal force	CE						
Vehicular braking load	BR						
Pedestrian live load	PL						
Live load Surcharge	LS						
Transient water load	WA_T		1.30	1.00	1.00	1.05	1.20
Uniform temperature	TU	Max	1.20	1.20	0.00	1.05	1.30
Creep	CR	Min	0.50	1.00			
Shrinkage	SH						
Temperature gradient	TG		1.20	0.50	0.00	0.00	1.30
Settlement	SE[e]		γ_{SE}	γ_{SE}	γ_{SE}	γ_{SE}	γ_{SE}
Earthquake loads	EQ[b,c]		0.00	0.00	1.00	1.05	1.20
Vehicular collision	CT		0.00	0.00	0.00	1.05	1.20
Vessel collision	CV[c,d]						
Ice load	IC[c]						
Frictions	FR		1.00	1.00	1.00	1.00	1.00
Support loss	SL[c]		0.00	0.00	0.00	1.00	0.00

[a] Load factor is for at-rest earth pressures.

[b] EQ used for Extreme Event I considers most severe earthquake anticipated on project-specific basis

[c] EQ, CV (sinking ship or falling anchor), IC, SL, CT, and CV are considered, one at a time, in Extreme Event II. A design earthquake effect less severe than Extreme Event I may be used for EQ in Extreme Event II.

[d] This load case is used to check a rare event for the simultaneous combination of loads shown.

[e] γ_{SE} is considered on project-specific basis.

[f] For Strength I case, the load factor on vehicular-type loads shall be increased from 1.5 to 1.75 for design components where live load is the dominant transient load and may be omitted for components where this causes greater loads.

Table 9.5 Load factors and combinations for AASFTO LRFD construction loads

		Limit state	
Loading		Strength I	Service I
Dead load	DC	1.20	1.00
Permanent water load	WA$_P$		
Dead load—surfacing and utilities	DW	1.00	1.00
Locked-in effects from construction	EL		
Construction load	CL	1.20	1.00
Temporary water load	WA$_T$	1.20	1.00
Earthquake load	EQ[a]	1.20	1.00
Wind load	WS	1.40	0.3

[a]   Construction Strength 1 combinations shall consider a smaller earthquake to be a static load, typically having a return period of 5 to 10 years.

Rare or infrequent combination (stress limitation)

Characteristic values of permanent actions and the dominant variable action, together with the combination values of other variable actions, given as

$$\Sigma G_{jk} + P_k + Q_{kl} + \Sigma \ \psi_{0i} Q_{ki}$$

Frequent combination (crack width checks)

Characteristic values of permanent actions, together with the frequent values of the dominant variable action and the quasi-permanent values of other variable actions, given as

$$\Sigma G_{jk} + P_k + \psi_{1l} Q_{kl} + \Sigma \ \psi_{2i} Q_{ki}$$

Quasi-permanent combination (stress limitation)

Characteristic values of permanent actions, together with the quasi-permanent values of the variable actions, given as

$$\Sigma G_{jk} + P_k + \Sigma \ \psi_{2i} Q_{ki}$$

Accidental combination (stress limitation)

Characteristic values of permanent actions, together with one accidental action and the frequent value of the dominant variable action and the quasi-permanent values of other variable actions, given as

$$\Sigma G_{jk} + P_k + A_k + \psi_{1l} Q_{kl} + \Sigma \ \psi_{2i} Q_{ki}$$

Seismic combination (stress limitation)

Characteristic values of permanent actions and seismic action, together with the quasi-permanent values of the variable actions, given as

$$\Sigma G_{jk} + P_k + A_{Ek} + \Sigma \, \psi_{2i} Q_{ki} \in$$

where "+" implies "to be combined with"
Σ implies "the combined effect of"
G_{jk} = characteristic value of permanent actions
P_k = characteristic value of prestressing actions
Q_{kl} = characteristic value of dominant variable actions
Q_{ki} = characteristic value of other variable actions
A_k = characteristic value of accidental actions
A_d = design value of accidental actions
A_{Ek} = characteristic value of seismic actions
A_{Ed} = design value of seismic actions
γ = partial factor
ψ = combination factor (suffix "0i" denotes combination value, "1i" denotes frequent value, "2i" denotes quasi-permanent value of the variable actions)
γ_l = importance factor

Ultimate limit state combinations

For the ultimate limit state, the following combinations of actions are considered.

Temporary condition (combination I) and in-service condition (combination IIa)

Design values of permanent actions and the dominant variable action and the design combination values of other variable actions, given as

$$\Sigma \, \gamma_{Gj} G_{kj} + \gamma_p P_k + \gamma_{Ql} Q_{kl} + \Sigma \, \gamma_{Qi} \psi_{0i} Q_{ki}$$

In-service condition (combination IIb)

Design values of the permanent actions, given as

$$\Sigma \, \gamma_{Gj} G_{kj} + \gamma_p P_k$$

Accidental condition (combination III)

Design values of permanent actions, together with the design frequent values of the dominant variable action and the design quasi-permanent values of other variable actions and the design value of one accidental action, given as

$$\Sigma \, \gamma_{GAj} G_{kj} + \gamma_{pA} P_k + A_d + \gamma_{QAl} \psi_{1l} Q_{kl} + \Sigma \, \gamma_{QAi} \psi_{2i} Q_{ki}$$

Seismic condition (combination III)

Characteristic values of permanent actions, together with the quasi-permanent values of other variable actions and the design value of one seismic action, given as

$$\Sigma G_{kj} + P_k + \lambda_l A_{Ed} + \Sigma \, \psi_{2i} Q_{ki}$$

Notations are as defined for the serviceability limit state.

Partial safety factors and combination factors

The partial safety factors and combination factors recommended are shown in Tables 9.6, 9.7, and 9.8.

STRUCTURAL ANALYSIS

Analysis of the tunnel structure behavior under the imposed loading conditions is required in order to design the structural components with the necessary structural capacity. This is a relatively straightforward process and

Table 9.6 Eurocode combination factors for serviceability limit state (SLS) and ultimate limit state (ULS)

Load	$\psi 0$	$\psi 1$	$\psi 2$
Earth pressure (surcharge)	0.7	0.5	0.3
Road and rail traffic loading	0.6 (0.8)[a]	0.75 concentrated 0.4 uniformly distributed	0
Wind	0.6	0.4	0
Water[b]	0.8	0.55	0.3
Temperature	0.7	0.5	0.2
Wave & current	0.8	0.7	0.6
Temporary construction loads	1.0	0.7	0.6

[a] The $\psi 0$ factor for the infrequent combination for road load.
[b] The ψ factors for the water load are used with water level variation.

Table 9.7 Eurocode loads and partial load factors for serviceability limit state

	Serviceability limit state		
	Load combinations[a]		
	I	*II*	*III*[b]
Load	*Temporary*	*In-service*	*Accidental/ seismic*
Permanent Loads (G)			
Self-weight of structure	1.0	1.0	1.0
Ballast concrete	1.0	1.0	1.0
Road pavement, furniture		1.0	1.0
Hydrostatic load (MWL)	1.0	1.0	1.0
Earth pressure	1.0	1.0	1.0
Settlements	1.0	1.0	1.0
Prestressing, creep, and shrinkage	1.0	1.0	1.0
Variable Loads (Q)			
Earth pressure (surcharge)	1.0	1.0	1.0
Road traffic		1.0	1.0
Wind		1.0	1.0
Water level variation	1.0	1.0	1.0
Temperature	1.0	1.0	1.0
Wave and current loads	1.0		
Temporary construction loads	1.0		
Accidental Loads (A): One of the following			
Explosion			1.0
Collision from road traffic			1.0
Sunken ship			1.0
Falling anchor			1.0
Earthquake			1.0
Tunnel flooding			1.0
Extreme high water and waves			1.0

[a] For all loads, the factors given in the column are to be multiplied by the corresponding load in all load combinations. If no value is given, the load shall not be taken into account.

[b] These combinations are considered for stress limitations of structural concrete only. Steel reinforcement stress is limited to the yield stress, f_{yk}. Crack width and deflections are not considered.

not dissimilar to any other type of structure whereby the conventional load deformation analysis should be carried out. Analysis should be undertaken of the transverse behavior of the tunnel—that is, how the cross section of the tunnel carries loads and deforms—and of the longitudinal behavior of the tunnel—that is, along the long axis of the tunnel. Consideration must

Table 9.8 Eurocode loads and partial load factors for ultimate limit state

	Ultimate limit state			
	Load combinations[a]			
	I	II[a]	II[b]	III
Load	Temporary	In-service		Accidental/ seismic
Permanent Loads (G)				
Self-weight of structure	1.2	0.9/1.25	1.0/1.35	1.0
Ballast concrete	1.2	0.9/1.25	1.0/1.35	1.0
Road pavement, furniture		0.9/1.25	1.0/1.35	1.0
Hydrostatic load (MWL)	1.2	1.0/1.25	0.9/1.35	1.0
Earth pressure	1.2	1.0/1.25	0.9/1.4	1.0
Settlements	1.2	1.0/1.25	0.9/1.4	1.0
Prestressing, creep and shrinkage[b]	1.0	1.0	1.0	1.0
Variable Loads (Q)				
Earth pressure (surcharge)	1.35	1.5		1.0
Road traffic		1.5		1.0
Wind	1.35	1.5		1.0
Water level variation	1.2	1.2		1.0
Temperature				
Wave and current loads	1.35			
Temporary construction loads	1.35			
Accidental Loads (A): One of the following actions				
Explosion				1.0
Collision from road traffic				1.0
Sunken ship				1.0
Falling anchor				1.0
Earthquake				1.0
Tunnel flooding				1.0
Extreme high water and waves				1.0

[a] For all loads, the factors given in the column are to be multiplied by the corresponding load in all load combinations. If no value is given, the load shall not be taken into account.
[b] Effects of creep and shrinkage only included if these create an unfavorable effect for the element under consideration.

also be given to the behavior of the structure in the temporary condition during construction, as well as in the permanent condition.

The process of design is quite different for concrete and steel tunnels, but in analysis terms, they are both relatively simple structures and, in the majority of cases, a simple two-dimensional plane frame analysis model on elastic supports will be suitable to design the structure. For transverse analysis, a plane frame supported on springs with a stiffness that accurately represents the compressive behavior of the substrata will be adequate, provided some sensitivity analysis is performed for likely variations in the sub-grade over the tunnel width. Spring stiffness may be based on the modulus of sub-grade reaction, provided the designer is comfortable with that approach.

Sufficient supports at close spacing should be applied to enable the optimizing of the design and the full understanding of the variation in bending moment and shear in the structural slab. Similarly, sufficiently close nodal points along the structural members in the plane frame model should be used to enable the accurate understanding of the moments and shears close to the supports and corners, and throughout the length of each element of the structure. Sensitivity analysis should include, for upper and lower bound soil properties and variation in soil properties across the width of the tunnel, particularly if the tunnel is a wide, multiple tube tunnel section. If the tunnel is to be placed directly on a gravel bed foundation layer, the variation in the thickness of the gravel bed may influence the analysis, particularly if the underlying soil is stiffer than the gravel layer. In soft soil materials, this is likely to have no significant impact. Dredging tolerances and methods may lead to a variation in thickness across the width of the tunnel. The tolerance on the finished level of gravel should also be modeled, but this can be applied as an imposed deformation at the supports in the frame analysis prior to their release to determine the correct soil–structure interaction.

The stages of construction should be modeled in the plane frame analysis to understand if there are any critical conditions other than the permanent condition. As mentioned earlier in this chapter this is often the case when a concrete tunnel element is constructed in a purpose-built casting basin that has a soft foundation. As the tunnel is not subject to uplift forces due to hydrostatic pressure, the amount of ground settlement that occurs during construction can be an important load condition. This could impact either the transverse or the longitudinal design of a tunnel element. Ideally, the foundation of the casting basin will be designed so that it is not a governing factor in the design, but this may not always be possible or cost-effective. It will also be the case for steel tunnels that have a more distinct staging in their construction, requiring the analysis of the partially complete structure under different temporary load conditions.

Support conditions for the purposes of structural design can normally be modeled as elastic springs. Nonlinear stiffnesses can be applied if the designer feels this is important. However, any nonlinear behavior of the foundation is likely to be of greatest significance for the longitudinal behavior of the tunnel—in particular, the settlement behavior, rather than the transverse behavior of the tunnel. It is common to undertake a separate longitudinal settlement analysis of the structure using specifically developed geotechnical software, and if this is done, then ideally there should be some agreement in the magnitude of deflections between the structural analysis model and the settlement model.

Lateral spring supports can be applied to the structure walls to model the interaction between the tunnel structure and the surrounding fill. This may be important if there is uneven surcharge loading on either side of the tunnel that could generate a sway condition. However, this adds a degree of complexity to the modeling and will require some subjective assessment of the appropriate stiffness parameters to build into the model. Some sensitivity analyses are usually required to understand the difference in loading between assumptions of passive earth pressure and at-rest earth pressures acting on the walls of the tunnel. Rather than trying to model the soil–structure interaction accurately in this way, it may be better to move to a finite element continuum model, which will give a better representation of the tunnel behavior. Time-dependent nonlinear behavior of soils, and the stages of construction, will also be easier to model using FE analysis techniques.

The geometry of the tunnel cross section should be modeled in detail. There are features such as haunches between the walls and slabs that can be very beneficial in design for reducing peak bending moments and shears and it can be beneficial to optimize these features to minimize reinforcement in the tunnel. The stiffness of the structure will typically be represented by uncracked concrete sections. The elastic modulus for the concrete will need to be considered in accordance with design codes, which will typically factor the modulus for short- and long-term behaviors, appropriate to the type of loading being considered. This is an important aspect of the design, as there are many short-term accidental loads to be considered and the appropriate structure stiffness and soil stiffness need to be represented in the modeling.

There is a growing tendency to model structures with more sophisticated software, such as finite element analysis packages and to model in 3D rather than 2D. This is useful where the tunnel may be particularly wide, and distortions due to the combination of transverse and longitudinal effects may be important. However, a typical four-lane road tunnel or twin-track rail tunnel will be relatively narrow compared to the length of the tunnel, and separate transverse and longitudinal analyses are perfectly sufficient. Where modern FE analysis has benefit is in modeling the soil and tunnel together, to get a good understanding of soil–structure interaction and how the soil stiffness modifies the bending moments and shear forces compared to an above-ground structure. While two-dimensional plane frame models can be used for this, programs such as PLAXIS and FLAC aid understanding and are now becoming more standard. These do depend on having a good knowledge of ground conditions and the engineering soil properties to create a representative model, but they can be used to model both structure and ground deformations such that a single analysis can be used for both the structural design and foundation design.

Longitudinal modeling of the tunnel should take account of the whole tunnel and approach tunnels to ensure the complete load effects are considered from varying soil conditions and varying backfill levels. The tunnel elements are normally modeled as simple beams on elastic foundations. The rotational and longitudinal stiffness of the rubber gaskets in the immersion joint can be modeled using the stiffness properties defined in the force-compression curves supplied by the gasket manufacturer. However, the analysis should consider the different stiffnesses that may arise, depending on how compressed the joints are. If they are compressed due to temperature rise in the structure, then the joints could be very stiff. If the joints are open due to temperature reductions, then they could have a relatively low stiffness (Figure 9.15). If the tunnel is a segmental concrete tunnel, then the rotational capacity of the joints should be modeled. Again, this should take account of nonlinear stiffnesses, because as the joint opens, compression will arise in the slab opposite to the side of the joint that is opening. This gives some restraint to rotation that needs to be modeled by a nonlinear rotational stiffness. The variation in stiffness is shown in Figure 9.16.

For steel tunnels, the analysis model should correctly represent the composite section. There will be a need to run various analysis models to represent the stages of construction. These will include the fabrication sequence of the shell, to determine the level of bracing and support required as the shell is constructed. This may require a number of analyses to look at various partially completed configurations. It will also require an analysis of the lifting conditions for individual elements of the steel shell to ensure that handling does not overstress the steel plate or sections through distortion or buckling. The steel shell alone will then need to be analyzed and then

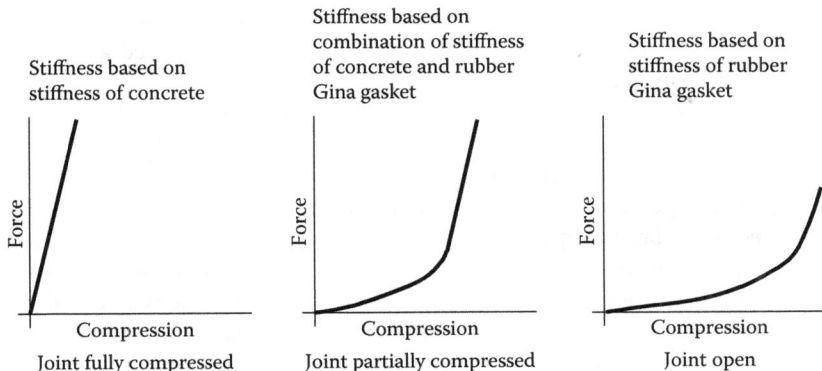

Stiffness based on stiffness of concrete

Stiffness based on combination of stiffness of concrete and rubber Gina gasket

Stiffness based on stiffness of rubber Gina gasket

Force

Compression
Joint fully compressed

Force

Compression
Joint partially compressed

Force

Compression
Joint open

Figure 9.15 Immersion joint rotational stiffness for a longitudinal analytical model.

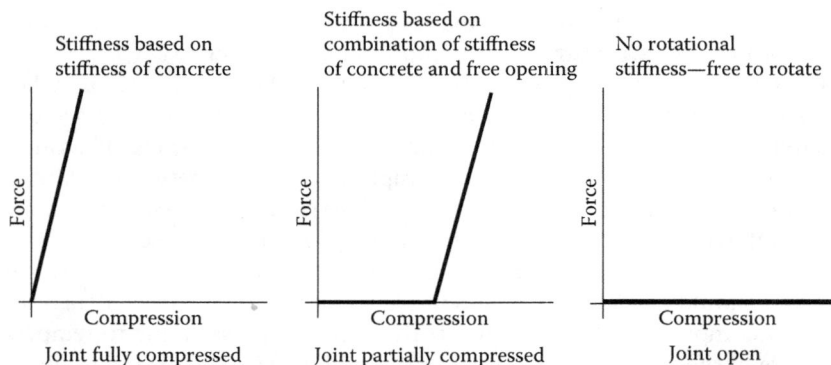

Figure 9.16 Segment joint rotational stiffness for a longitudinal analytical model.

the gradual addition of concrete ballast can take place. Initially, this is usually just the keel concrete placed prior to launching the tunnel element. The condition when the keel concrete is placed and the tunnel element is set on its launching sleds should also be analyzed. Launching loads need to be applied to the element modeled as a beam. Dynamic loading as well as static loads must be assessed in this instance.

The construction sequence from here on is important to the remaining analysis work required. There may be further concrete ballast added prior to towing the tunnel element to the site, or the element may be towed high in the water. If this is not known, the designer may have to undertake a number of analyses to ensure either can be accommodated or he may have to restrict the contractor to build the elements in a prescribed manner. For the tow condition, wave loading should be applied to the shell acting as a beam to assess the risk of buckling or overstressing. Again, stiffening may need to be designed to cater to the loading during tow. If the steel shells are transported using semi-submersible barges, then the tow loads clearly do not occur, but some handling loads may need to be considered as the shell is transferred on and off the barge.

The outfitting of the steel shell will impose a number of further loading scenarios. The inner structural concrete is placed as the element is floating, which gradually reduces the freeboard of the element. Each concrete placement will change the load distribution along the tunnel element and the element needs to be analyzed both longitudinally and transversely. In the transverse condition, it is important to look at the sections adjacent to the last concreted section as they may be drawn lower into the water and have increased hydrostatic pressure acting on the external plate, but with no concrete yet behind the plate to provide stiffening. The designer will need to define the sequence of concreting based on the analysis, and provide this to the contractor to ensure the steel shell is not overstressed. The designer should analyze both local and global effects to ensure all parts of the shell have adequate resistance. However, for a double shell tunnel, there is a three-dimensional distribution of load through the shell. The diaphragm that spans between the form plate and inner shell acts as a stiff ring, utilizing an effective width of the form plate and inner shell. The hydrostatic load acting on the shell between the diaphragms is transmitted to some degree to the diaphragms by stiffener plates welded to the inside of the shell. Some assessment is necessary as to how much load is carried by the shell, stiffeners, and the diaphragm. Nowadays, this would lend itself to a sophisticated three-dimensional frame analysis or FE analysis, but as the double shell tunnel form has not been constructed for some years, this is unlikely to be a subject of further development in immersed tunneling. When the internal concrete has been placed, a steel tunnel element must be analyzed similar to a concrete tunnel element as it undergoes the immersion process.

Japanese steel tunnels are constructed as a true composite structure and would be more straightforward to model in an analysis. Nevertheless, they still require that the various stages of construction be analyzed as described and there are some specific design procedures to follow that relate to the detailed features of the construction.

Although all upward hydrostatic forces and downward weight of the structure and imposed loadings will be modeled in a structural analysis, the determination of the buoyant behavior of a tunnel element should not be modeled as part of the structural analysis. This should always be carried out separately to obtain clarity of the buoyancy characteristics at each of the stages of construction described earlier.

STRUCTURAL DESIGN

The transverse design of a reinforced concrete tunnel structure is relatively conventional and will be governed by the design codes once the key durability parameters, such as cover depth and permissible flexural crack width, are defined and the strengths of materials are selected. There are some important aspects to consider however, such as remembering to include the benefit of the axial compression in walls and slabs that arises due to the constant hydrostatic pressure that is applied to the exterior of the tunnel structure in the permanent condition. This increases the bending capacity of the section.

Flexural cracking is controlled in the usual way for reinforced concrete structures, to ensure the required level of durability for the structure. Because of the marine environment and the highly corrosive environment within a tunnel, it is common to impose tight limits of flexural crack widths, particularly in the water-retaining parts of the structure and external surface of the structure. Typically, maximum crack widths on the external face will be limited to 0.15–0.2 mm, and on the inside faces, between 0.2 mm and 0.25 mm. National design codes may also give guidance on this aspect, though it is not common for immersed tunnel structures to be specifically dealt with.

For reinforced concrete tunnels, through-section cracking should not be permitted, so the common design procedure for determining the size and spacing of reinforcement bars to control this phenomena arising from early age thermal and shrinkage effects is not applicable to the watertight perimeter of the tunnel. The required approach to dealing with early age cracks and shrinkage cracks is described in Chapter 11. Minimum reinforcement requirements of the design codes will apply, however.

When considering the longitudinal design of monolithic tunnel elements acting as beams in bending, the designer should ensure that the either the top or bottom slab does not go into full tension under the peak bending moments as this could result in cracks through the full depth of the slab. Because the overall tunnel section is large, the ratio of section modulus to applied bending moments is large, and so, it is often the case that tensile stresses are low, and may be within the tensile capacity of the concrete, resulting in no flexural cracking occurring. However, if tension is of a magnitude such that cracking will occur, or has a moderate to high risk of occurring, it is likely to extend the full depth of the roof or base slab. In this case, the structure should be prestressed to prevent it. The likelihood of cracking extending through the full depth of slabs is one of the reasons an external membrane is applied to monolithic tunnel elements.

If the whole length of the tunnel structure is made fully continuous, then thermal expansion and contraction due to seasonal temperature and air temperature fluctuations can cause the tunnel structure to be placed into tension. Care needs to be taken in selecting the interval and type of jointing to prevent this. Steel tunnels are often made continuous, but the ductile nature of the steel shell means that, in theory at least, this can cope with movements and remain watertight. Concrete is not so ductile and will crack, and therefore cannot be designed in the same way. It should be checked that this condition does not arise as part of the design process.

Longitudinal compression waves can be induced by earthquakes that can cause a tunnel structure to go through a rapid cycle of compression and tension. The tension will be across the whole section and cause instantaneous cracking. However, provided the section remains in its elastic range, there will be no permanent deformation and cracking will not be severe. Although cracks may not fully close once opened, there will not necessarily be a leakage problem. However, this is a situation that will require inspection and potential remedial works, such as grouting to seal the visible cracks.

The strength of the tunnel concrete must be known for the purposes of design, but it is not usually necessary to select a high strength for this type of structure. The tunnel walls and slabs are often sized for the purposes of obtaining good floating characteristics and achieving the required factor of safety against uplift in the permanent condition. This means they are not specifically made slender for the purposes of efficient structural design. This has the benefit that reinforcement can be designed reasonably comfortably without the need for using the larger bar sizes or multiple layers of bars. This also means there is no demand for high concrete strengths as the sections are relatively thick, and strengths in the order of C40 or C45 are generally adequate. Larger tunnels with wide tubes containing three of four traffic lanes need to be looked at more carefully and there starts to be benefit in using larger reinforcement bar sizes and increased concrete strengths.

In some circumstances, the construction method may require high concrete strength at an early age, in which case a higher characteristic strength may become necessary. This was the case for the Øresund Tunnel, which used a construction method that required the tunnel segments to be slid on temporary bearings only a few days after casting the concrete. The structure was very sensitive to unevenness in the sliding surface and the stresses this could introduce at an early age. Therefore, a minimum strength was needed before the sliding operation could take place and this resulted in a higher characteristic strength concrete becoming necessary than would otherwise have been needed.

When detailing the reinforcement for the walls and slabs of the immersed tunnel, the designer should attempt to minimize the presence of shear links in the structure. This is particularly important in the walls of the tunnel where, historically, there have been some incidences of water piping along the shear link where concrete has shrunk or settled, leaving a small void beneath the horizontal bar of the link. Where good concrete mix design is used along with sound concreting methods and practice, this can be easily avoided, but if the designer can avoid the need for links, then the risk is immediately eliminated. For smaller tunnels either carrying railways or just one or two lanes of road traffic in each tube, this is considerably easier. For tunnels with wider spans carrying 3 or 4 lanes of traffic, it will be difficult to avoid placing shear links in the slabs. As already mentioned, usually, wall thicknesses are not slimmed down to the minimum for structural adequacy because of governing factors such as buoyancy. This may help to avoid the need for shear links in the walls. For the same reasons through-wall formwork ties are often not permitted during construction.

Optimizing the tunnel cross section by introducing haunches between the walls and slabs can improve load paths and minimize the shear reinforcement needed. Shear enhancement should also be taken into account for slabs and walls close to supports. Even though the structure comprises walls and slab rather than beams, the dominant load effects are permanent and uniformly distributed, so it is reasonable to use the shear enhancement rules within the design codes. The recognized approach of smoothing peak bending moments at supports should also be taken. There are numerous design texts that explain this principle and the methods to follow, which are not repeated here.

Design for shear is an ultimate limit state (ULS) condition whereby shear cracking is intended to be prevented. Designers should be aware that if shear cracks do occur close to supports, then the hydrostatic pressure acting in the shear crack will prevent the normal strut-and-tie action developing, which is the basis of the normal shear enhancement design rules close to supports. Some project specifications have precluded the use of shear enhancement for this reason. However, Det Norske Veritas (DNV) guidance for offshore concrete structures is that this effect may be neglected for water depths of less than 100 m, which would cover all the immersed tunnels designed and built so far.

In the longitudinal direction, reinforcement detailing is very conventional. If the tunnel is a segmental construction, the longitudinal bending in the structure will be very small and the minimum reinforcement requirements of the design codes will most likely be the governing criterion for the design. If, however, the tunnel is constructed using monolithic tunnel elements, the longitudinal bending moments may be significant and will require substantially greater quantity than the minimum provision. In this instance, it can become important to consider the most efficient way to group or layer bars, particularly where the transverse reinforcement is at its most dense over walls and at mid-spans in slabs.

In detailing the reinforcement around the tunnel element joints, it is important to ensure that structural continuity is achieved between the structural reinforcement and the shear connectors cast into the concrete at the back of steel plates associated with the immersion joints. Good overlap between the closure bars for the main wall and slab rebar cages and the shear studs should be achieved. The detailing should also follow good practice for half-joints at the immersion joints, with diagonal bars in accordance with the detailing codes. Where waterstops are cast into the faces of segment joints, they tend to be close to the exterior face of the tunnel, so that they may avoid features such as the shear keys. There is a risk of pressures from the waterstop grouting causing the edge of the concrete to spall and render the waterstop ineffective. It is important to ensure the outer mat of reinforcement extends around the waterstop to prevent this happening.

Shear keys

Differential settlement may occur between tunnel elements if they are left to settle independently of one another. A small amount of differential movement can be tolerated, but any significant amount of vertical or lateral relative shift could cause damage to the watertight components in the tunnel joints, resulting in joint leakage. Even if the differential movements are predicted to be small by the designer, it is not good practice to leave the tunnel elements independent of each other as settlement behavior can be unpredictable and present a risk to the tunnel integrity. It is therefore necessary to provide some form of shear restraint between the tunnel elements. Shear keys are provided in the tunnel joints to resist transverse and vertical differential movement between the tunnel elements, and for concrete segmental tunnels, between the individual segments. They can also play a part in the temporary condition and be used as the temporary supports in a so-called nose/chin arrangement for one element to rest on the previous element during the immersion process. This approach is particularly suitable when steelwork shear keys are used, but it is not an essential requirement. If sufficient alternative temporary support is provided, the permanent shear keys can be installed at a later date.

Shear keys may not need to be installed immediately after a tunnel element is immersed. Indeed, there may be some advantage in allowing a percentage of the predicted settlements to occur before forming the keys. This can enable a lower loading to be used in the design of the permanent shear keys. However, this approach is too high-risk to be adopted in the long term, so the designer should judge when the best time to form the keys will be.

Shear keys for the concrete tunnels are located in the space on the inside of the Omega seal. There is often only a limited width of concrete available in which to construct the shear key and, as a result, it is quite common for the walls to be widened locally to increase the width of wall available for the shear key construction. Historically, they have also been located in the base slab of the tunnel. For some steel tunnels, shear transfer has been achieved by making the tunnel continuous across the joint. More recently, with a change to flexible immersion joints, the shear keys can be constructed in the permanent inner reinforced concrete lining on the inside of the steel shell. The following types of solution that are commonly used are described as follows:

- Steel shear keys
- Dowel shear keys
- Discrete shear keys in walls
- Half-joints

Steel shear keys

Steel shear keys may be deep I-sections or box sections mounted on a base plate secured to the end of the tunnel element. Steel shear key assemblies have been used on a number of tunnels, notably in Hong Kong, for example, the Western Immersed Tunnel carrying the MTR Airport Railway. Here, stiffened box-type keys were bolted to the end faces of the concrete tunnel elements. To resist the vertical movement in both directions, a three-part shear key is required. Two keys were fixed to the secondary end of an element and a single key to the primary end of the next tunnel element.

The top key of the secondary end is initially omitted, allowing the tunnel element being immersed to locate its primary end shear key onto the lower secondary end key. The top key is then installed later on, in accordance with the designer's assumptions on settlement.

As a three part key is needed, the overall depth of an individual steel key can be no more than approximately one-third of the tunnel total height. In fact, the height is a little less than a third of the total height because of the space needed for immersion joint components in the top and bottom slabs. The combination of shallow depth and high eccentric loading means that a high level of tension arises at the extreme fiber of the shear key due to the moments applied. This tension needs to be transferred back into the tunnel element structure and can require a significant quantity of horizontal tensile reinforcement and high tensile bars, such as Macalloy bars, to fix the base plate of the shear key to the concrete structure (Figure 9.17).

Lateral eccentric loading is an important consideration in the design of the keys. If there is a horizontal misalignment between the ends of the tunnel elements or if there is any unevenness or out of tolerance on the bearing surfaces, this could cause eccentric loading. The shear keys should be designed for an element of torsion to account for the possibility of these effects. The use of bearings in the shear key assembly will mitigate uneven loading. The type of bearings selected needs to be free of any long-term maintenance requirements. Access for inspection and possible replacement has to be considered carefully in the design. It is possible for the temporary condition to design on the basis of there being no bearings, but to use packing or shim plates between the keys to get even bearing pressures on the contact surfaces. Bearings should be introduced for the permanent arrangement, however.

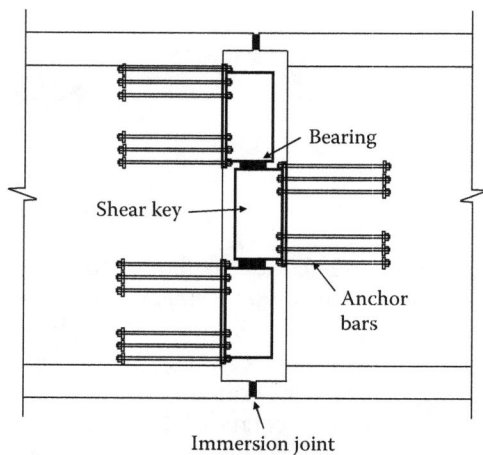

Figure 9.17 Steel shear key arrangement at immersion joint.

Dowel shear keys

Dowels cast into the second stage infill concrete placed above the immersion joint can offer a suitable shear connection. This is a solution that was originally used in tunnels built in the Netherlands. The dowels are commonly steel circular hollow sections filled with concrete that are fully embedded into one side of the immersion joint, but sit within a sleeve on the opposite side, which allows movement and joint rotation to occur. The arrangement is shown in Figure 9.18.

This type of shear connection has a limited shear capacity compared to discrete shear keys in the tunnel walls. The reason for this is the limited depth of concrete in which the dowels can be located. It is normal for the ballast concrete to be replaced by structural concrete over the joint to provide a greater depth, but even with this, there is limited room above and below the dowel. Dowel diameters are typically 200–250 mm and the depth of concrete will be in the order of 800–1500, depending on the depth of ballast concrete required in the tunnel. At the lower end of this range, there remains only 250–300 mm of concrete above and below the shear dowels in which to provide the shear resistance. As the shear transfer is concentrated at the dowels, loads can be high and the shear stresses can demand high levels of reinforcement to prevent shear failure of the concrete. Because the dowels need a reasonable depth of concrete in which to be fixed, the only place they can be located is in the base of the tunnel within the ballast concrete depth. This limits the shear capacity of the joint as a whole.

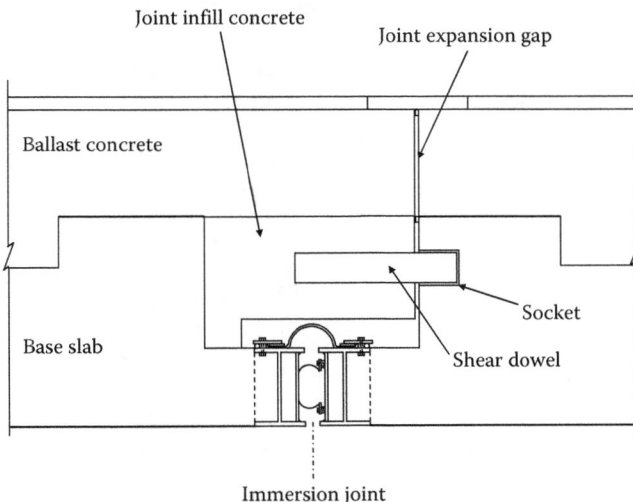

Figure 9.18 Dowel type of shear key.

An unusual dowel joint arrangement was used on the Bilbao Metro Tunnel, which utilized a dowel in the space between the Gina and Omega seals, welded to the immersion joint end frame steelwork on one side of the joint, and located within a bearing block on the opposite side of the joint. This arrangement has certain advantages in that the assembly is on the far side of the Omega seal, and hence does not obstruct access, and can be utilized in the full tunnel perimeter, if desired. However, on the converse, the shear keys cannot be accessed or inspected, which could be viewed as a clear disadvantage.

Half-joint arrangement at segment joints

The half-joint arrangement shear key has been a popular form of shear connection for concrete segmental tunnels that has been widely used in the Netherlands and elsewhere in Europe on tunnels such as the Medway Tunnel in the United Kingdom and the Jack Lynch Tunnel in Ireland. The typical arrangement is shown in Figure 9.19. The end of each segment is finished as a half-joint and effectively provides a spigot and socket type connection between the tunnel segments. The section thicknesses for external walls are typically around 1 m, and so, this enables approximately 0.5 m thickness of section to be utilized in resisting the shear forces.

This is an attractive solution from the point of view of simplicity of detailing. A difficult aspect of the design, however, is to assess the proportion of the half-joint that is effective in transferring shear. Because of shear lag effects, for wide spans of up to and in excess of 10 m, large parts of the half-joint are likely to be ineffective, as shear transfer is concentrated in the zones close to the walls of the section. This restricts the capacity of the joint.

As the joint is match-cast, there are no issues with tolerances on bearing surfaces. However, some care over the geometry is needed to allow the joint

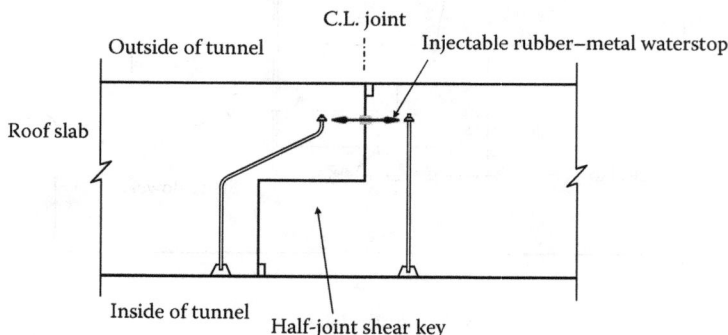

Figure 9.19 Half-joint type of shear key.

to articulate correctly. This may require slight slopes to be formed on the bearing surface or some compressible filler on the vertical faces to allow rotations to occur. Otherwise, an angular concrete-to-concrete interface may cause the joint to lock up and resist rotation, with the possible result of damage to the concrete.

The injectable waterstop is cast into the outer part of the half-joint. Reinforcement detailing around the waterstop should ensure the external concrete is not left unreinforced or this could be at risk of spalling as the joint rotates. Correct detailing for the half-joint is needed to avoid shear cracking. Because of the shear lag effects and the limited capacity of the joints, this type of solution is now less favored and discrete shear keys in the walls are the most common solution being adopted by designers.

Discrete keys in walls

The solution that is currently most favored with designers is to locate the shear keys within the walls and slabs of the tunnel. This approach can be applied at immersion joints and within segment joints for concrete tunnels and has been used on the majority of concrete tunnels built since the Øresund Tunnel in the late 1990s, including the high-speed line tunnels in the Netherlands, the Bjørvika Tunnel in Norway, the Limerick Tunnel in Ireland, and the Busan-Geoje Tunnel in South Korea. It is an attractive solution because it means the shear connections are in the most appropriate locations for shear transfer through the structure: in the walls. The typical arrangement is shown in Figure 9.20. There are no issues of shear lag in the slabs leading to lengths of slab having ineffective shear transfer, as would be the case with the half-joint solution. Shear lag is considered only in as much as determining the relative stiffness of the walls, taking into account a short length of roof and base slab above and below each wall, to determine the spread of shear force between the individual keys. This type of key has probably the greatest capacity of any of the shear key types described.

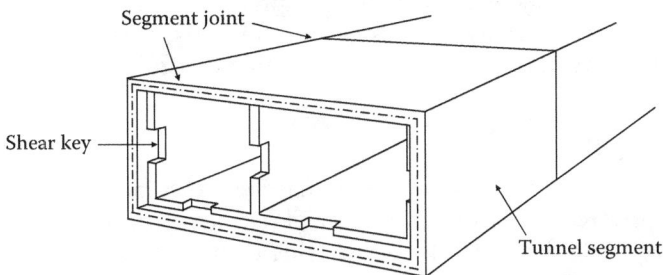

Figure 9.20 Discrete shear key solution.

In designing the key itself, the following aspects must be considered:

- Local bearing stresses at the contact surfaces of the shear keys
- Concentration of stresses due to rotation, due to opening of the joints, and arising from friction on the concrete surfaces
- Global shear through the body of the key
- Corbel behavior using strut-and-tie models to transfer tension back into the tunnel element
- Shear in the wall of the tunnel element behind the shear key due to transfer of load back into the tunnel structure

The quantity of reinforcement in the shear keys just below the bearing surface can be very high and particular care needs to be taken with the detailing of the reinforcement to ensure the keys are buildable and concrete can flow through the reinforcement. Small-diameter bars are needed to ensure there is reinforcement close to the full perimeter of the key, that bearing areas are adequately reinforced, and that there are no unreinforced corners that may be susceptible to spalling under concentrated loads.

Some tunnels have used single shear keys in each wall or slab. Others have used a number of keys, for example, the Aktio-Preveza Tunnel in Greece. For simplicity, the single shear key offers easier construction and there is no loss in capacity. Good detailing to ensure joint articulation is required. This means rounding of the leading edges so that the joint can rotate. To limit friction on the contact faces, a slip membrane can be used. It is difficult to obtain proprietary systems that offer a sufficiently high compression capacity, but provided they have sufficient bearing capacity for the service loads, it is better to put the best product in rather than omit this, as friction forces can be surprisingly high. Alternatively, stainless steel plates can be used at the bearing interface to improve performance.

Because shear keys in segment joints are cast into recesses in the end of previously cast segments, the effect of thermal expansion and contraction should be considered. The recess can cause lateral restraint to the expansion of the tunnel segment and cause early age cracking through the shear key. It is therefore recommended to install a compressible layer against the vertical faces in outer wall shear key recesses and, similarly, to the horizontal faces in the shear keys in the roof and base slab that control lateral movement.

Prestressing

Reinforced concrete tunnel elements may be either permanently or temporarily prestressed. Monolithic tunnel elements may be permanently prestressed if the load effects require this to provide adequate structural capacity to resist the applied bending moments. Segmental tunnel elements must be temporarily prestressed to clamp the tunnel segments together for the purpose of floating, transporting, and immersing the element. The prestress in this instance is then generally cut, but occasionally, for example in seismic regions, the prestress is designed to be left in place permanently. Some designers believe the temporary prestress does not have to be cut at all. Left in place, it provides additional longitudinal strength and if the joints do rotate excessively, the prestress will yield and the situation will be similar to having cut them in the first place. However, this introduces a degree of uncertainty into the design and it is preferable to cut the prestress so that the structural action is known. The decision on the approach must be made at the outset of design. It will not necessarily be possible to leave prestress uncut if it has been designed on the basis that it will be cut. Joint stiffness, settlement profiles and, hence, loading on the tunnel structure will be different in either circumstance.

Different criteria apply for the design of permanent and temporary prestressing systems. The methods to be followed for designing permanent prestress are largely as defined in structural design codes. Generally, the highest class of structure is chosen from the point of view of durability. This will stipulate that no tensile stresses will be permitted in the service condition, to ensure there is no flexural cracking that could lead to water ingress and corrosion of the prestressing cables. The loadings to be considered in the design should include both temporary loads due to the construction stages and processes, and the permanent condition. Arguably, some tension could be permitted in the structure in the temporary condition if this is more severe, provided there is none in the permanent condition once the tunnel is complete.

Conventional design software for the design of prestressed beams can be applied to immersed tunnel elements. The usual design process of evaluating losses in the prestress force due to friction and wobble over the length of the prestress cables must be followed. In this respect, immersed tunnels are somewhat simpler structures to design prestress for, compared to a bridge for example, as cables are generally straight and located in the roof and base slabs of the tunnel. As there is little curvature to the alignment of the cables, the friction losses are generally lower. The main reasons for curving cables would be to avoid features in the tunnel, such as the low point drainage sump, or cast-in items such as the anchorage assemblies for pieces of temporary works equipment that will be mounted externally on the tunnel.

Cable anchorages are usually located in the end faces of the roof and base slabs at the immersion joint, on the inside of the Omega seal. As there will usually be infill concrete placed in this area after the tunnel elements are immersed, the anchorages can be cast into this concrete and, therefore, be provided with a level of corrosion protection. Due to the exposure conditions for an immersed tunnel and the potential consequences should prestress cables corrode, the cables should always be grouted to offer the maximum corrosion protection.

Temporary prestress is used for segmentally constructed concrete tunnels. In this situation, the function of the prestress is only to create a monolithic structure for floating, towing, and immersion. The design loading for the prestress will derive from wave and current loads and the dynamic impact of these. If the tunnel is located in sheltered conditions, then the design loading condition will just be the imbalance of loads applied to the tunnel element during the immersion process, from the various items of equipment mounted on and in the tunnel element. There will be three conditions to consider here: the floating condition, where the element is supported by the hydrostatic uplift and has discrete loads applied from bulkheads, ballast tanks, survey, and control towers; the immersion condition, where the element is supported at approximately the 1/5th points along its length and has the same loads with the exception of increased ballasting to achieve the necessary overweight for immersion; and the immersed condition, with the tunnel element on temporary supports, which are often wider than the 1/5th points and with potentially further ballast load applied. These three conditions will provide differing bending moment envelopes to derive maximum tensile stresses to then design the prestress. The objective of the design is to preserve the integrity of the tunnel element as a monolithic structure during these stages of handling. To achieve this, there should always be a compression force across the segment joint interface under the most extreme conditions, that is, compressive stress at any point in the joint always greater than zero, and a minimum compression stress of not less than 1 N/mm^2 at all points of the segment joint interface under service conditions.

The design process is very simple in that the tensile stress is calculated at the extreme fiber of the tunnel cross section due to the bending moments arising from the applied temporary loads, assuming the tunnel element is monolithic. The quantity of prestress is simply that required to provide the equivalent compression to nullify the tension for the extreme condition, or to ensure a minimum compression is achieved at the joint position for the service condition. As with a permanent prestress design, the prestressing force that should be applied to the tunnel cross section to give the compression force needs to be calculated, taking into account all losses due to friction, wobble, and locking off the anchorages.

There are a number of detailing requirements for the prestressing that are specific to immersed tunnels. These include special slip-ducting at segment joints where it is important to allow the prestess cable to strain, should the segment joint open slightly. This typically extends a meter or so either side of the joint. Also, at the segment joint, there needs to be a facility for cutting the prestressing tendons once the tunnel elements have been immersed and backfilled and the prestress is no longer needed. These can either be pockets in the roof and base slabs close to the joints, or the joint can be provided with a slight widening of the joint gap to allow a saw cut to be made at the joint interface. This recess would need to be repaired with grout once the prestress has been cut. If saw cutting is planned, it is advantageous to have the tendons grouped to enable the cutting process to be as quick and efficient as possible. Tendons are located towards the inside face of the tunnel structure to enable easier cutting and to ensure they are on the inside of the segment joint waterstops. The cutting sequence should be designed to release the prestress as evenly as possible along the length of the tunnel. Care should be taken to avoid any blocking of the cutting recesses with temporary equipment, such as ballast tanks, that would prevent cutting being carried out, and the cutting operation needs to be timed with other finishing works to allow free progressive access through the tunnel.

There is a practical limit to the anchorage size if it is to be located at the end face of the tunnel element by the immersion joint. The available depth typically limits anchors to something around a 19-strand tendon. This enables the anchor to be located with reasonable clearance from the inside face of the tunnel. This is also important for ensuring the standard spiral bursting reinforcement can be fitted without infringing on the cover depth to the inside face. This is not always possible and it is fairly common for the spiral reinforcement to be replaced with orthogonal bars in the form of closed links. This creates congestion in the main reinforcement mats, however, and needs careful detailing.

The location of the anchors and the general tendon alignment will be dictated by other features in the tunnel. At the end of the tunnel element, there can be little space available as there are many cast-in items, such as bulkhead support brackets and anchorages for marine equipment such as bollards.

The grouting of the prestress cables is conventional, although the distances can be quite long for immersed tunnel elements, possibly up to 180 m, as seen on the Øresund and Busan Tunnels. Regular grout vents are needed, but with some pre-testing of grout mixes to ensure complete filling is achieved, the grouting is relatively straightforward. Well-sealed coupling sleeves for the ductwork are needed at the segment joints to prevent grout leakage into the segment joint. Pre-testing with water is common to check the watertightness of the ductwork.

Special features

Two special features that are common in most immersed tunnels are the drainage sump and technical rooms. These have particular structural design and detailing considerations.

Drainage sumps

To collect the surface water or spilled liquids in the tunnel, a drainage sump is required in the low point of the tunnel. For longer tunnels, there may be more than one such sump. The sump is generally sized to cater for a ruptured tanker or for fire hydrants running for one to two hours within the tunnel and most highway immersed tunnel sumps have a storage capacity of around 50 m^3, but requirements will vary according to the nature of the road and the operational philosophy for the tunnel. The challenge with an immersed tunnel is finding space to locate this sizeable void. It is undesirable to have protrusions below the bottom of a tunnel element as this creates difficulty in construction. The base of the casting basin or dry dock would need to be shaped to cater for any features below the element and the dredged trench would have to be shaped similarly. If the tunnel foundation is to be gravel, then this would need to be shaped to support the tunnel element correctly. This adds complexity and cost to the construction of the tunnel. The draft of the tunnel when floating will also be greater and thus the water depth for access to the casting facility and the towing route need to be increased. To avoid this unnecessary complexity, it is common to form the sump within the depth of the ballast concrete and structural base slab. This also has its challenges, but overall, is considered to be the lower-cost, more manageable approach to design.

In a highway tunnel, the drainage carrier pipes usually are located in the ballast concrete beneath the road surface. In order to form a low point for the collection of liquids, the base slab can be locally thinned so that the top surface steps down. Similarly, the ballast concrete that sits above the structure base slab is omitted at this location and a thinner walled structure can be formed to create space for equipment and liquid storage and to support access covers in the road surface. The same approach can be used in a railway tunnel; however, there is typically less depth of ballast concrete and drainage may be provided using surface channels. This is because drainage inflows are likely to be less, which is useful because it means a sump can still be created in the reduced depth of ballast and structure.

In order not to compromise the watertightness of the structure, it is common to ensure a minimum thickness of structural slab of 500 mm is preserved beneath the sump. Sump chambers tend to be quite long, typically 10–15 m along the length of the tunnel. This length is needed to generate sufficient storage volume. However, a thin slab may have insufficient transverse bending capacity if it extends over such a long length and it is usually necessary to stiffen the slab with transverse beams. This has the effect of dividing the sump chamber into a series of parallel chambers in between the stiffening beams. These can be interconnected with pipes to create the needed storage volume. The stiffener beams also serve as the support structure for the frames for access covers, which will be needed to open up the sump chamber for periodic inspection and cleaning. A typical drainage sump arrangement is shown in Figure 9.21.

Durability is of some concern for the reinforced concrete forming the sump chamber as there is potential for aggressive materials to be stored.

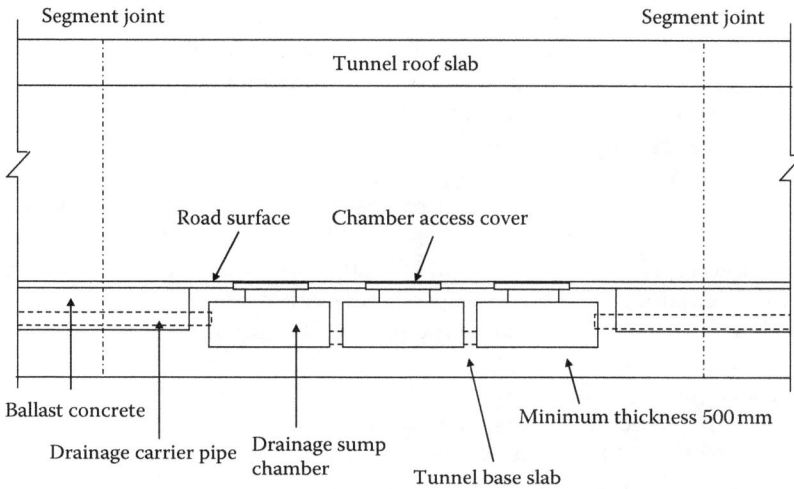

Figure 9.21 Typical drainage sump arrangement.

Many sumps are lined or painted in order to protect the concrete. Crack width and cover depth need to be selected with this aggressive environment in mind.

Technical rooms

Longer tunnels may require technical rooms along their length to house mechanical and electrical plant, switchgear, and transformers for the power supply to the tunnel. This can be problematic in that there is very little space in which to house them. If there is an emergency escape tube and service duct alongside, separate from the main highway tube, then the solution can be to utilize this space. This means the escape route will be impassable, but if there is a single room at the low point of the tunnel, then the direction of escape would normally be away from the room, which may be an acceptable solution. This is still quite a narrow space in which to fit the equipment and enable access for maintenance. It is feasible to do this however, and this approach was taken in the Øresund Tunnel. In this instance, the walls separating the escape gallery from the highway tubes were thinned to maximize the horizontal width in the room. This resulted in a room that is 2.0 m wide and is just sufficient to house the equipment while still allowing access through the room to each piece of equipment for maintenance personnel.

The Bjørvika Tunnel in Oslo, Norway provided technical rooms to the side of the roadway in a specially widened tunnel element. In this tunnel, there was no central gallery and one element was to be widened in any case to allow a lay-by to be created within the tunnel. It was therefore relatively easy to combine the technical room into the same tunnel element. This was quite an unusual circumstance, but the solution of adding width to the tunnel was certainly easier than adding depth.

The tunnel proposed for the Fehmarnbelt link is planned to have special deeper elements at intervals along the tunnel. This is needed as there are multiple locations where technical rooms will be needed and blocking the escape corridor at multiple locations would be undesirable. As the tunnel is 19 km long, there will be several special tunnel elements that are deeper than the remainder. This will undoubtedly cause some difficult construction challenges, but is seen to be the preferred approach at the early stages of planning.

For a tunnel element where ballast is removed or structural sections are thinned, the structural design needs particular attention. Internal walls were thinned to 300 mm on the Øresund Tunnel, which made construction very difficult. In retrospect, some pre-cast wall units installed as a secondary construction exercise may have been a better solution than building the walls in situ.

The buoyant behavior of the tunnel element must be carefully assessed as it is likely to be lighter and float higher in the water. It may also need a greater amount of ballast, and so, may need special provisions, such as external ballast, to achieve the required factor of safety against uplift. If it is part of a long tunnel element, a short length where the construction is different may not be that significant, but for shorter tunnel elements around the 100 m range, the impact of a sump or technical room could change the behavior significantly.

SETTLEMENT MODELING

As well as analytical models for structural analysis, it is important that settlement behavior is understood. Analysis using software such as OASYS Pdisp, or similar settlement modeling software, is ideal for this task as it can take account of multiple soil layers with varying time-dependent compressibility characteristics and a series of construction stages. Add to this the sensitivity aspects of variations in soil properties or the tunnel bedding layer and multiple scenarios for settlement behavior can be assessed.

Global settlement is important to understand as the final road or railway alignment will be dependent on this. Normally, however, an immersed tunnel can tolerate a considerable amount of global settlement, provided it is relatively uniform and the tunnel joints have been designed to accommodate it. Differential settlements on the other hand can cause significant problems and need to be thoroughly investigated and catered for in the design.

An accurate modeling of longitudinal variation in soil conditions and foundations must be carried out. The area where this is most critical is in the tunnel approach ramps, cut-and-cover tunnels, and the ends of the immersed tunnel. In this area, the tunnel founding depth is changing, often going from soft material to more competent material as the tunnel gets deeper. However, the foundation solution also typically changes between the approach tunnel and the immersed tunnels, and frequently, there is an abrupt change in foundation stiffness, and sometimes structural stiffness, at the interface. This can cause the immersed tunnel to "hang" from the end of the cut-and-cover tunnel, causing increased foundation loading on the cut-and-cover tunnel and uneven settlement behavior. A number of solutions may need to be investigated by analysis as often a transition foundation solution is needed to smooth out the settlement profile satisfactorily. For high-speed rail tunnels, this is vitally important as a very tight tolerance is usually specified on the alignment. While much of the settlement will occur early during construction and can be accommodated, the subsequent long-term settlement will be problematic for the railway.

A particular phenomenon that occurs with immersed tunnels is a downdrag effect from the tunnel backfill. The submerged weight of the backfill will exert a greater pressure on the underlying ground than the tunnel structure itself, which is only negatively buoyant by a small amount, according to the final factor safety against uplift that is achieved for the completed condition. This means the backfill may settle more than the tunnel structure, but the influence of the backfill settling can cause the downdrag of the tunnel, through friction and by the pressure bulb created below the tunnel at each side. This can be particularly apparent where ground improvement has been necessary to strengthen the underlying soils. The ground improvement may need to extend to some degree beneath the backfill also to prevent this behavior. This emphasizes the need to consider the transverse settlement profile as well as the longitudinal settlement profile.

SEISMIC DESIGN

A number of immersed tunnels have been built in highly seismic regions. These include the BART Tunnel in San Francisco, many tunnels in Japan, the Aktio-Preveza Tunnel in western Greece and, most recently, the Bosphorus Tunnel in Istanbul and the Busan Tunnel in South Korea. Immersed tunnels are very capable of withstanding the effects of earthquakes and this has been demonstrated now a number of times. Both the BART Tunnel and the Preveza Tunnel have been subject to relatively severe earthquakes and both have performed very well with no damage occurring. In Japan, the Osaka Port Tunnel was under construction during the Kobe earthquake of 1995. This, too, was unaffected, whereas surrounding infrastructure was severely damaged.

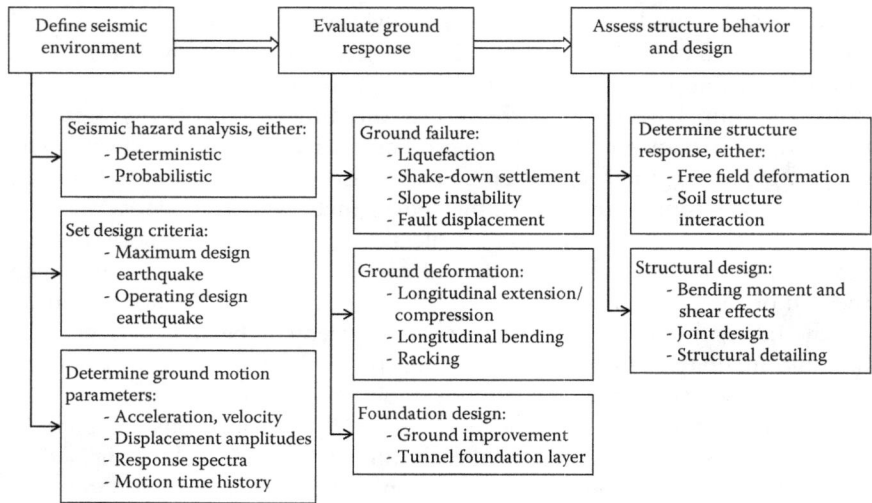

Define seismic environment	Evaluate ground response	Assess structure behavior and design

Seismic hazard analysis, either:
- Deterministic
- Probabilistic

Set design criteria:
- Maximum design earthquake
- Operating design earthquake

Determine ground motion parameters:
- Acceleration, velocity
- Displacement amplitudes
- Response spectra
- Motion time history

Ground failure:
- Liquefaction
- Shake-down settlement
- Slope instability
- Fault displacement

Ground deformation:
- Longitudinal extension/ compression
- Longitudinal bending
- Racking

Foundation design:
- Ground improvement
- Tunnel foundation layer

Determine structure response, either:
- Free field deformation
- Soil structure interaction

Structural design:
- Bending moment and shear effects
- Joint design
- Structural detailing

Figure 9.22 Steps for seismic design. (Developed from Hashash 2001.)

The design of an immersed tunnel in a seismic region requires the involvement of both specialist seismic engineers and analysts, combined with experienced immersed tunnel designers, as design often requires iteration, taking into account various possibilities and solutions. Seismic analysis and design is a specialized topic, but because it is often important to immersed tunnel designs and even their feasibility, we have set out the main design principles. The steps to follow for seismic design are summarized in Figure 9.22.

Seismic design is quite different from the other aspects of tunnel design in that loadings cannot be determined with exact precision and they must be dealt with in terms of risk and probability. The loadings to the structure are transient and reversing and occurring at a high frequency. Due to their onerous nature, the levels of acceptable damage must be defined rather than simply ensuring the tunnel structure can adequately resist all loads.

The design of tunnel structures under seismic loading is very different from that of above-ground structures. Buildings and structures above ground are subject to shaking. The movement of the ground surface causes a corresponding movement at the base of the building and the inertia of the building causes the deformation to occur. The building design then looks at stiffness, structural adequacy, and articulation to mitigate either the movement or the loadings arising from inertia effects. Buried structures move with the ground and the inertia of the soil is generally greater than the inertia of the tunnel, so the inertia of the tunnel is no longer of particular relevance. It may be relevant at the tunnel portals if there are above ground service buildings or ventilation shafts, but not for the long buried tunnel structure in between.

Earthquakes give rise to pressure waves that pass through the ground, causing deformation. These are known as body and surface waves. Primary (P) waves and secondary (S) waves are body waves; Raleigh (R) waves and Love (L) waves are surface waves. R waves create a rolling motion of particles in the surface stratum and L waves cause a horizontal shear motion that is greatest at the surface and dissipates with depth. The body waves, however, are the most significant in design terms and the principal effects of these waves on an immersed tunnel structure, which must be considered in its design, are discussed next.

Longitudinal compression caused by P waves

The P wave is a pressure wave that causes compression at the wave front and can cause compression–extension behavior as is travels along the length of the tunnel. When the wave direction is along the tunnel alignment, this behavior is sometimes referred to as worming. It will cause compression and tension loads in the structure, which have to be designed for, and it will cause the opening and closing of the tunnel joints. When the P wave is at an angle to the tunnel alignment, there may be lateral compression on the tunnel. When the angle of incidence is between 0° and 90° to the tunnel alignment, there is a time lag that must be considered, as the P wave will hit the structure at one end before the other end. This can also cause horizontal bending due to tension and compression zones arising on opposite sides of the structure (Figure 9.23).

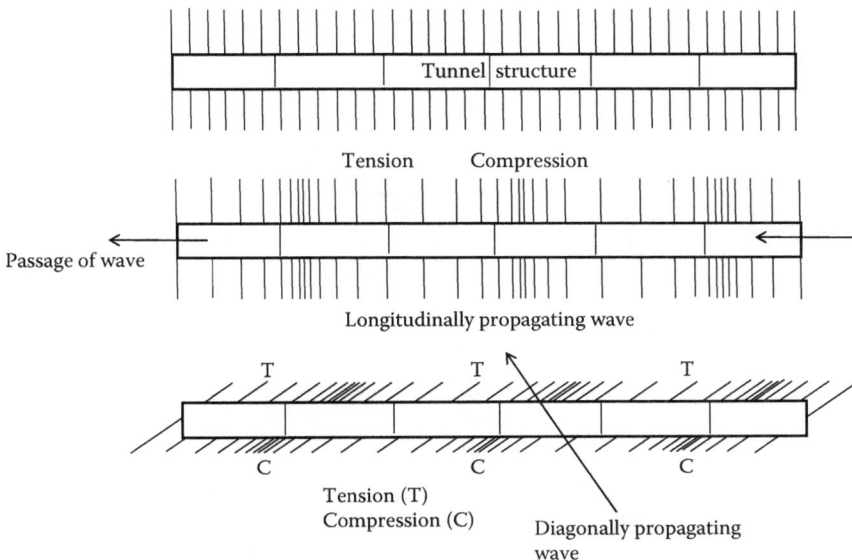

Figure 9.23 Seismic longitudinal compression wave.

Horizontal and vertical ground displacement caused by S waves

S waves are sinusoidal and travel at half the velocity of P waves; they have both a vertical and horizontal component. They dictate the peak ground acceleration (PGA) and peak ground velocities (PGV) that are used in design. PGV is a more relevant parameter for underground structures and PGA for above-ground structures. The tunnel structure will distort according to the direction of travel of the waves and their angle of incidence on the tunnel. This behavior is sometimes referred to as snaking. It will cause vertical and horizontal bending moments in the structure, and where there are joints, will cause joints to open and close due to the curvature of the structure.

Racking

This is the result of ground shaking and the inertia in the ground stratigraphy, causing a horizontal distortion of the ground mass—in effect, a shear wave passing vertically up through the ground. In a circular section tunnel, this will cause ovaling, and in a rectangular tunnel, this will cause racking whereby the rectangular section is distorted to a trapezoidal shape, as seen in Figure 9.24. This will cause large bending moments and shear forces in the structural walls, particularly at the stiff wall–slab connections.

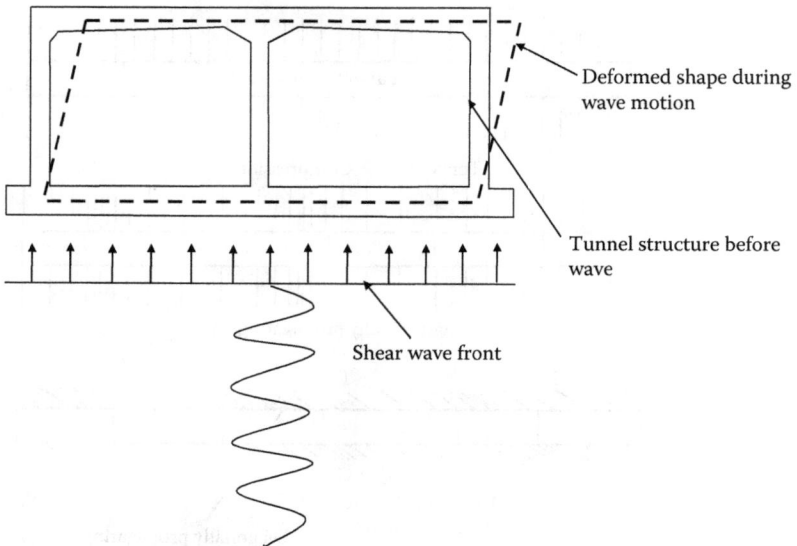

Deformed shape during wave motion

Tunnel structure before wave

Shear wave front

Figure 9.24 Racking behavior due to vertically propagating shear wave. (Adapted from Hashash 2001.)

Liquefaction

This is an effect on sandy soils and is the process whereby the cyclical pressure waves passing through the soil repeatedly compress the soil faster than the water held within the soil structure matrix can escape. The pore water pressure builds up to the point where it exceeds the contact pressure between the soil particles and the soil matrix loses all strength and behaves as a liquid. Structures founded on liquefiable soils run the risk of sinking. An immersed tunnel is already buried, but depending on the net pressures on the soils beneath the tunnel, may be susceptible to either floating or sinking.

Fault displacement

Shear movement in the bedrock is almost impossible to design for and it might not be prudent to construct a tunnel with any such risk. However, immersed tunnels would rarely be built in the bedrock and are more likely to be in the overlying soft deposits. A shear movement may be mitigated with the increasing depth of soft material between the tunnel and bedrock, and so, it might be possible to design for this circumstance.

Slope instability

Ground shaking or liquefaction may have deleterious effects on soil slopes around the tunnel portals or on the inclined banks of the watercourse at each side of the waterway crossing or even on the underwater slopes. The design of slopes may need to take this into consideration.

Damage criteria

Before commencing any design work, it is necessary to define the performance criteria that a tunnel should achieve in an earthquake event. Two design levels are commonly used.

Level I: Service life event

This is a seismic event of a magnitude that could be expected to occur over the design life of the tunnel. This is often expressed in terms of an event that has a probability of exceedance in the order of 40–50% over the design life. The precise definition will depend on the design codes that apply to the project, for example, the UBC code or ASCE-07. This event may also be referred to as an operating design earthquake (ODE). Under loading from the level 1 earthquake, the structure should always behave elastically and suffer no permanent distortions. Yielding of reinforcement should not be allowed to occur as this will lead to residual cracking causing leakage of the tunnel structure. Tunnel joints should remain watertight at all times. Minor damage may be tolerated, but this would be to the internal fittings and equipment and should be readily repairable. No interruption of service should be expected other than that necessary for undertaking inspections and minor repairs, which could be carried out during off-peak routine maintenance closures.

Level 2: Maximum credible event

This is a large-magnitude event that would have a low probability of being exceeded in the design life of the structure. The probability of exceedance in this instance is generally less than 5% and often in the range of 2–3%. Again, the precise criteria will need to be developed according to local practice and applicable design codes. This is also sometimes known as the maximum considered earthquake (MCE) or maximum design earthquake (MDE). Under this loading, temporary disruption to the operation of the tunnel might be expected, but damage should not be sufficient to prevent the tunnel being brought back into operation within a few months of the earthquake. Some level of repair may be necessary to the structure and internal equipment. Some plastic behavior of the structure may be permitted, provided sufficient ductility is available to prevent any catastrophic collapse and the structure is repairable.

Design load effects

A site-specific seismic hazard analysis should always be undertaken for the particular site where a tunnel is to be built. Indeed, this is mandatory in some national codes for highly seismic zones. This enables a quantitative estimation to be made of the ground motions. The hazard assessment may be either deterministic or probabilistic. A deterministic approach considers a particular earthquake based on local history and evidence. A probabilistic approach considers multiple events in the region and, hence, establishes the seismic event for an appropriate return period being considered. Both methods will make reference to international databases for seismic source information and time history records.

The deterministic approach considers all earthquake sources capable of producing ground motions at the tunnel site, the earthquake potential, and the distance from the site. Distances to both the epicenters and actual geological fault zones are considered, along with their predicted impacts. The most significant earthquake is then selected and defined by size and distance from the site. The characteristics of the seismic hazard may be defined by PGA, PGV, peak displacements, or by response spectra, usually obtained from design codes. Alternatively, ground motion time histories may be used. This approach defines a worst-case scenario.

The probabilistic approach enables a better assessment to be made of the frequency at which a particular earthquake event may arise, and therefore ties in well to the level 1 and level 2 damage criteria. This procedure entails a similar review and characterization of potential seismic sources that could give rise to ground motion at the site. The level of uncertainty associated with the various sources and events is also assessed. The frequency and magnitudes of earthquakes are assessed for each source and ground motion parameters are determined for each potential source and magnitude. The probability that any particular ground motion parameter could be exceeded is calculated for any given site and time interval. Model accelerograms are created using a selection of time history information, rather than using specific site time histories directly. Site-specific time histories tend to need adjustment in any case to represent seismic events at depth.

The probabilistic seismic hazard assessment is considered the most appropriate method to determine the ground motion characteristics, particularly when trying to define the level 1 design parameters. However, there are circumstances where it might be combined with the deterministic approach— for example, probabilistic methods could determine the level 1 criteria—but the deterministic approach could be used to provide the upper bound condition for the level 2 design parameters. For the Bosphorus immersed tunnel in Istanbul, the highest possible magnitude earthquake has a high probability of occurring during the design life of the tunnel. In this instance, the level 1 and level 2 criteria are more or less equal and there is a highly dominant source, the North Anatolian Fault, which is well documented. Therefore, a deterministic approach is more appropriate and a single design basis earthquake was defined for the design of the tunnel.

For either method of selecting the earthquake characteristics, the ground motions can be defined either by design response spectra or by accelerogram time histories. Design response spectra are either developed from records of ground motions from earthquake events, or by using codified approach as set out by the Uniform Building Code or Eurocode. The developed response spectra aim to capture and formularize the free field motions in terms of displacement, velocity, and acceleration against natural frequency. Recorded events have a jagged nature, with peaks and valleys in the spectra chart, whereas the design spectra will smooth these out and aim to envelope the available records to ensure they are all accounted for. They therefore take account of a number of time histories. In some respects, this could be considered conservative, but could equally be viewed as a suitably safe design approach. Assumptions need to be made as to the degree of damping of the ground and, typically, a range of spectra may be developed to encompass the likely conditions.

Response spectra by their nature produce load effects that correspond to the maximum peak values of ground displacement. Some consider this to be a drawback as the maximum response in the structure may occur at a different instant in the time cycle of the earthquake loading. Equally, others consider the approach to be generally conservative and more suitable because it takes account of a variety of time histories.

The alternative approach to design response spectra is inputting ground motion time histories into the analysis directly. Time histories (accelerograms) represent ground acceleration versus time and can be based on actual records of seismic events, for which large databases exist, or can be developed synthetically. Some consider the time history approach to be nonconservative because the time histories can be very specific in terms of the response they cause in the structure and there is a risk that some potential responses could be overlooked. Synthetically developed time histories can overcome this to a degree. There are codified approaches to developing accelerograms, such as in Eurocode, which can be used. These will effectively generate a time history that fits with a site-specific response spectrum. If there is a great deal of data for a specific site and the source is equally specific, then the use of accelerograms is a very good approach. The amount of available data for the site should, however, be considered when deciding on the design approach.

Longitudinal analysis

The longitudinal design of the tunnel structure requires an analysis of the ground deformations to determine the applied loadings. A simplified upper abound approach is to look at the free field deformations caused by the shear waves and impacting at an angle to the axis of the tunnel. As this is an upper bound solution, it can be used as a check on detailed calculations and to get a feel for expected behavior, but it could be a highly conservative approach if used for final design. The method of determining the axial ground deformation and the transverse ground deformations was described by Newmark and Keusel in the late 1960s, according to the angle of incidence of the propagating wave compared to the axis of the tunnel. The method assumes that the tunnel structure deforms entirely with the ground and takes no account of soil–structure interaction affects. It therefore lends itself to highly flexible structures. Whilst segmental tunnels are flexible, they will most likely have a joint restraint system that causes a degree of stiffness in their response to the ground. For tunnels constructed with monolithic tunnel elements, the flexibility to follow the ground is provided only at the immersion joints. However, the monolithic element itself will be rather stiff in relation to the ground, particularly if it is a wide rectangular box structure. Therefore, the interaction between the soil and the structure is likely to make a significant difference to the behavior. Hence, this type of analysis is not recommended other than to get an initial feeling for the behavior of the tunnel.

A more suitable method of analysis for the structure is a dynamic analysis using a ground motion input. Numerous pieces of design software exist in the marketplace to enable such an analysis to be carried out. However, some care is needed in selecting the best approach in order to get meaningful design load effects and ground deformations. Linear or nonlinear analysis can be carried out.

The dynamic mass spring model has been the most popular method of modeling immersed tunnel structures. This takes account of the soil–structure interaction by defining the connectivity between the structure and the soils as a series of complex springs. The tunnel structure is linked through the springs to a series of masses representing the ground. The ground masses have stiffness parameters defined to model the interaction between them. Motion is introduced to the masses and the structure response can be seen, based on how the connectivity is defined. This type of model is often known as a Winkler model. The tunnel structure can be modeled as a series of elements or segments, each with its own mass spring. The rotational and compression stiffness characteristics of the joints between tunnel elements and between tunnel segments can be modeled. The stratigraphy between the bedrock and the tunnel can be modeled in the soil mass characteristics with springs and dashpots to represent the variation in ground inertia with depth. This was illustrated by Kiyomiya in a number of papers and the principle described can be seen in Figure 9.25.

The FE continuum model is now becoming the preferred method of analysis, where the soil mass can be modeled in three dimensions, the structure can be modeled in detail, including the joint characteristics, the ground geometry can be modeled, and the various ground motions applied either to the soil mass or to the bedrock and nonlinear analyses undertaken. There are many sophisticated pieces of software that can be used for this, usually as time step analyses, but they still need much care when defining critical parameters. With careful definition of the soil and material properties and the interface conditions between the soil and the structure, the mass-spring approach is replicated with much a greater degree of accuracy and in a continuous manner rather than as a series of discrete masses. Nonlinear hysteretic soil models can be used to accurately define the ground characteristics. These can be based on experimental data or recorded data. This approach is recommended to get the best understanding of the behavior of the structure and the joints under seismic load conditions.

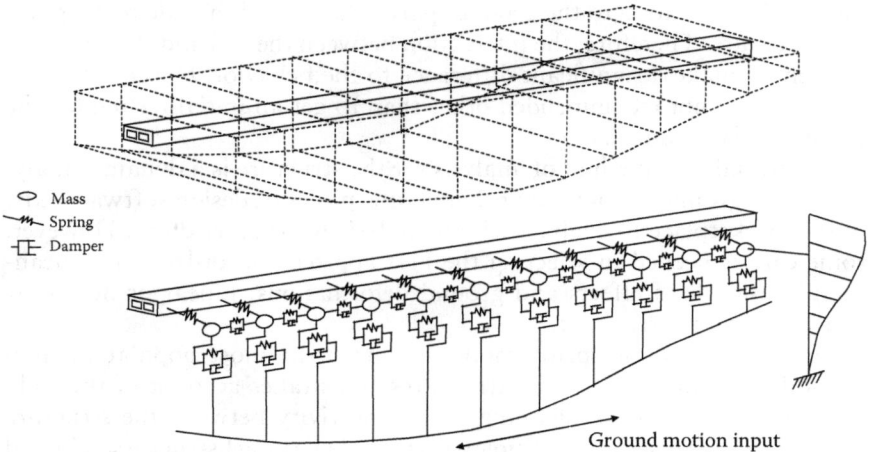

Figure 9.25 Mass spring model. (Adapted from Kiyomya 1995.)

Important elements of seismic modeling

There are a number of important aspects of modeling to take account of that are specific to immersed tunnels.

Joints: Segment, immersion, and terminal

Modeling the joint characteristics accurately is vitally important. They will determine the longitudinal stiffness of the structure and its ability to articulate and deform to follow the ground deformation. The Gina gasket at the tunnel immersion joints should be modeled as a ring with an assigned compression stiffness, which will automatically generate a vertical and horizontal rotational stiffness (Figure 9.26). The tunnel could be modeled as a line beam with a simple hinge with the compression and rotational stiffnesses defined, but experience shows that while this is reasonable for the articulation, it is more difficult to interpret in terms of the joint opening and closing. Restraint mechanisms also should be modeled and if the tunnel cross section is not symmetrical vertically, then these may not be uniform, and so, the full joint plane should be modeled. Shear restraint also needs to be modeled and the limited articulation at tunnel segment joints should be carefully considered. The detailing of shear keys in many tunnels is such that joints would need to open before they can rotate and many analyses are over-optimistic in expecting full rotational ability. To reflect these characteristics, a complex set of stiffness parameters may need to be defined. Terminal joints may have their own different characteristics with, for example, increased movement capacity or ground support.

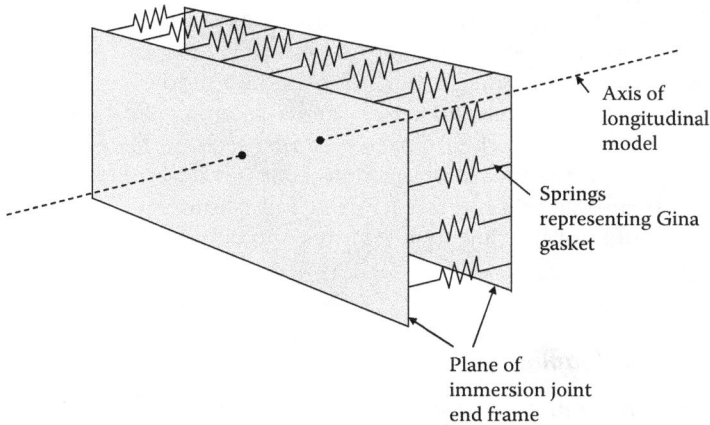

Figure 9.26 Idealization of Gina seal at immersion joint for dynamic seismic analysis.

Interface with soils

Immersed tunnels are shallow structures. Under longitudinal compression waves, the longitudinal movement may cause the soil to slide relative to the tunnel, due to the restraint within the tunnel structure. The friction between the tunnel and the surrounding soils should be defined in the model to get the correct interaction effects. The correct modeling of the soil mass around the tunnel is also important. If the cover is very shallow over the tunnel, the cover may simply move with the tunnel and shear against the surrounding soils. Backfill to an immersed tunnel is relatively extensive and may be quite different to the natural soils. This must also be modeled correctly to get the true behavior of the tunnel. Similarly, ground improvement beneath the tunnel should be accurately represented as it may create stiffer blocks of soil that may also have a modifying effect of the deformations.

Geometry of ground

The ground motion will be modified above the bedrock by the overlying soft deposits in which the tunnel is located. It is very important to model the site geometry and soil stratigraphy accurately to understand this modified behavior and obtain the correct ground deformation characteristics at the level of the tunnel.

Geometry of the tunnel, tunnel backfill and foundation, and tunnel approaches

The structure response will vary according to the surrounding backfill, overburden, and any ground improvement or ground support within the tunnel foundation. It is important to model these features accurately as they may give rise to accentuated load affects at critical points. For example, the transition between the cut-and-cover tunnel and immersed tunnel often results in a hard/soft transition in the foundation. This could be significant in that the cut-and-cover tunnel section may be semi-rigid in the ground compared to the immersed tunnel section. Shear forces due to the imposed deformations may become concentrated and, therefore, significantly greater at this interface. Modeling a uniform, idealized length of tunnel would not pick up such features and could be a nonconservative approach.

Time phase of loading

Long structures will experience seismic pressure waves at different times along their length. The direction of the source must be known and modeled to take account of these effects. If this is not known, a worst-case scenario can usually be determined in the modeling to give upper bound solutions.

Racking analysis

The transverse racking behavior of the tunnel can be modeled in a number of ways, each of which has a different level of accuracy. These are described in the following subsections.

Free field deformation approach

In this instance, the lateral shear deformation of the tunnel is assumed to follow that of the ground. This is only appropriate if there is a very low shaking intensity or the distortions are very small, as would be the case in very stiff ground. This is therefore not very appropriate for an immersed tunnel.

Equivalent static analysis

A number of approaches have been used, including the following:

- Forces applied to the structure model directly or through springs representing soil
- Earth pressure coefficient methods, such as the Mononobe–Okabe methods
- Hydrodynamic methods, such as Westergaard theory
- Simplified plane frame analysis approach by Wang

These methods require a great many assumptions to determine the loading or to model the soil–structure interaction. They are therefore prone to inaccuracy. The Mononobe–Okabe method is not recommended, except perhaps for simple retaining structures. The last of the methods listed, by Wang, is the best method to use. In this instance, the free field motions are determined using a SHAKE analysis or similar. In addition, the relative stiffness of the soil and structure need to be determined to establish a flexibility ratio. Racking coefficients can then be determined using published graphs and this enables equivalent loads to be derived, to apply to a plane frame analysis model, as a concentrated force applied at the roof slab of the tunnel (to determine bending moments) and a triangular distribution applied to the side of the tunnel (for shear forces).

Seismic coefficient methods

Seismic coefficient methods are simple methods by which static loads are applied, representing the effects of acceleration under seismic induced motion. They are not unreasonable, but are likely to be conservative and are better used to assess the impact on fixtures and fittings within the tunnel or elements of the tunnel that can be accelerated by the motion of the main tunnel structure. The soil–structure interaction is not accounted for, and so, this can only be an upper bound solution to cross-check against more detailed analyses.

Imposed deformations method

Imposed deformation methods can be used that are either free field assessments using two-dimensional plane frame models, which do not take account of soil structure interaction, or can be carried out using 2D FE models, where the tunnel and soil are modeled. As noted previously, the free field approach is not recommended. The displacements can be applied at the soil boundaries to obtain the correct soil structure interaction (SSI) behavior such that the free field deformations are correctly modified and limited due to the structure stiffness. This can give quite reasonable results.

2D dynamic analysis

The most appropriate method of analysis is a two-dimensional time history dynamic analysis using FE modeling techniques. This can either be carried out using a fully developed FE continuum model that represents both tunnel and ground and enables the ground motion to be input at the bedrock level, or by determining free field motions, using a SHAKE analysis or similar and applying the shear deformations at the level of the tunnel to a simpler FE model to obtain the racking distortions. Racking coefficients can then be derived to apply to the design of the structure.

Joint design

The tunnel structure can generally cope very well with the seismic load effects. Of greater concern is the ability of the joints to cope with the ground movements while remaining watertight. A number of different approaches are taken with regard to joint design for seismic load effects and the first decision the designer must make is whether to make the tunnel a continuous structure or to retain the flexibility at the immersion and segment joints. It is becoming less common to make the tunnel fully continuous, although this approach has been taken with some steel tunnels. If this approach is taken, it will be important to ensure the immersion joint does not represent a weak spot along the length of the tunnel or the onerous seismic loading may concentrate forces at the joint and cause damage.

By far, the most common approach is to allow flexibility at the immersion joint as this is healthy for the structure. This movement allows forces to be released, and so limits the impact on the structure. However, to allow flexibility, the risk of leakage at the joint must be managed. Typically, a seismic immersion joint will contain the rubber Gina gaskets that are used for the immersion process, plus a secondary seal such as the Omega gasket, but will have a restraining mechanism to prevent the joint from opening to a point where leakage can occur. The types of joint that facilitate this are described in Chapter 10 and are either the restrained immersion joint, the crown seal joint, or the steel bellows joint. The latter two arrangements have been tried in Japan, but on only a small number of tunnels. The restrained joint is the most common approach. This has restraining bars or prestressed tendons installed across the joint that are not initially tensioned. As the joint opens, tension is introduced into the system and the bars or prestress tendons become active. They may permit a small amount of strain, but no more than the Gina gasket can tolerate before losing its watertight seal.

To determine the amount of restraint required, the joint stiffness must be modeled in the dynamic analysis with the compression characteristics of the Gina seal and the tension characteristics of the restraining bars/tendons. The forces are large and to restrain the opening, a large number of bars may be needed. The dynamic analysis is somewhat iterative to determine the right number of restraining bars. Initially, the analysis should assume no restraint to determine the magnitude of the problem to be solved. An initial assessment of the restraint should be made and then the model should be amended to reflect this. It may take several iterations before the number of restraining bars or tendons is finalized.

Special seismic Gina gaskets have been developed that have a greater capacity to accommodate seismic loading and the large potential movements. This may not obviate the need for a restraint system, but may reduce the number and size of restraints that need to be installed. Gina gasket suppliers can advise on the possibilities and work with designers to optimize the solution. In low-level seismic areas, the Gina gaskets working on their own, without additional mechanisms, may have sufficient capacity to accommodate the seismic movements that might occur without leaking.

Shear keys must be designed with a high level of robustness for the dynamic forces. The rapid cyclical loading and unloading may introduce local high stresses in concrete surfaces, resulting in spalling or damage. It is therefore recommended that either steel shear keys are used or that the bearing surfaces have steel plates cast in that can better tolerate the dynamic loading.

In segmental concrete tunnels, it is common to cut the prestress installed for floating, towing, and immersing the tunnel element. However, in seismic regions, it may be preferable to keep the prestress as a permanent feature to give additional security that segment joints will not open in the event of an earthquake. In such circumstances, the increased stiffness that the prestress brings to the structure needs to be modeled in the dynamic analysis as the tunnel no longer has the advantage of articulation at the joints. This philosophy was followed for the Busan Tunnel. In addition, a more robust inner seal was applied to the segment joints, in that instance a small Omega gasket.

Features may be needed in the joints to relieve pressure buildup, should water enter a joint past the outer seal as the joint opens, and then compressed rapidly as the joint closes. This could damage the seals and a pressure relief system to return the water to the outside of the tunnel through a valved outlet can be used. Alternatively, a compressible component within the joint space can be used.

Liquefaction

The impact of liquefaction occurring in the soils surrounding an immersed tunnel is that uncontrolled movement of the structure may occur. Movement may also occur due to shake-down settlement effects, where the granular material naturally consolidates due to the loss of its structure. This is akin to a jar of ball bearings arranging themselves in the most condensed way. Additionally, there is the question of the behavior of the tunnel when it is effectively floating again within a very dense liquid. In this respect, it is possible to assess the likely movement by undertaking some buoyancy assessment of the structure and its backfill depending on the relative weight

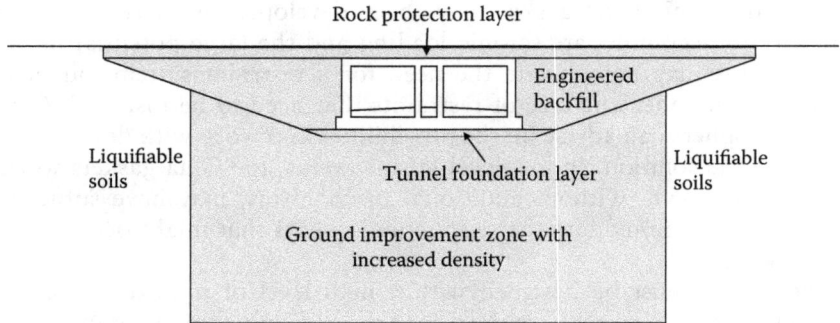

Figure 9.27 Backfill zone of influence for liquefaction assessment.

of soil and tunnel. If the tunnel alone were to be considered, then it is most likely that the tunnel will float upwards as the weight of displaced dense liquid is likely to be higher than the weight of the tunnel. However, this assumes uniformity in the ground conditions and that the tunnel is free to float to the surface. The backfill around and above the tunnel will be selected engineering fill and should not be susceptible to liquefaction, so the impact must be considered on both the tunnel structure and the backfill materials, and also any ground improvement. This creates a large block of soil and structure to consider, as shown in Figure 9.27. This becomes rather too complex to assess as the extent to which the backfill acts as a block with the tunnel is not easy to determine.

Although the potential movement may be of interest in understanding the risk, as it is difficult to quantify the precise behavior of the ground, there is a basic requirement to prevent liquefaction from occurring. Any uncontrolled movement of the structure is undesirable and could lead to overstressing of the structure or leakage of the tunnel joints. An assessment should be made of the extent of potential liquefaction in the ground supporting the tunnel and ground treatment should be designed in order to prevent any liquefaction occurring that could cause instability. Methods of ground improvement are discussed in Chapter 12.

An assessment of the liquefaction potential of soils can be made using recognized methods such as Youd and Idriss (2001) and Moss (2003). Potentially liquefiable soils are generally silty sands that can behave in a granular fashion once the strength has been lost, but have sufficient silt content to prevent relief of the pore water pressure buildup. Some silt materials may also be susceptible and it is important to have good geotechnical investigation data, such as cone penetration testing (CPT) or standard penetration testing (SPT), as well as recovered samples for grading and testing from the potentially liquefiable zones in order to make a thorough assessment.

A further impact of liquefaction could be to cause submarine landslips, where material in the shallow foreshore areas flows down into the deeper channel and increases the overburden on the tunnel at this location. This could give rise to increased settlements, but is something that can be assessed by making some reasonable assumptions on the likely increase that could occur and designing accordingly.

Chapter 10

Joints

There are many types of joints involved in the construction of an immersed tunnel:

- Immersion joints between the tunnel elements
- Segment joints between the segments of a concrete element
- Terminal joints at the ends of the immersed section of the tunnel
- Closure joints, which are the last joints formed in the immersed part of the tunnel
- Seismic joints, which are variations to all the above joints in seismic regions
- Construction joints between concrete pours

This chapter explains the different types of structural joints used and describes the main features and aspects of their detailing and construction that need to be considered. It deals with the joints in the main tunnel structure; joint features that are located within the ballast concrete are described in Chapter 15, Finishing Works, and although the shear keys are described to a degree in this chapter, they are described more fully in Chapter 9, Design Principles.

The objective of any joint in an immersed tunnel (other than the construction joints) is to allow the articulation of the structure while creating a barrier to water ingress. Joints should be designed and detailed to be 100% watertight and this should always be the objective. The one thing all joint types have in common is the need for high standards of workmanship if they are to be effective and trouble-free during the lifetime of the tunnel. However, this does not mean that leakage does not occur; there are many examples of immersed tunnels that have suffered from water ingress. In most instances, this does not cause any serviceability problems for the tunnel and the leakage can usually be contained or remedial measures can be implemented to correct the problem. With forethought and good detailing, it is possible to create effectively watertight tunnels that meet the serviceability and durability requirements of tunnel owners and that do not have onerous maintenance needs.

IMMERSION JOINTS

Immersion joints are a key part of the immersed tube tunnel concept; without them, the tunnel cannot be constructed. They are the connections formed between the individual tunnel elements that make up the tunnel. They play an intrinsic role in the construction process and they contain sealing gaskets that provide temporary watertightness during the construction phase and permanent watertightness for the completed tunnel. The principle of how the joint works is applicable to all immersed tunnels, whether constructed in concrete, steel, or a steel/concrete composite construction.

The procedure for forming the connection between two tunnel elements during the immersion process is shown in Figure 10.1. The tunnel element is fitted with a continuous rubber gasket around the external perimeter of the tunnel at what is called the primary end of the element. The gasket is most commonly a type called a Gina gasket. Although other types are

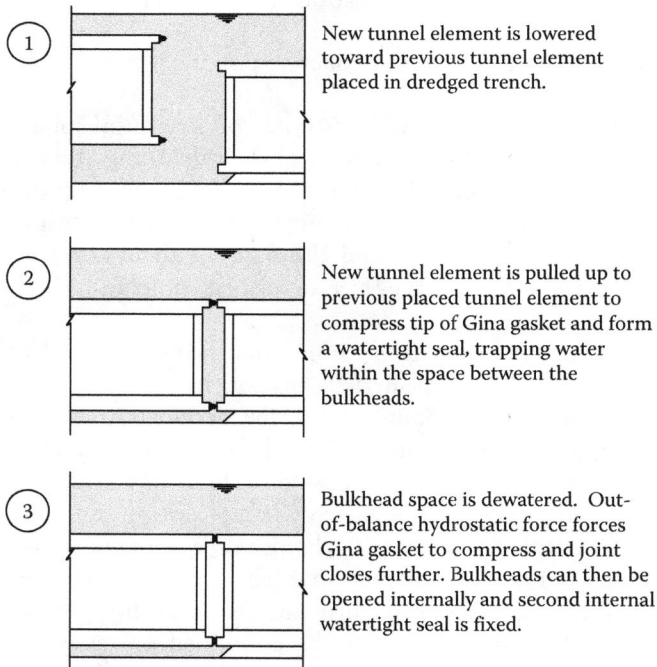

1. New tunnel element is lowered toward previous tunnel element placed in dredged trench.

2. New tunnel element is pulled up to previous placed tunnel element to compress tip of Gina gasket and form a watertight seal, trapping water within the space between the bulkheads.

3. Bulkhead space is dewatered. Out-of-balance hydrostatic force forces Gina gasket to compress and joint closes further. Bulkheads can then be opened internally and second internal watertight seal is fixed.

Figure 10.1 Formation of immersion joints.

occasionally used, this is the most common, and therefore, we will use this name throughout the discussion on joints. The Gina gasket is formed by a composite of two parts: a lower stiffness nose and a higher stiffness body. In the early immersion stages, when the elements are pulled together, the soft nose makes contact with the end face of the opposite tunnel element and provides the initial seal that allows the joint space to be dewatered. The ends of the elements are fitted with bulkheads, which make the elements water-tight and are set back a little from the end of the tunnel element. During the initial pulling together, water is trapped in the space between the bulkheads of the two tunnel elements. When this water is removed, the pressure in the space between the bulkheads reduces, but the full water pressure is still exerted on the other end of the tunnel element. This out-of-balance hydro-static force causes both the tunnel element to move closer to the previously immersed element and the Gina gasket to compress further. The compression load in the gasket is now taken by the stiffer main body of the Gina.

The finishing works to the joint include the installation of a second seal, an Omega seal, on the inside of the Gina gasket and the construction of shear connectors to ensure that the tunnel elements cannot displace hori-zontally or vertically relative to each other. There may also be specific fea-tures to control the future movement of the tunnel under loading, such as during seismic events.

Concrete tunnels have the typical arrangement of Gina and Omega seals, shown in Figure 10.2. This has been adopted as a robust and safe solution since the 1960s and has only changed marginally over the years.

Steel tunnels are also now adopting this combination of sealing gaskets, but historically have featured a pair of small rubber gaskets to form a tem-porary seal during the immersion process, and then, a welded connection across the joints. This is shown in Figure 10.3. These do not have the same flexibility, but have ductility, which has proved successful in allowing some movement to occur without leakage arising within the tunnel. Some recent tunnels in Japan have used different solutions because of their seismic load-ing. These are discussed later.

Design of gaskets

The Gina and Omega seals are often referred to as the primary and second-ary seals. Some use this terminology simply to indicate which is installed first and which second. However, some projects have used these terms to designate which seal provides the primary barrier to watertightness and which is secondary.

Historically, the initial seal—the Gina (or an equivalent)—was a tem-porary measure only. The secondary seal would be the permanent seal, which would be designed for the desired life of the tunnel. In this context, the Omega seal is occasionally referred to as the primary seal as it is the

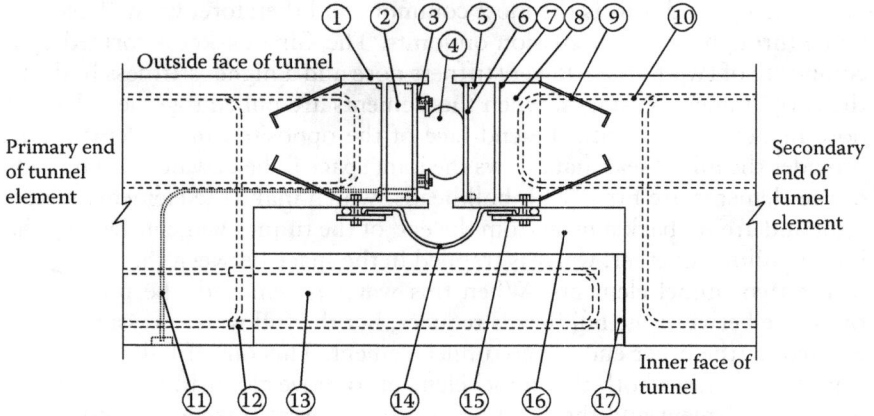

Figure 10.2 Components of immersion joints—concrete tunnel (section through roof slab).

Item no	Component
1	Immersion joint end frame
2	Grout infill to void
3	Gina gasket clamping system—spacer, clamping bar, and securing bolt
4	Gina gasket
5	Welded spacer block to position the counterplate
6	Welded counterplate
7	Grout injection tube (passes through stiffener plate)
8	Stiffener plate
9	Anchor bar to secure the frame to concrete (can use shear studs)
10	Structural reinforcement within concrete slab
11	Pressure test pipework
12	Couplers on reinforcement for connection infill section bars
13	Infill concrete, cast after Omega seal is fixed
14	Omega seal
15	Omega seal clamping system—clamping bar, spacer, and torqued bolt
16	Void
17	Compressible joint filler and surface seal

primary barrier to water ingress. However, it is common now for both seals to be designed for the full life of the tunnel and so the words "primary" and "secondary" within this book refer to the Gina and Omega, respectively. The rubber materials of the gaskets should be durable for the design life of the tunnel. They are made from styrene-butadiene (SBR) rubbers that have a large natural rubber component. Although this could deteriorate with exposure to ultraviolet (UV) light and oxygen, there is no UV light present in the environment of the joint and the oxygen supply is very limited, particularly for the Gina gasket. They can therefore be expected to continue to perform almost indefinitely. Supplier testing for accelerated aging is generally available, along with resistance tests for chemicals and abrasion to demonstrate their durability. The Gina seal is mounted onto the immersion joint counterplate with a simple retaining bar arrangement, as shown in Figure 10.4.

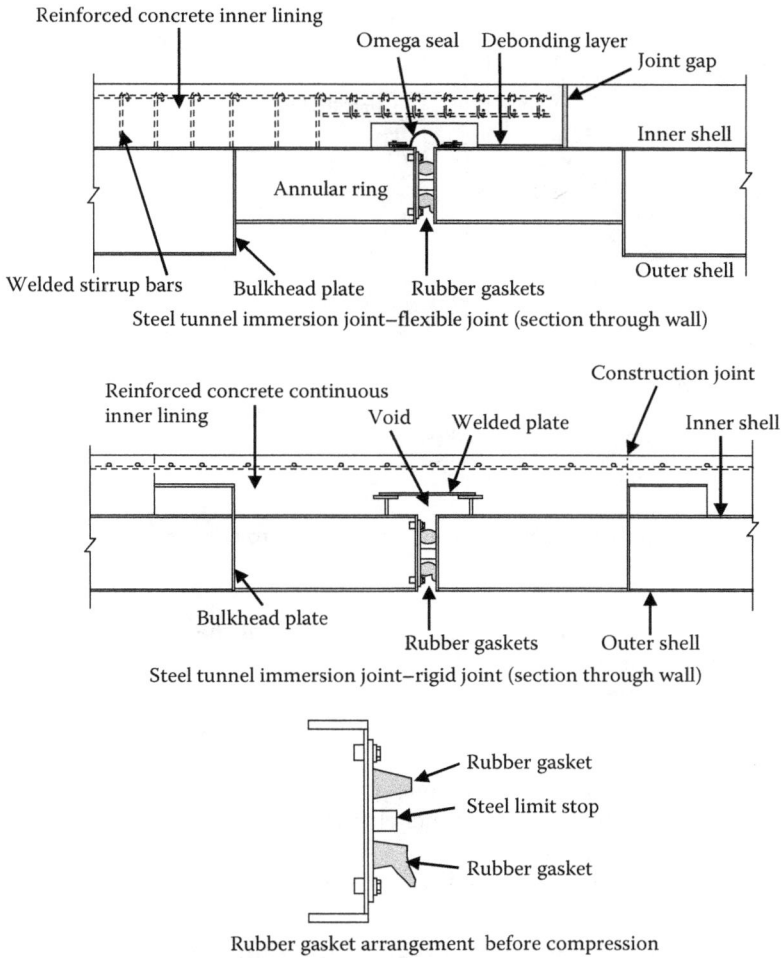

Steel tunnel immersion joint–flexible joint (section through wall)

Steel tunnel immersion joint–rigid joint (section through wall)

Rubber gasket arrangement before compression

Figure 10.3 Components of immersion joints—steel tunnel.

A number of Gina gaskets are available off-the-shelf. In selecting the appropriate Gina gasket, the following need to be considered with regard to its movement capacity:

- Force–compression characteristics
- Modification of the seal characteristics over time due to creep and relaxation
- Initial and overall compression
- Effect of local tolerances
- Residual compression capacity required once all other effects have been taken into account

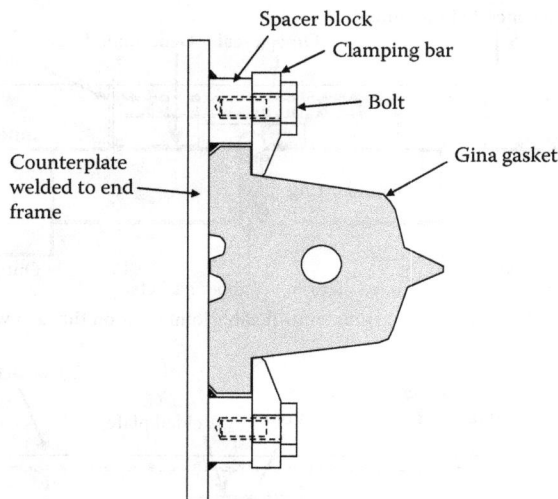

Figure 10.4 Gina gasket mounted on counterplate prior to compression.

- Change in sealing pressure due to opening of the joint due to
  - Settlement
  - Shrinkage
- Temperature when the joint is sealed and subsequent variation causing movement
- Backfill friction restraining movement
- Accuracy of placing the tunnel element
- Realignment of the tunnel
- Seismic movements, if applicable

The information that can be obtained from the gasket supplier should give the first two of these. The remainder will depend on the tunnel geometry and the methods of construction. It is therefore very important to engage the contractor in these discussions, or make reasonable allowance for these effects, based on practical experience.

The sealing design criterion to be used for the Gina gasket is often expressed as a factor of safety on the water pressure that exists on the outside of the seal. This factor of safety is often set at 2.5 and is easily achievable for shallow tunnels. The sealing pressure caused by the gasket compression needs to exceed the applied external water pressure by this factor. This approach is quite conservative, but can be accommodated by the gaskets in most circumstances without leading to any additional cost or complexity in the design or construction. However, it becomes more difficult to apply this safety factor for deeper tunnels, where the effective depth of water that the seal is being designed for becomes quite unrealistic and onerous. In such circumstances, a more rigorous approach should be applied to assess the safety factor, using a risk-based approach and, provided this is done thoroughly, lower safety factors can be considered.

The steps in designing a Gina gasket are given below and illustrated on the force–compression curve in Figure 10.5:

1. Calculate the hydrostatic force to be applied through the Gina. Note that this is the force removed at the immersion joint in question, not simply the hydrostatic pressure applied at the opposite end of the element. The force is calculated as the pressure acting on the area within the Gina gasket divided by the length of the gasket, to arrive at a force per meter value. Select a Gina seal based on the force–compression curve so that the initial compression is at around the mid-point of the curve. Read the compression that corresponds to the force per meter compression.

2. Check that the maximum compression—taking into account tolerances, realignment, temperature, and shrinkage—does not cause the joint to fully close and that some reserve compression capacity is maintained. This reserve is normally an additional 10 mm margin for movement.

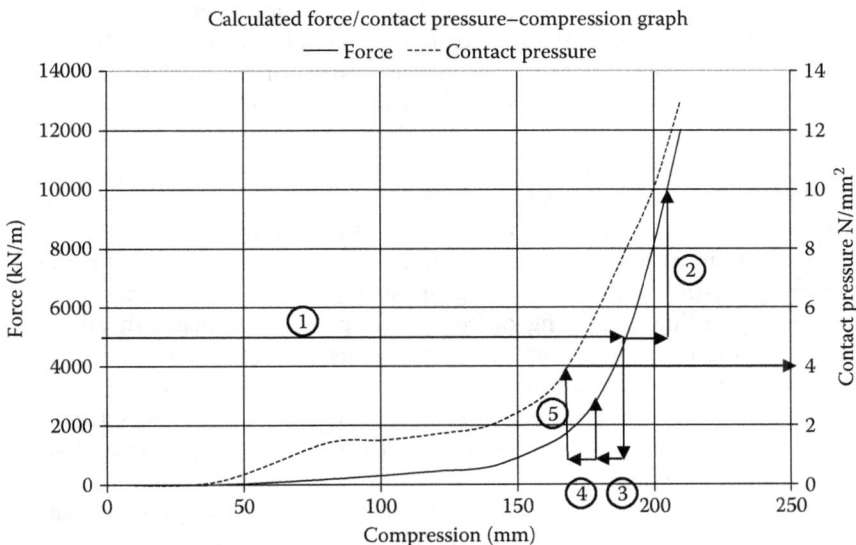

Figure 10.5 Sample Gina force–compression curve.

3. Check that the minimum compression—taking into account the opening of the joint due to tolerances, realignment, temperature, and shrinkage—does not result in a compression in the flat portion of the force–compression curve. This would be indicative of the gasket relying on the soft nose for sealing, rather than on the body of the gasket.
4. From the minimum compression value, deduct an equivalent compression that corresponds to the relaxation of the rubber material over its life (typically 45%–50% of its elasticity).
5. For the corresponding sealing pressure, check it exceeds the external water pressure by the required safety factor, with all effects taken into account.

It should be noted that once the immersion process is complete, the geometry of the joint gap is more or less fixed. Subsequent movements can occur due to settlement, shrinkage, and temperature effects, but the creep/relaxation behavior of the rubber materials will not cause a change in the joint gap, but simply a change in the sealing pressure.

The season during which immersion takes place can have an effect on the subsequent positive and negative temperature range that should be considered when assessing the sealing behavior and this should be taken into account in design. This is covered in more detail in Chapter 9.

For the condition during immersion where the tunnel element is first drawn up to the preceding tunnel element, the sealing behavior under low compression forces must be considered. At this stage, the joint has not been dewatered, and so, the sealing pressure is applied by winching or by a pulling jack. The force–compression characteristics of the seal at these low compression forces can be used to determine the winch cable loads or the size of pulling jack required to achieve the initial seal.

The Omega seal is fixed in place shortly after the immersion process. A suitable seal should be selected to accommodate the predicted movement range at the immersion joint over the life of the tunnel. Movements along the axis of the tunnel will occur due to temperature change, shrinkage, and creep, resulting in the opening or closing of the immersion joint and the corresponding stretching or compressing of the Omega. In addition, some opening or closing may occur due to settlement effects. A degree of shear movement at the immersion joint may occur before shear keys are introduced and the relative positions of adjacent tunnel elements are fixed. The selected seal must be able to deal with both the elongation and shear movements combined.

To provide an effective seal, the Omega must be compressed against the steel end frame. The flange of the Omega is clamped to the end frame by a bar arrangement and the clamping bolt is torqued until a pressure is reached that gives a factor of safety of 2.5 against the acting external water pressure. Allowance should be made in the design for creep in the rubber materials to ensure that sufficient pressure remains applied in the long term.

Design of supporting steelwork

Gina seals are clamped to a steel frame that is cast in to the end of the tunnel element and is continuous around the perimeter of the tunnel. A similar frame is cast into the secondary end of each element, and this provides the sealing surface for the Gina gasket. The frames are relatively complex and their construction requires great accuracy and must provide durability and watertightness.

It is a common debate during the tendering phases of a project and in value engineering exercises as to whether the steel frame supporting the Gina gasket can be omitted or simplified. There are a number of outline designs that have proposed alternatives, such as mounting the seals directly onto concrete and using timber or synthetic materials for the mounting frame. However, the risk that a new system may not provide such a high degree of watertightness has generally prevented much innovation in the joint arrangement.

If the frame was omitted, then water paths through the concrete behind the seals become much shortened and any defects in the concrete will have a much greater impact on the watertightness. There is also the issue of surface tolerance, which cannot be achieved to the same degree of accuracy with concrete as is achievable for a steel frame with a secondary welded counterplate.

Simplifying the steelwork has been tried, most notably on the Øresund Tunnel, where a single plate was used without the secondary counterplate. The construction of the tunnel elements for that project was undertaken in factory-like conditions with a high degree of control on tolerances; so, for that project this approach was suitable. However, it was not without its difficulties. The onsite welding of the sections of the end frame, after they had been cast into the concrete, created water paths at the back of the welds due to plate distortions and fabrication inaccuracies. So, care should be taken if considering a simplification of the arrangement.

It is clear that the immersion joint assembly is a very expensive part of the construction, but ultimately, the cost savings that may be perceived from optimizing the design of the joints are likely to be spent on rectifying leakage or dealing with construction difficulties. Therefore, unless there are very specific project circumstances, most project owners prefer the tried and tested solution.

Off-the-shelf steel sections can be used to fabricate the end frame, though they tend to be at the heavier end of the range of sections that are available. Care needs to be taken to ensure that the flanges of the beams are wide enough to allow the rocking bar of the Omega clamp to bear on the steelwork, but not too wide to clash with the flange of the opposite frame when the immersion joint compresses. Equally, the height of the frame needs to be sufficient to accommodate the Gina clamping bars and allow enough working room to tighten up the clamping bolts. Because of these constraints on geometry, it is more common for the I-sections to be fabricated as welded plate girders.

The counterplate welds do not need to be designed to carry the compressive force from the Gina as the void behind the counterplate is grouted. Compressive forces should, therefore, be transferred directly to the grout. However, a proportion of the loading may need to be assumed for their design for the event that grouting of the void is not fully achieved.

The frame will need stiffeners on its rear face for the temporary stage when it is being assembled, to avoid buckling during lifting and while it is temporarily supported, prior to casting it into the concrete. The various lengths making up the frame should be welded with full penetration butt welds, with appropriate testing of the welds to ensure they are watertight and without defects. If any welding is carried out after the frame is cast in, measures to seal the water path behind the frame are required as the heat from welding may cause the separation of the plate from the concrete surface. The surface tolerance on the counterplate at the welds will need to be agreed upon with the supplier of the Gina seal, but typically will require there to be no steps, and limits will be imposed on angular deviations.

Clamping forces on the Gina and Omega seals are internal forces within the steel items and can be designed for quite readily. The arrangements of the clamping systems for both the Gina and Omega are well established, and recommended arrangements are provided by some seal suppliers. These can be readily adapted for the particular geometry and circumstances of any project.

There are some alternative ways of detailing the clamping arrangements for the Omega seal. Threaded studs can be welded to the frame and nuts used to fix and clamp the flange of the Omega seal. Alternatively, the frame can be drilled and cap nuts welded to the rear side of the frame to receive a bolt. This has an advantage in that there are no protruding studs during the transportation of tunnel elements that may risk being damaged. Either method can be satisfactory.

The end frame steelwork must be secured to the tunnel concrete and the usual practice is to use shear studs or anchor bars welded at regular centers to the rear flanges of the I-section. Detailing requires extreme care if there are any requirements to ensure electrical isolation between the frame and the main reinforcement cage of the reinforced concrete. This is usually necessary to avoid creating galvanic cells that lead to metal corrosion.

Tolerances

Tolerances on the finished surface of the immersion joint counterplate are highly important. There are limitations on the accuracy that the joints can be constructed to and these must be taken into account in the design of the Gina and Omega gaskets. Variations in the planarity of the counterplate on which the gaskets are mounted or surface irregularities will reduce the range of compression and expansion over which the seals can provide the required sealing pressure. Typically, there will be specifications for the planarity and surface tolerance of the counterplate, which will be of the order of a few millimeters (Figure 10.6).

The tolerances that must be considered are

- Vertical planarity of the counterplates—both due to the initial construction accuracy and due to settlement.
- Horizontal planarity due to the inaccuracy of placing. These can be positive (+ve) or negative (–ve) differences from the intended plane of the surface.
- Localized surface irregularities—for example, steps at welds. Again, these can be positive or negative differences from the intended position of the surface.

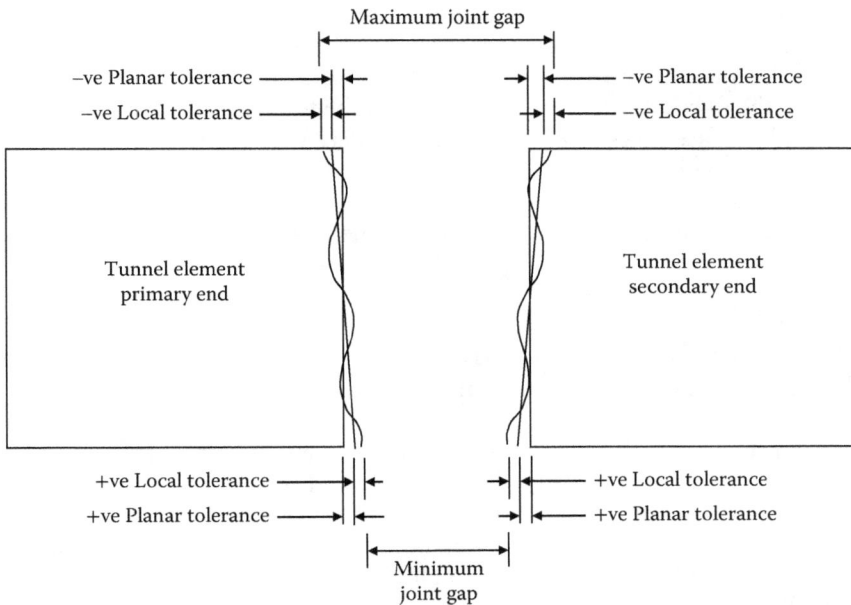

Figure 10.6 End frame tolerances.

The problems of horizontal planarity appear during the placement of the tunnel. During the immersion process, when a tunnel element is being pulled up to an already placed tunnel element, it will try to orientate itself at 90° to the plane of the immersion joint, to achieve a uniform compression in the Gina seal. If either of the end frames is not set at 90° to the line of the tunnel element, the element will naturally position itself off of its intended alignment. This can normally be corrected by forcing the tunnel element back to the intended alignment and locking it in place with backfill. In practice, this can be achieved by simply lifting the free end and moving it slightly to correct the alignment or by using jacks within the immersion joint to move the element back to its intended position. However, the Gina gasket will be over-compressed at one side of the element and under-compressed at the opposite side, reducing its possible movement range thereafter.

When elements are being constructed, it is possible to account for some out-of-tolerance by undertaking as-built surveys of the elements within the casting basin or dry dock. If there is any out-of-tolerance detected, it can be compensated for by offsetting the plane of the end frame counterplate in the adjacent tunnel element.

All of these tolerance issues must be considered when designing the Gina gasket, to ensure it retains sufficient sealing pressure under all circumstances. An assumption on tolerances needs to be made in the design of the Gina seal, but once designed and selected, this will dictate the accuracy of construction required and should be reflected in the specifications for construction of the end frame and in the immersion procedures. At the stage of choosing the tolerances to allow, it is important for the designer to either consult with the team which will be responsible for placing the tunnel, or to have good knowledge of the likely construction methods and the tolerances that could result.

Local surface tolerances must be considered in terms of the possible reduction in sealing pressure in the Gina gasket that could lead to leakage. The Gina gasket will be able to cope with many variations, but sharp steps in the surface or abrupt deviations in the line could lead to leakage.

Detailing

The careful detailing of the joint is an essential part of its success in ensuring that watertightness is achieved. Particular features to be aware of include:

- In sizing the steelwork to accommodate compression of the Gina gasket, ensure that opposite flanges of the end frame do not clash or that the Gina clamping bars do not touch when the joint is fully compressed.
- Provide pipework from the void between the Gina and Omega seals to the inside face of the tunnel to allow for pressure testing of the joint.
- Maintain the pipework used for joint pressure testing as access points for maintenance inspections and for checking water ingress. Should

any leakage occur, this can provide useful information on the source of the leakage and the quantities of water entering the tunnel.

- Fix injectible grout pipes behind the end frame in the areas where there is highest risk for poor compaction of concrete. At least two lines should be installed and grouted as a matter of course to ensure that potential water paths around the back of the frame are blocked.
- Electrical isolation between the end frame and the reinforcement within the concrete is difficult to achieve. It will be necessary to have shear connectors attached to the back of the end frame and stiffener plates for the temporary stage of fabricating and assembling the frame. Careful spacing of U-bar reinforcement is needed to prevent clashing with these features, and there may be some benefit in coating the rear of the frame with a high build epoxy or providing insulation caps on shear connectors.
- Space needs to be left above the Omega seal to allow for compression and outward deflection of the bulb. Space is always tight in this area and it is easy to forget that the seal needs space around it.
- The joint will include shear keys to prevent vertical or lateral displacement at the joint. The shear keys must always cater for the longitudinal movement that is needed at the immersion joint and feature compressible material to allow the movement to occur.

Durability

Durability is covered in Chapter 11, but there are some key points worth summarizing here. The watertightness of the seal assembly depends not only on the rubber gaskets, but also on the supporting steelwork. It is important to ensure this steelwork will perform for the intended service life of the structure. For steelwork, this means there must be some corrosion protection applied. The environment around the steelwork is relatively inert. The steel exposed to the marine environment on the outside of the immersion joint will be in a saline environment, but buried within the backfill of the tunnel. The renewal of oxygen will therefore be limited and the joint is rather well protected. This does not mean corrosion will not occur, but it will have an impact on the rate of corrosion, which will be much less than in a splash zone or where water is moving across the surface of the steel. On the inside of the tunnel, the steelwork should be considered to be in a damp environment.

In a shallow buried environment, there may be other corrosion causes; for example, bacterial corrosion such as sulfate reducing bacteria (SRB), particularly if the tunnel is in a harbor environment with polluted sediments in the vicinity. This gives rise to pitting corrosion at accelerated rates compared to general corrosion in seawater.

It is common to coat the steelwork with a protective coating. This will have a limited design life, but will ensure that corrosion does not commence during construction, when floating tunnel elements may be moored in the splash zone for several months. It is important to ensure the immersion joint steel work is not accidentally connected to the internal reinforcement as this can cause the formation of a corrosion cell and cause rapid corrosion. In this situation, coating of the steel may do more harm than good as corrosion will be concentrated at defects in the coating and be greatly accelerated. This is not a reason to omit the coating. It emphasizes the need to avoid contact between the end frame and the reinforcement and to provide a good quality coating.

Corrosion of the main frame supporting the Gina seal is a threat to watertightness if the corrosion progresses beneath the gasket. This is quite an unlikely event, but if the steel were to deteriorate, the compression locked into the gasket would mean it would always push into any void. Coatings will delay the initiation of any such corrosion. In the longer term, when coatings have deteriorated, the steel will corrode typically at rates of around 0.1–0.3 mm per year, though corrosion rates for crevice corrosion are more variable and may well be higher. Sacrificial thicknesses could be considered, but this is a somewhat unreliable approach. It is relatively simple and cost-effective, however, to use cathodic protection and attach sacrificial anodes to the immersion joint steel work to limit long-term corrosion. If SRB corrosion is a risk, then sacrificial thickness and coatings are appropriate as cathodic protection will not protect against this. It is therefore common for a combination of measures to be used to deal with the various causes of corrosion.

The outer clamping bar to retain the Gina gasket does not warrant any protection over and above that applied to the general steel frame. It has fulfilled its task once the tunnel has been constructed and serves no purpose in the long-term performance of the tunnel. The gasket is locked into position by the compression across the joint and does not need the retaining bars to keep it in place.

The space between the Gina and Omegas seals is completely locked and there should be little or no corrosion. If subsequent inspection suggests that water is getting into the space between the two seals, causing a corrosion risk, then it is possible to treat the void with a corrosion inhibitor on a periodic basis. This can be inserted through the pipe installed for pressure testing and inspection.

The inside components of the immersion joint are susceptible to corrosion, particularly at the bottom, as they are exposed to the internal atmosphere in the tunnel, although they are in a void that is below the slab supporting the road or railway. The main items of concern in this area are the clamping components for the Omega seal. At the base of the tunnel, they are in the lowest point in a void that is susceptible to collecting water and could be considered to be regularly damp, and potentially filling and emptying. Should the clamping bolts or the clamping bars corrode, there is a risk of loss of compression on the Omega seal flange and, hence, loss of watertightness of the seal.

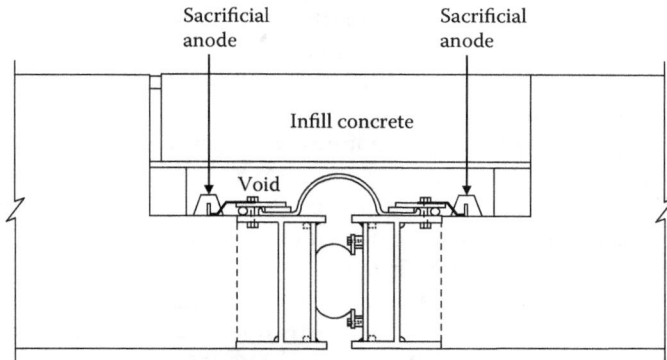

Figure 10.7 Omega protection detail.

These components need to be selected with steel grades to minimize the risk of corrosion, such as stainless steel, or be coated or protected. Protection is sometimes as simple as applying grease or wrapping the components in protective tape. A better alternative is to attach sacrificial anodes, as shown in Figure 10.7. Given the limited space, anodes in the form of ribbons are useful, and these can be connected along the length of the clamping system.

Trial castings

Trial castings are common for all civil engineering projects where specific finishes are required or unusual techniques are being used. It would be unusual for site operatives to have direct experience in building immersion joints, so the benefit of undertaking trial castings cannot be overstated. Ideally, the casting in of a section of end frame and the subsequent grouting operation should be trialed and cores taken to verify that the concreting and grouting procedures have successfully filled all voids behind the end frame. Any unusual operations, such as welding once the frames are already cast in, should also be trialed. It is also very informative to undertake a trial fixing of a length of Gina and Omega seals. People carrying out the installation, particularly of an Omega, are often surprised at how difficult it is to form a watertight seal at the corners of the tunnel, even when they are doing it in the workshop, let alone under field conditions.

Seal installation

Gina gaskets can either be made to length in the production factory or can be vulcanized on-site to form the complete ring required to run around the perimeter of the tunnel. The supply, either whole or in parts, is largely dependent on the size of the tunnel and, therefore, the practicalities of shipping a complete joint. Final lengths can be adjusted to suit as-built dimensions, if necessary. A frame is used to suspend the gasket during lifting both to maintain its shape as it is lowered into position and to enable easy fixing to the end frame.

The fitting of the Omega seal requires the erection of scaffolding in the joint gap. The seal is preformed as a complete loop and vulcanized based on as-built dimensions, with preformed corner pieces. The initial clamping of the seal at the corners is the most difficult area. Special clamping pieces are needed to ensure even pressure is applied right into the corner. Difficulties can arise if there is any misalignment of one tunnel element relative to the other. The Omega accommodates this quite readily along the straight lengths, but at corners, this can make it difficult to fix the seal and make it watertight.

The weight of the clamping bars for the Omega seal needs to be considered from a practical and safety point of view as they have to be lifted to the roof of the tunnel for installation. It can be advantageous to store the clamping bars in the tunnel element prior to floating and immersion so that they are close by and difficult transportation through many bulkheads can be avoided.

The sequence of clamping the Gina and Omega seals needs to be developed to ensure that the seal is evenly spread and no distortion occurs. This requires progressive fixing around the perimeter of the tunnel, and the advice of seal suppliers should be sought.

Once the seals are fixed, it is usual to undertake a water pressure test to verify the watertightness. The space between the Gina and Omega is filled with water and pressurized. The pressure is then monitored for 24 to 48 hours. There will be some initial drop off in pressure as the rubber relaxes, and this should be corrected at the start of the test. Water should be drained from the space on completion of the test.

Protection of seals

Temporary protection of the Gina gaskets is required once they have been clamped to the steel end frame of the tunnel. Typically, a simple timber box is provided along the exposed top edge of the tunnel and the top parts of the walls. This is to protect the gasket from damage during the transportation and maneuvering of the tunnel elements. During these operations, there are numerous winch cables running from the tunnel element to moorings and tugs, which can easily cause damage to the seals. This protection should not be overlooked as it is a difficult and expensive process to replace a Gina gasket once the tunnel element is floating. Any bolt threads used for the clamping systems need to be protected during immersion. Greasing them is normally sufficient.

After the tunnel elements are immersed and the joint gap has been cleaned of seawater and sediment, the inside face of the compressed Gina gasket should be protected until such time that the Omega seal is being installed. Whilst the rubber is a relatively tough material, and the compression within the gasket is so high that the seal cannot be displaced, there will be equipment and scaffolding brought into the joint gap to clean the end frame ahead of mounting the Omega seal, and there will be various materials used that could damage the rubber, such as cleaning solvents and fuel for generators. The Gina gasket should therefore be well protected during this period.

The same applies to the Omega seal. Once this is in place, the temporary bulkheads are removed and greater access through the tunnel can be achieved. The Omega is less robust than the Gina and the joint area should be boarded over to prevent damage during operations to remove the bulkheads and complete the joints. It is also important to be very careful not to spill oils, petrol, or diesel in this area, which could damage the rubber materials. There can be quite a delay between the installation of the Omega and the final finishing of the ballast concrete and road/railway, so this protection is necessary.

Some tunnels have mounted external protection plates on the roof and to the sides of the tunnel to prevent backfill or debris from entering the joint gap. These are fitted by divers after the immersion process is complete. This prevents fine-grained fill material entering the gap and preventing closure if the tunnel expands. Many designers consider this to be a minimal risk and assume the material will be displaced due to the high forces that will be applied, but this should be considered according to the geometry of the Gina and end frame, and the type of backfill proposed.

Temporary works

There are a number of temporary works items that need to be accommodated within the immersion joint area. These often affect the permanent work design as they require space to be provided for them. These include:

- *Bulkheads*: The bulkheads are mounted at or near the end of the tunnel element and, therefore, may have fixing details that coincide with the immersion joint.
- *Prestressing*: This may be permanent or temporary, but the anchorages will typically be located in recesses at the immersion joint. Consideration is needed on how the anchorages are cast in and covered.
- *Alignment jacks*: These may be installed to correct horizontal inaccuracies in the placement of the elements. They may be located on the inside of the joint. If so, it is important to consider the sequence

of construction and whether they will be removed before the Omega seals need to be fixed, as they may cause an obstruction.

- *Pull-up jacks*: These provide the pulling force that joins the elements together initially. They can be mounted externally, but some projects have utilized a jack that uses the center wall of the tunnel between the bulkheads.
- *Temporary shear keys*: These may be used to land the end of a tunnel element onto the previously placed tunnel element. They therefore occupy the space between the bulkheads at the internal walls, and potentially at the external wall also, on the inside of the joint seals.

Joint finishing

When the tunnel elements have been placed and the Omega seals fixed, the internal part of the immersion joints can be completed. This comprises a section of infill concrete that protects the Omega seal. This infill is usually made continuous with the concrete of the tunnel element on one side, leaving a gap for expansion and contraction at the opposite side. The concrete in the base of the tunnel then supports the ballast concrete layer and the road or railway on top.

On high-speed rail tunnels, this infill layer can be used as a rocking slab to take out any differential settlement that occurs and ensure that a smooth alignment across the joint is achieved. This type of arrangement was used on the high-speed link tunnels under the Oude Maas and Dortsche Kil canals in the Netherlands.

In theory, Omega seals can be replaced but, in practice, this would be difficult because of the shear keys, infill concrete, and the road or railway. The Bjørvika tunnel in Oslo has made some provision for this, however, and the infill concrete was made accessible by constructing the overlying ballast concrete such that it could be removed if there was ever a need to replace the Omega gasket. The concrete was precast in short lengths so that they could be removed in sections without extensive breakout requiring repair. Similar infill is required to the walls and roof of the tunnel. Often, this is done with reinforced concrete but, in some instances, simple cover plates can be fixed. Whatever solution is chosen, the infill must be capable of resisting impact at road level and should be resistant to fire. This infill is also commonly used to provide shear restraint across the joint both horizontally and vertically.

Joint leakage

The question most tunnel owners and operators are keen to understand is whether immersed tunnel joints leak and, if they do, what can be expected for maintenance and whether this causes any serviceability or durability

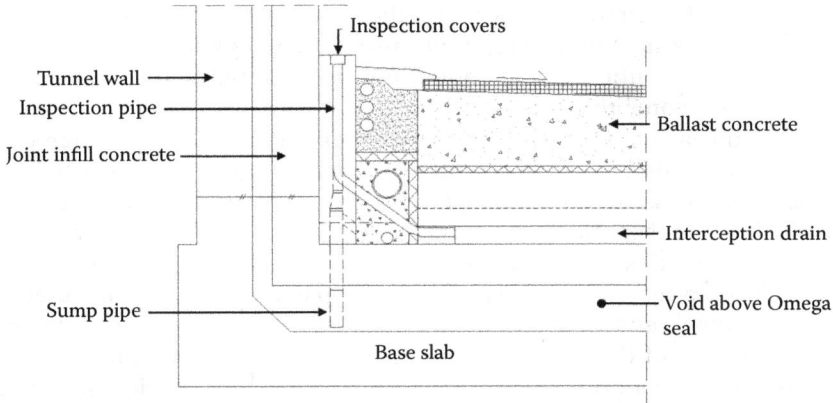

Figure 10.8 Joint inspection detail—part section through immersion joint.

problems. The short answer to this question is yes, they do leak occasionally, but this is not normally sufficient to create an onerous maintenance requirement nor will it compromise the serviceability and durability of the tunnel.

If water is collecting in or around the joints, it can come from several sources. It is unlikely that the Gina is leaking as this will have been tested fully during the construction process and proved effective. Nothing is likely to have happened since construction to affect its watertightness. It is more likely that water is seeping through small voids in the concrete behind the end frames or that the water is from inside the tunnel. This could come from leaking pipes or water moving through the joints in the ballast concrete and collecting at the immersion joints. Inspection facilities can be introduced into the joint, as shown in Figure 10.8. Such details enable any water collecting at the joint to be identified and removed.

SEGMENT JOINTS

Segment joints are the joints within a segmental concrete tunnel that separate a tunnel element into a number of discrete segments. The joints are match-cast joints that have no continuity reinforcement through them and are de-bonded to allow articulation and the opening and closing of the joint. Rotation is permitted, but vertical and horizontal displacement of one segment relative to another is prevented or restricted by shear connections or by the shape of the joint itself. Segment joints are typically spaced at 20–25 m through a tunnel element to permit the structure to articulate in the event of settlement occurring, such that longitudinal bending moments in the structure are limited and reinforcement quantities can be minimized. This length also aids with the control of early age thermal and shrinkage cracking. Segment joints are also sometimes known as expansion joints or dilatation joints, but should not be confused with construction joints within monolithic tunnel elements.

The segment joint incorporates a watertight seal across the match-cast faces. This is usually a grout injectable waterstop that has a rubber bulb in the joint gap to allow for joint opening without tearing of the waterstop, rubber flanges either side of the central bulb to provide a long waterpath, and metal plates that extend further into the concrete, at the end of which are injectible foam sponges. This type of waterstop is almost universally used in concrete tunnels and has been found to be highly reliable, provided the initial installation of the waterstop during concreting is carried out correctly and the subsequent grouting process is undertaken carefully. It is shown in Figure 10.9, which illustrates the common segment joint components.

Grouting is a process that is carried out as a matter of course on the completion of concreting, before flooding of the casting basin or dry dock in which the elements are constructed. The grouting facility is not intended as a remedial measure should leakage occur; the grouting process is a quality procedure for the initial construction to ensure there are no waterpaths

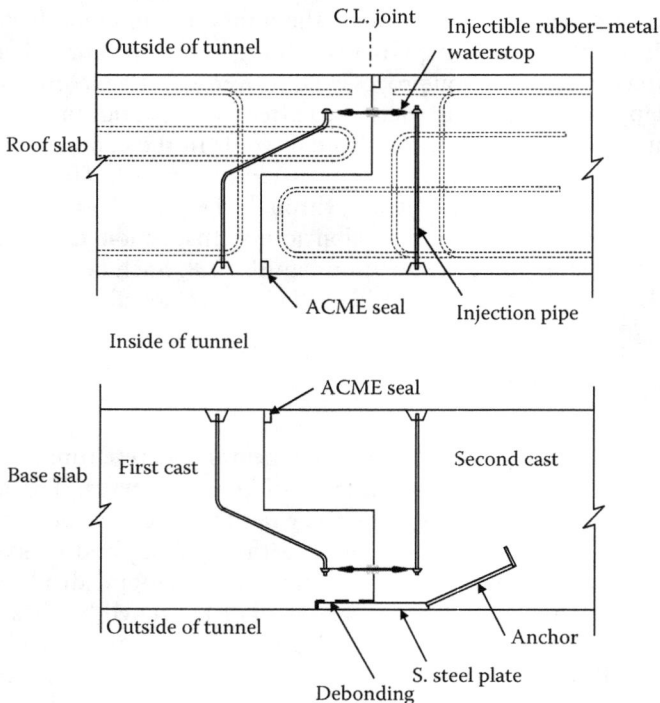

Figure 10.9 Typical segment joint.

around the seal due to trapped air or honeycombing during concreting. If such a facility is required for possible future remedial grouting, a second injection system should be provided. This would typically be an injection tube system.

It is now common to require a second barrier to water ingress in addition to the grout injectible waterstop. This may take a number of forms, but is generally a different type of barrier to the injectable waterstop. Hydrophilic seals have been used on a number of tunnels; these are set into the match cast joint and have a delay on them to prevent them expanding during the initial concreting. This type of seal is located on the inside of the injectible waterstop as a second line of defense. However, careful selection and detailing of this seal is required to ensure that it will function under the full range of joint movements that may be encountered. The main worry with this type of seal is its effectiveness when the tunnel is at its coldest and the joints are at their widest. If thermal movements are combined with the joint opening due to settlement, the seal may not be able to plug the joint gap until a significant amount of seepage has occurred.

An alternative approach is to apply a waterstop externally around the perimeter of the tunnel. Rear guard waterstops that are commonly used for in situ construction of retaining structures and cut-and-cover tunnels can be used, but the difficulty comes with the roof slab. For the base and walls, the waterstop can be cast into the concrete satisfactorily, but for the roof, this is less practical. Extremely careful workmanship is required to ensure the concrete is fully compacted beneath the waterstop and all air is expelled, but the risk would always remain that some voids could arise that create leakage paths. A good alternative is to use a post fixed system across the roof that is similar to the traditional waterstop. This could be a bonded membrane or a clamped system. The Limerick Tunnel in Ireland and the Bjørvika Tunnel in Olso, Norway, have both recently used such systems. Grouting of the joint gap is not desirable as, even with flexible, low-strength grouting materials such as polyurethane, their effectiveness in sealing a joint that opens after grouting has to be questioned.

Solutions with double injectible waterstops have been considered, but there is normally insufficient space in the end of the wall to position two seals with enough cover depth and space between the seals. This is particularly problematic because of the need to provide either discrete shear keys or a continuous half-joint around the perimeter of the tunnel. Some projects have considered the use of Omega seals as a secondary seal at segment joints, but this is a very expensive approach and, similar to the double injectible seal idea, the Omega takes up a lot of the section thickness and generally could not be fitted together with the shear key features and the injectible seal.

If an external waterstop is not provided at the joint, seals are required at the outside surface of the joint around the tunnel perimeter to prevent soil from entering the joint gap and preventing it from opening and closing freely. This is normally achieved with a Neoprene or HDPE ACME-type seal, which is compressed and glued into a recess. This can be applied to the walls and roof after casting of the segments, but for the base slab, an alternative arrangement is needed. A stainless steel plate is used, cast into one-tunnel segments using an anchor or shear stud to hold it in place. The opposite side of the plate beneath the second cast concrete has a de-bonding material applied so that it can slide against the concrete as the joint opens and closes.

Expected joint movement range

It is wise to consider the movements of the complete set of joints, including immersion joints and fixed points, such as piled cut-and-cover approach structures, to understand how movements vary along the tunnel. For example, greater rotations, and hence joint opening, may be expected at the connections to the cut-and-cover tunnels if there is a significant change in the foundation type.

Typically, the injectible waterstop in segment joints can accommodate quite large movements, but the permissible movement will reduce with increasing water depth. Information on the capacity of waterstops to accommodate movement for different water pressures is generally available from the suppliers of the waterstops in the form of design curves, to aid the selection of the appropriate product. As an example, a stretching of the center bulb of up to 80 mm, combined with a shear movement of the same magnitude, could be accommodated in 20–30 m water depth. This well exceeds the movement that would ever be likely at a segment joint.

Detailing

The location of the grout injectable waterstop needs careful consideration. It is usually positioned toward the outer surface of the tunnel. There should be enough space to fix a layer of reinforcement between the outer surface and the waterstop to ensure that grouting pressures do not cause the outer nib of concrete to shear.

Injection tubes should come to the internal surface and have injection nozzles with a small recess around them to enable a good repair once the grouting has been carried out. Grout tubes can be angled to avoid the surface of the half-joint or shear keys, as necessary.

The waterstops will also need to be positioned to allow prestressing ducts to pass through the concrete sections, with sufficient cover between the two to ensure concrete can be well compacted. It should be remembered that prestress duct couplers will be needed at the joints, and will cause additional width as a result. Prestress ducts may also have a wrap of compressible material around them to allow differential expansion and contraction of one segment relative to the other during curing, to prevent the prestress ducts from being damaged.

Construction considerations

Segment joints look relatively simple, but there are several aspects of construction that, if not treated carefully, can lead to maintenance issues in the tunnel:

- The waterstop grout injection process should be trialed in advance of grouting the permanent works, to ensure the grouting procedure will fill all likely voids. The type of grout should be compatible with the injection tubes and sponge to ensure it will travel and fill voids. The frequency of injection points along the seal will also depend on the grout type and injection process. The normal procedure of progressive grouting around the perimeter and ensuring overlap between grout injection points should be followed. This sounds simple but frequently lengths of waterstop have not been fully grouted.
- The grouting operations need to be carried out carefully so that the grouting of the waterstops and the grouting of the prestressing cables do not interfere with each other. There have been instances where the waterstop grout has entered the prestressing ducts, preventing the cables being tensioned.
- Good-quality concreting around injectable waterstops is important and the manufacturer's advice should be followed. The waterstops are quite wide, so it is particularly important to lift the edges of waterstops as concreting proceeds and fully compact the concrete beneath them. The waterstop needs to be tied off onto the reinforcement to fix it in position. They should not be nailed to the formwork as this defeats the object of a waterstop, but it has been done.
- De-bonding of the match-cast joint is important to ensure that the joint can release and function as designed. The use of a de-bonding agent is sufficient to achieve this.

TERMINAL JOINTS

Terminal joints are the joints located at the ends of the immersed part of the tunnel, though they are not necessarily the last joints to be constructed. Generally, they join the immersed tunnel to the adjacent cut-and-cover tunnel or to a ventilation building.

Often, terminal joints are very similar, or indeed identical, to the normal immersion joints between the tunnel elements, but their form and function depends on the sequence of construction. For example, if the first immersed tunnel element is placed against a cut-and-cover tunnel or ventilation building, then the form of joint can be the same as the joints between the immersed tunnel elements. The immersed tunnel is placed and joined to the cut-and-cover section in exactly the same way as it would be joined to another immersed tunnel element. If the closure joint is provided somewhere along the length of the tunnel between two immersed tunnel elements, then the terminal joint at the other end of the immersed tunnel can also be the same as a normal immersion joint. Construction then proceeds from both ends towards the closure joint somewhere in the middle of the tunnel.

A variation on this is that the construction of the immersed tunnel starts against one cut-and-cover tunnel and continues toward the other. The last tunnel element then has to be placed against the other cut-and-cover tunnel. In this case, the second terminal joint has to take a different form, which must allow for fabrication and placing tolerances. A simple form as illustrated in Figure 8.7 within the chapter that covers approach structures, from the Medway Tunnel, shows the end of the immersed tunnel element being exposed after placing so that the connection to it can be constructed in the dry. This can be achieved by placing the element so that its end lies within a temporary canal, possibly the exit from a casting basin.

During construction of the land connection, it is not only water exclusion that has to be considered. It is essential that the compression is maintained in the seals between the immersed tunnel elements. If this compression were relaxed, there is a possibility that the joints could open sufficiently to leak. In any event, the flexibility of the immersion joints would be compromised. To prevent this happening, the end of the last immersed tunnel element must not be allowed to move when the hydrostatic pressure is removed from one end by the dewatering operation. This might be done, if space permits, by using the weight of the backfill placed around the end of the immersed tunnel. If it is sufficient to generate ground friction that gives an adequate safety factor against sliding, this can be enough. However, often, some additional restraint may be needed, which can be provided in a number of ways:

1. By propping against the base slab of the approach structure; the slab may need to be piled to provide sufficient resistance and would clearly have to be constructed in advance of the immersed tunnel. To some degree, this may negate the advantage of placing the immersed tunnel first.
2. By installing temporary anchor piles which can be propped against.
3. Using a combination of earthwork and temporary piling.

In the case of canal-type construction, the end of the element can be propped against the concrete floor of the canal. The forces involved are large, requiring substantial props to the side closure panels as well. This canal can then be dewatered to allow construction of the adjacent tunnel or ventilation building. The joint to the adjacent construction can be a relatively simple in situ watertight joint.

An important aspect of this type of joint is the detailing needed around the end of the tunnel element to provide a barrier to water entering the dewatered excavation for the approach structure. This excavation may be open for some time, and so, it is essential to have a robust solution for this. The sides can be sealed with sheet or tubular steel piles, and a steel sheet pile wall can be erected on top of the element. This, coupled with sealing around the perimeter, is usually sufficient to make the construction watertight. Beneath the tunnel, a common solution is to place a rubber gasket on top of a sheet pile cut-off wall such that the tunnel element, when immersed, rests partially on the gasket and seals the waterpath.

For many tunnels, the terminal joint is required to act in a particular way. This can be to

- Enable increased rotation at the joint compared to regular immersion joints.
- Provide support to the end of the immersed tunnel.
- Accommodate large seismic displacements.
- Provide a transition to a completely different form of construction, for example, bored tunnel.

The terminal joint may have to be provided with some sort of shear key to prevent differential movement between the sections. This depends on the form of construction and the amount of shear that has to be transferred between the two. The loadings on the cut-and-cover section are usually greater than the immersed tunnel and the form of construction is often different to reflect this—for example, the cut-and-cover tunnel could be constructed with diaphragm walls. These would be stiffer and would give different settlement characteristics from the immersed tunnel, which is more flexible, resulting in a shear transfer between the two if the immersed tunnel is "held up" on the cut-and-cover tunnel. If this is restrained by shear connections, either large shear keys or a cill structure—extending from the base slab of the approach structure on which the immersed tunnel can bear—will be required.

In extreme cases, if the ground under the immersed tunnel is particularly poor, these shear forces may be very large. It might be necessary to remove a section of the poor ground under the immersed tunnel and replace it with a better granular material before the tunnel element is placed to improve the foundation characteristics and reduce the shear forces. This ground improvement might also be achieved by the installation of stone columns. Another option is to support the end of the immersed tunnel element with piles and a crosshead. In less severe cases, it is possible to set the end of the immersed tunnel high against the adjacent cut-and-cover and then let it settle before the shear connection is made. This relies on the slip characteristics of the Gina gaskets, which, in theory, can sustain up to around 150 mm of shear movement within the gasket itself and then the whole gasket could slide down the sealing plate. However, reliance on these principles alone—where the tunnel element is moving in an uncontrolled (albeit predicted) way—is high risk and most designers would rather rely on more positive action.

Connection to bored tunnels

There have been a few instances where immersed tunnels are connected directly to bored tunnels. This type of connection can be made quite effectively, but requires many safety features and some unusual techniques. This technique was used successfully on the Marmaray immersed tunnel in Istanbul. The connection here was well out into the Bosphorus and was made using a modified immersed tunnel element rather than a separate caisson. This had the following advantages:

- A large shaft for TBM removal was not needed.
- Marine operations and construction operations were kept consistent for both the general tunnel elements and the special end elements.
- Alignment transitions were largely contained within a single structure.

As described in Chapter 9, the elements at each end of the tunnel were special elements, with one end widened to form a reception chamber into which the tunnel boring machines (TBMs) could be driven. After being placed, the terminal elements were surrounded with a weak concrete mix that extended back to the competent rock face.

The TBM drive through the concrete plug is made following a rigorous survey of the connection element and final adjustments of the TBM alignment. The speed of the TBM drive is restricted for the drive through the concrete plug. Grout probes on the cutting face of the TBM ensure any potential cracks or voids in the concrete plug, in advance of the drive, are pre-treated to further ensure potential water paths are sealed off. Similarly, the annulus against the excavated face is grouted as the precast concrete tunnel linings are put in place.

The TBM is driven into the reception chamber until the skin plate is within the collar wall and past the clamping plate. The clamping plate will comprise a steel-braced rubber plate fixed to the perimeter of the collar wall, fitted with an inflatable grout tube to the rear. Once the TBM is within the collar wall, the grout tubes are inflated to seal the clamp plates against the TBM skin plate. The area between the clamping plate and skin plate is then pressure-grouted using grout tubes cast within the element.

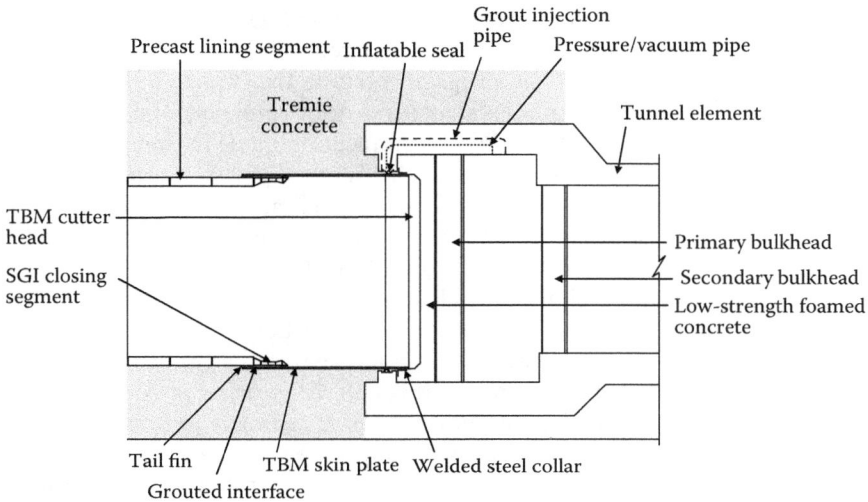

Figure 10.10 Immersed tunnel connection to bored tunnel.

The primary seal consists of steel closure brackets welded to the internal edge of the skin plate and collar wall. Back propping is provided whilst the TBM is dismantled and an in situ lining is cast against the skin plate from the end of the last bored tunnel ring to the collar wall. A spheroidal graphite iron (SGI) transition ring, connected to the skin plate, ties the last concrete tunnel ring, to ensure stability of the lining once the ram/thrust jack has been dismantled. The arrangement described is shown in Figure 10.10. This scheme shows that it is possible to make the connection to an immersed tunnel even under extreme conditions.

CLOSURE JOINTS

The closure joint is the final joint formed between the immersed tunnel and one of the approach structures, or between two tunnel elements. Typically, a space of between 0.5 to 1.5 m is left between the two structures and this is in-filled in one of a number of ways:

- Concrete tunnel: in situ joint
- Prestressed segment
- Terminal block
- V-wedge
- Steel tunnel: tremie joint
- Dry joint

The choice of how the closure joint is made will depend on a number of variables. Whether a wet or dry joint is selected will be dictated largely by program considerations. If the program permits the immersed tube to be placed first, before the approach structures are completed, the end of the immersed tunnel can be exposed and a connection made in the dry, similar to the terminal joint described earlier. This is the simplest approach for the structure and the joint can be quite conventional and little different from the other joints along the tunnel. Otherwise, a wet joint will need to be formed. This is one that is constructed underwater.

The choice of articulation at the closure joint will depend on the overall articulation of the tunnel, and in particular, on how the tunnel needs to behave at the point of the closure. It is possible to either build the joint to make the tunnel structure continuous through the joint, or provide a degree of flexibility. Provision of a combination of a segment joint and a construction joint is common practice, with injectible waterstops in both joints to provide watertightness. This arrangement is illustrated in Figure 10.11. Before the segment joint/construction joint arrangement was used, a double Omega joint detail was used for tunnels in the Netherlands, but this is no longer used.

Because the larger movement capacity of an immersion joint is lost with this arrangement, the movements at the immersion joints on either side of the closure joint due to thermal expansion need to be assessed to ensure they are not unduly increased beyond their capacity. If this occurs, then an alternative arrangement with greater flexibility may be required, or a shortening of the tunnel elements on either side of the closure joint can be used to control the joint movements to an acceptable level.

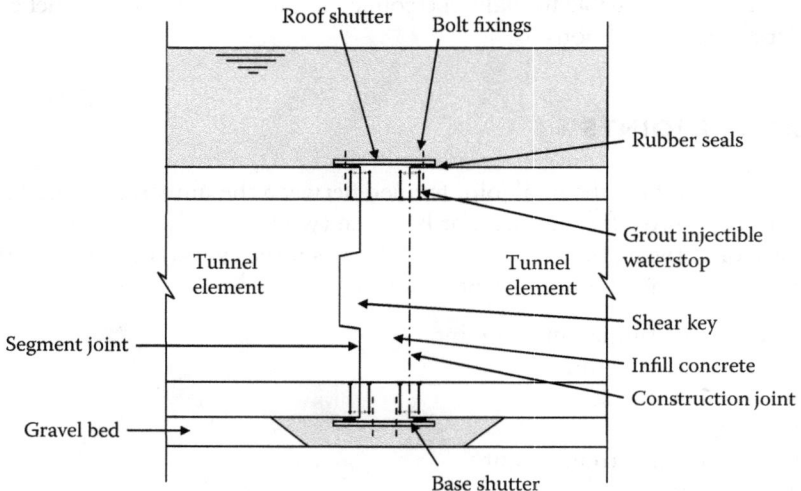

Figure 10.11 Øresund Tunnel closure joint.

The process of construction for a conventional closure joint is relatively simple:

- Tunnel elements are placed leaving a 0.5–1.5 m space between the ends of the two elements.
- The joint space is blocked out by wedges or jacks.
- Closure formwork panels are assembled around the external perimeter of the tunnel.
- The bulkhead space is dewatered.
- The concrete slabs and walls are formed by in situ concreting from within the tunnel.
- Temporary wedges or jack are removed and the load is transferred to the permanent concrete.

For this type of joint, it is essential to carry out the concreting work to as high a standard as the remainder of the immersed tunnel. Although there is sacrificial formwork around the tunnel perimeter, this is generally considered to be an additional benefit and not the basis of watertightness. Therefore, the concreting has to be carefully planned and executed. The conditions for this work are quite difficult as it has to be carried out within the bulkhead space. Whilst the closure panels are deemed to create a safe working environment for the periods of time required to erect the formwork and carry out the concreting, bulkheads should remain in place until the concreting is completed, to protect the remainder of the tunnel, should anything untoward occur.

The concreting work will generally be carried out in stages—base slab, walls, then roof—for practical purposes. This may be further complicated if there are temporary wedges bracing the tunnel elements apart, as the concreting sequence will have to work around these. This adds further operations as construction joints need to be carefully prepared to ensure they are watertight. Due to the construction joints between the concrete pours themselves, and the joint with the end of the tunnel element, there is a high risk of early age cracking occurring and it is common to require cooling pipes to be installed in the concrete pour to control the heat development during curing and to avoid cracking.

A special pumpable concrete mix will be required for the roof slab pour as access is only available from below, and careful planning is required to ensure that complete filling of the roof void is achieved, as the results are difficult to inspect. If a pumpable mix is proposed, it should be trialed to test the effectiveness of the method. Historically, sprayed concrete has been used as this can be relied on to fill the void satisfactorily, but now pumpable mixes are equally effective for forming the slab.

The closure panel formwork that is fixed around the perimeter of the tunnel is usually steel construction and has to be reasonably stiff to resist the hydrostatic pressures that will be acting when the joint is dewatered from the inside. They are typically sandwich panels with steel plates on either side of I-sections. As a result, the panels are very heavy and the craneage required for handling the panels may need to be quite large. Panels need to have rubber-sealing gaskets mounted on the face of the steel plate that faces the concrete. Overlapping details are needed where panels meet at 90° at wall-slab corners. The panels for the walls and roof of the tunnel can be simply bolted to the external face of the tunnel concrete. The panel beneath the tunnel needs to be placed in the bottom of the dredged trench ahead of the last tunnel element being immersed. It can then be pulled up to the underside of the tunnel and clamped in place using beams spanning the base slab on the inside of the tunnel.

The external sealing of the panels can be a time-consuming process and will be carried out by divers. Simplicity of detailing is needed to enable this to be done effectively. Therefore, the end of the tunnel needs to be simplified if, for example, there are toes on the base slab. These should be stopped short of the joint to allow a simple rectangular panel to be used.

If the tunnel rests on a gravel foundation, the length of the gravel bed beneath the closure joint is generally omitted. The percentage area of support that is lost is relatively small and is unlikely to affect the settlement to any significant degree. However, this should be assessed by the designer. If there is a need to fill the void, this can be done using grout injection sacks, which are either pulled through beneath the tunnel after the bottom closure panel has been fixed, or can be secured to the underside of the closure panel. This approach was adopted for the Fort Point Channel Tunnel in Boston. Grout injection could be used, but is more difficult due to the need to fix curtains around the panel to contain the grout.

With a pumped sand foundation, although this is generally installed quickly after the tunnel elements are placed on their temporary supports, it is possible to delay the completion of this process until the closure panels have been fixed to the tunnel, and thereby form a continuous foundation layer beneath the tunnel.

After concreting the internal structure, the outside temporary formwork panels are not removed. However, if the closure joint is required to accommodate movement, unless a sliding ability is built into the design of these panels, one side of the formwork needs to be released from the tunnel to enable it to slide. Otherwise, the effect will be to clamp the joint so that it is effectively rigid.

Before dewatering the closure joint, the tunnel elements need to be braced apart to lock in the compression force that has been built into the tunnel from the immersion process. This can be achieved by a simple concrete wedge block system, as shown in Figure 10.12, which is lowered into the joint space and has the ability to take up any tolerances on the width of the joint. The wedges would be lowered and braced against the walls of

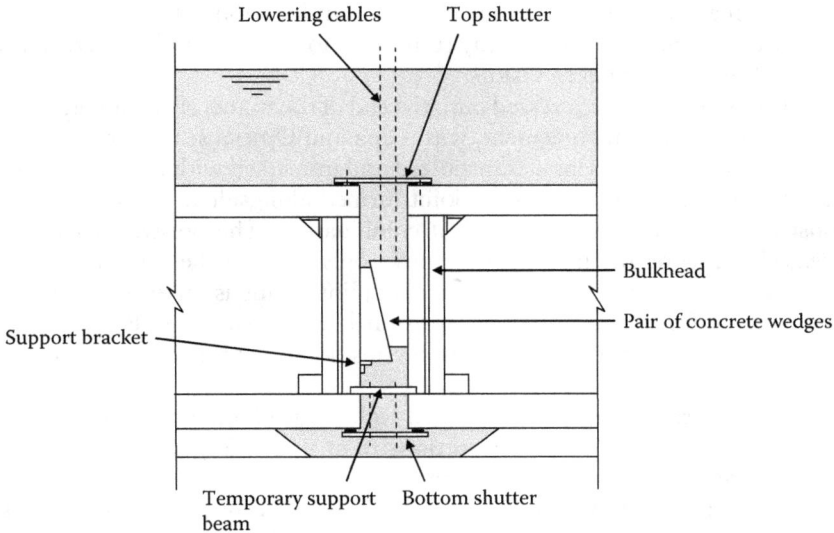

Figure 10.12 Wedge block restraining system.

the tunnel, either to the internal or external walls. Provided the wedges are symmetrical and prevent any eccentric loading, it does not matter whether internal or external walls are used. The main consideration will be the ease of the subsequent concreting operations and the sequence of works required to transfer the loading from the wedges to the permanent concrete.

Once placed, the wedges would be fixed in place by divers, though, in practice, friction would prevent them from slipping. After concreting the perimeter of the tunnel, the wedges would need to be dismantled or broken out from within the tunnel.

An alternative approach is to use hydraulic jacks to brace the tunnel elements apart. This has some advantages in that the jacks take up less room in the joint space and are easier to dismantle once the joint is complete. The jacks can be applied externally if desired to keep completely clear of the permanent works, using temporary concrete nibs or steelwork to brace against.

Prestressed segment

The reduced longitudinal flexibility associated with the normal closure joint may result in the movements in adjoining immersion joints nearing or exceeding their capacity. In such a case, a segment/construction joint solution may not be feasible and greater flexibility may be required. The options here are to either reduce the length of the tunnel elements close to the closure joint, to get a reasonably uniform behavior over the length of the tunnel, or to introduce an additional immersion joint within the length of the tunnel element close to the closure joint location. This approach has been used for the Airport Railway tunnel in Hong Kong, and more recently, on the Bjørvika Tunnel in Oslo.

A short segment is prestressed onto the end of the tunnel element with a pre-set immersion joint arrangement, with Gina and Omega seals, formed in the casting basin. The element is transported and immersed with the end segment clamped in place, and the closure joint formed alongside in the wet, using a construction joint and segment joint combination. The prestressing is then released, allowing the immersion joint to operate with the same movement range as the remainder of the immersion joints. This is an effective way of achieving uniform behavior along the tunnel. This is shown in Figure 10.13.

There are a number of construction considerations for this approach:

1. The short end segment will need to be braced against the tunnel element with rigid shear connections to ensure safety during the transportation of the tunnel element.
2. The segment must be constructed at a distance away from the remainder of the element in the casting basin to allow the immersion joint

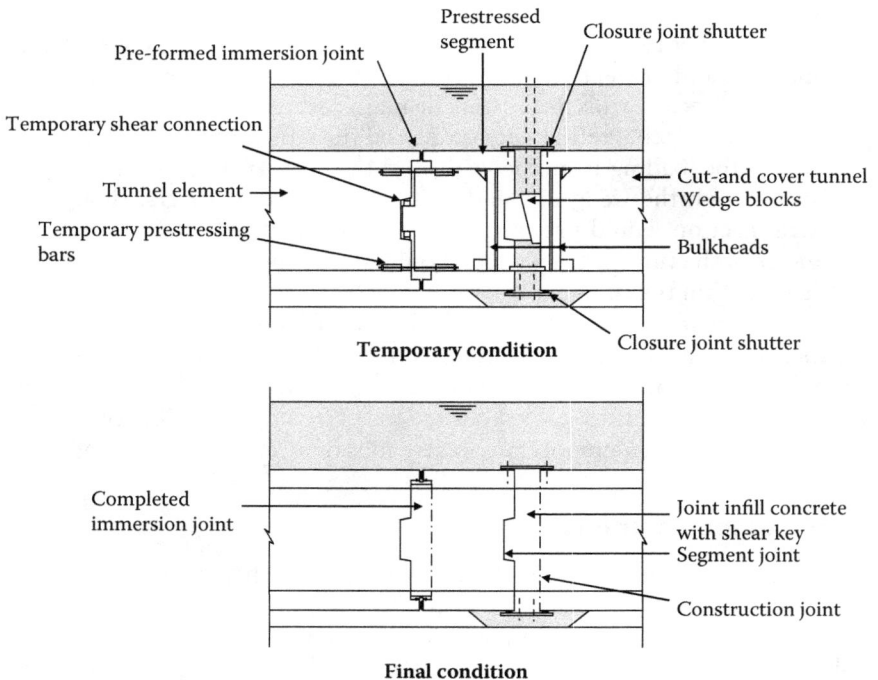

Figure 10.13 Prestressed segment closure joint.

to be formed and the Gina gasket to be mounted. The end segment, therefore, needs to be kept short and not too heavy, as it must be slid to close up the immersion joint. Typically, the length has been kept to around 10 m so that the sliding does not require complicated jacking or temporary supports.

3. The degree of pre-compression in the Gina should be set to match the equivalent compression that would be experienced due to the hydrostatic forces acting at the water depth where the joint will ultimately be located.

Terminal block

One closure joint solution developed in Japan is the terminal block. This is a closure system that relies on the closure joint being at the terminal joint position, that is, between the immersed tunnel and the approach structure (Figure 10.14). It is a relatively complex solution and has only been used on a few occasions.

A short segment, the terminal block, is constructed within a sleeve at the end of the approach structure while it is in the dry. Once the end of the approach is flooded and the last immersed tunnel element has been placed, the terminal block can be jacked out from the approach structure to meet the end of the immersed tunnel, with a Gina gasket providing the watertightness in the same way as a conventional immersion joint. By installing a bulkhead within the approach structure and within the terminal block, water can be allowed to flow to the rear of the terminal block and the hydrostatic force can be used to compress the Gina gasket fully as the bulkhead space between the terminal block and the immersed tunnel is dewatered. Once the joint is formed, the space behind the terminal block is grouted such that the side of the terminal block adjacent to the approach structure is made continuous.

The complexity of this approach lies with the tolerances of the initial terminal block construction to ensure the block slides freely within the sleeve. Guide rails can improve this. Another important consideration is how to achieve the water cutoff around the perimeter of the terminal block sufficiently to enable the bulkhead space between the terminal block and the approach structure to be dewatered. This is achieved by a combination of seals and grouting from the sleeve.

When the terminal block has been jacked outwards and the joint formed, the block needs to be held in position by jacks until such time as the space behind the terminal block is in-filled, otherwise the force transmitted by the Gina seal will cause movement back in the reverse direction.

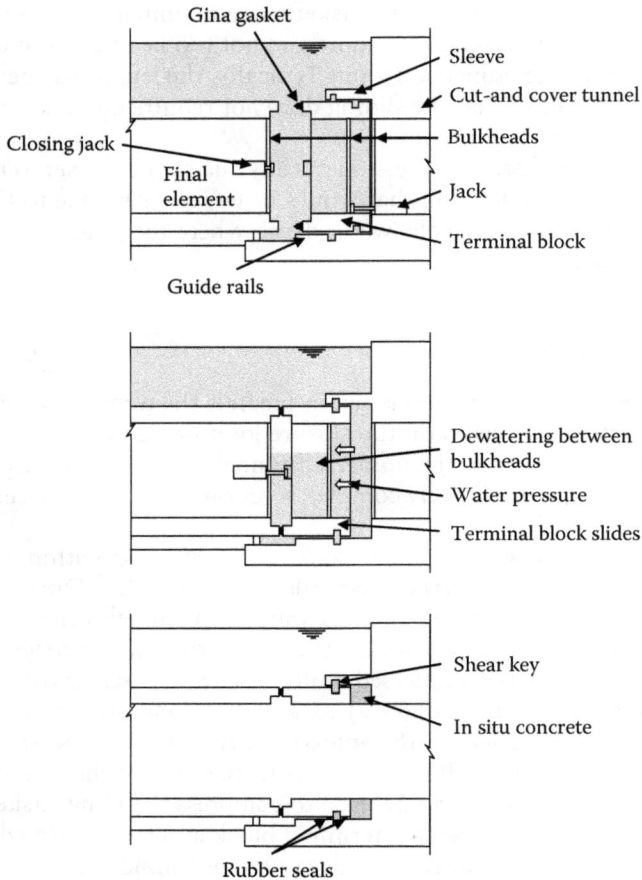

Figure 10.14 Terminal block closure joint.

V-wedge closure joint

The V-wedge joint is a relatively new innovation that has been used on the Naha Tunnel in Japan. The technique utilizes the difference in hydrostatic pressure between the roof slab and the base slab to maintain the stability of the segment once it is connected to the tunnel on either side. It therefore has inherent stability in addition to being physically connected to the remainder of the tunnel. The arrangement is shown in Figure 10.15.

The principle of the V-wedge is that it has sufficient capacity in the rubber gaskets and within its geometry to accommodate the tolerances in construction. There will be variation in the gap between the ends of the tunnel elements, the tolerances on the planes of the inclined faces, and there may be vertical or horizontal displacement between the elements. It is an elegant

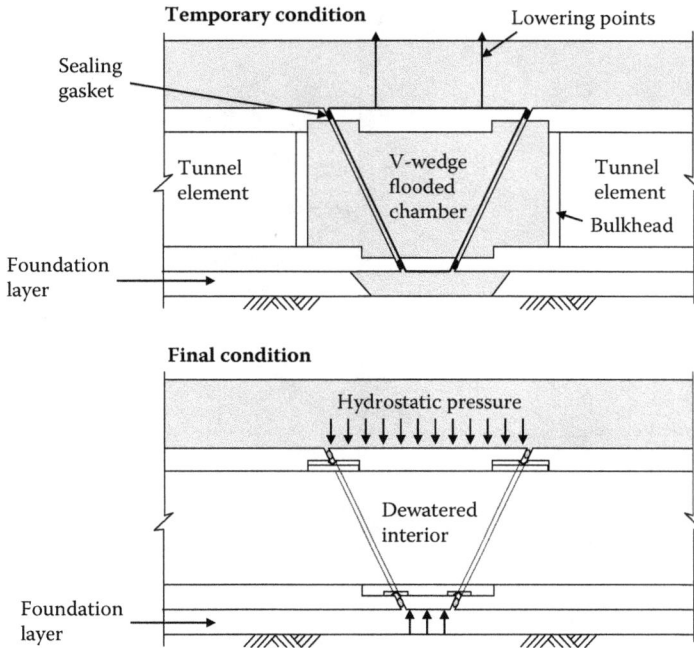

Figure 10.15 V-wedge closure joint.

structural solution in that it avoids the use of external underwater form-work and reduces diver activity. However, for the completed tunnel, it has no particular advantage over more conventional closures.

Steel tunnel tremie joint

For double shell steel tunnels, the common method of constructing the closure joint is to form an enclosure around the joint, which is filled with tremie concrete. The external shutter to the enclosure is formed by fixing vertical curved sheet piles down the sides of the tunnel connected to joint clutches mounted on the steel shell of the tunnel element on each side of the joint. The internal shutter for the tremie concrete is formed by circular hood plates placed by divers that overlap the inner steel shell, which pro-trudes from the end of each element. The hood plate provides a sufficiently watertight seal to allow dewatering. The inner steel shell is subsequently welded and grouted to provide a permanent watertight connection and the internal finishing concrete lining is cast within this. The typical arrangement is shown in Figure 10.16.

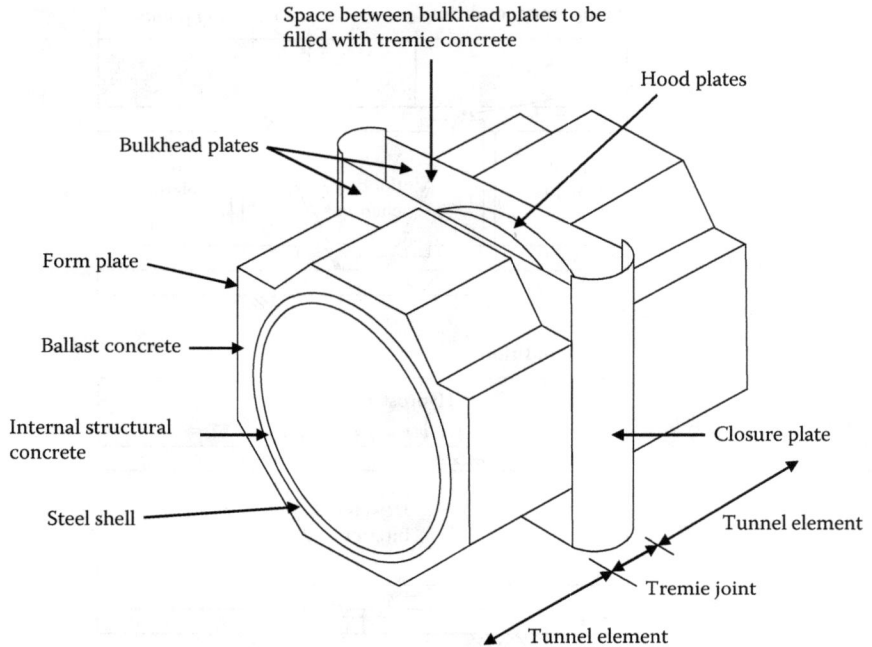

Figure 10.16 Steel tunnel closure joint.

This type of joint makes the tunnel continuous across the joint and so an assessment of the thermal expansion and contraction, and of movements from settlements would need to be carried out to make sure the tunnel can accommodate them satisfactorily.

CROWN SEAL JOINT

An innovative solution was developed for the Umeshima Tunnel constructed under Ohsaka Bay, Japan, which could accommodate a large, differential ground settlement. The joint is constructed immediately adjacent to a conventional immersion joint that is made rigid. It features a double seal that permits large deflections to occur and has prestress across the joint to prevent excessive opening. This is a one-off solution, but is a highly effective way of dealing with a large shear movement, if this is anticipated, and it may therefore offer a solution for future tunnels that need this type of facility. The joint arrangement is shown in Figure 10.17.

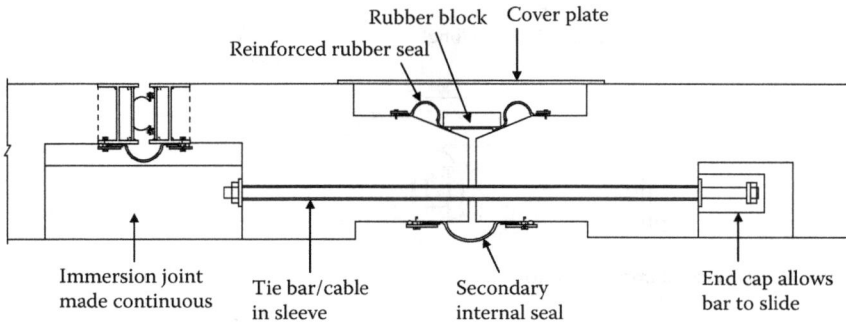

Rubber block   Cover plate
Reinforced rubber seal

Immersion joint made continuous   Tie bar/cable in sleeve   Secondary internal seal   End cap allows bar to slide

Figure 10.17 Crown seal joint.

SEISMIC JOINTS

Seismic joints are those that are designed to maintain the integrity and watertightness of the tunnel during a seismic event. In moderate seismic zones, it may be possible for the normal immersion joints and terminal joints to accommodate all movements that could occur. Under high levels of seismic loading, there may be a risk that joints could displace to the point that they lose watertightness and the tunnel is at risk of inundation. The methods of analysis and design of the structure and joints are described in Chapter 9. The analysis may show that seismic joints are needed at each immersion joint. Alternatively, there may just be a need for a single seismic joint at the connection to the approach structures at each end.

Seismic immersion joints

Where there is a risk of the joint opening to the extent that it loses its watertight seal, the accepted approach is to restrain the joint movement by means of high tensile bars or cables strapped across the joints. This is shown in Figure 10.18. Whether bars or cables are used, the principles of operation are the same. A sufficient length of cable is provided to allow them to strain and accommodate movement, but their stiffness is such that there is restraint to the joint opening and the degree of opening is controlled.

The cables or bars are connected up after the elements are immersed. They are installed ahead of immersion, but withdrawn into a recess, so that that the duct can be made watertight while the elements are afloat or being immersed. Once the bulkheads are removed, the bars are then pulled through and connected from each side to a coupler. Once connected, the slack within the cables is taken up, but no prestressing force is applied. Similarly bars are not tensioned, simply a notional torque applied to ensure that the clamping plates are firmly seated on their bearing surface.

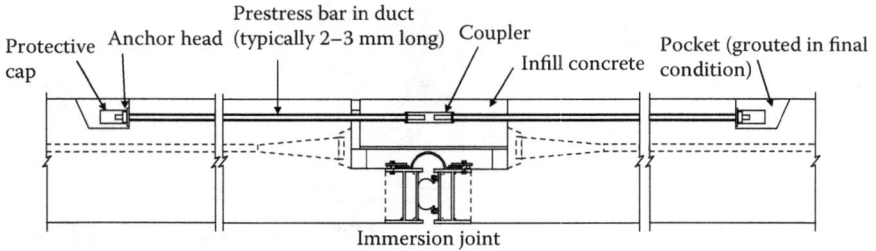

Figure 10.18 Seismic immersion joint.

During a seismic event, if the joint experiences compression, there is a risk of dynamic impact of the bar or coupler-bearing plate occurring once the joint rapidly opens and closes. Detailing of the bearing area and the concrete around the anchorage needs to take this into account by treating it as an anchorage zone to avoid damage occurring to the structure.

An alternative solution has been used in Japan that features a wave-shaped steel plate, sometimes called a bellows joint, which is welded to the tunnel element on either side of the joint on the inside of the Gina gasket. The principle of this arrangement is that if the tunnel joint opens and closes due to the seismic ground motion, the wave shape is able to stretch and compress without losing watertightness. Unofficial reports suggest that this form of joint was effective, but was not without its difficulties in construction. Other arrangements have been used for steel tunnels incorporating a high degree of vertical, lateral, and longitudinal movement capacity. One such arrangement is shown in Figure 10.19, along with the bellows joint.

Seismic terminal joints

Terminal joints have a part to play in the design of immersed tunnels in seismically active regions. The individual joints between the immersed tunnel elements are designed to cope with the movements that occur as the seismic waves pass through the site. But, these movements can build up and result in a larger displacement at the end joint between the immersed tunnel and the more stable cut-and-cover tunnel. The terminal joint is designed to accommodate this. Such a joint was provided for the 5 km long BART tunnel in San Francisco. Although the tunnel itself does not cross any active faults, it is located in a highly active seismic area. Thus, the terminal joint was also designed to accommodate seismic movements resulting from activity on any of the major faults in the surrounding area. The terminal joint between the last element to be placed and the ventilation building also allowed for any inaccuracies in placing the tunnel elements as well as tolerances for thermal movement of the completed tunnel.

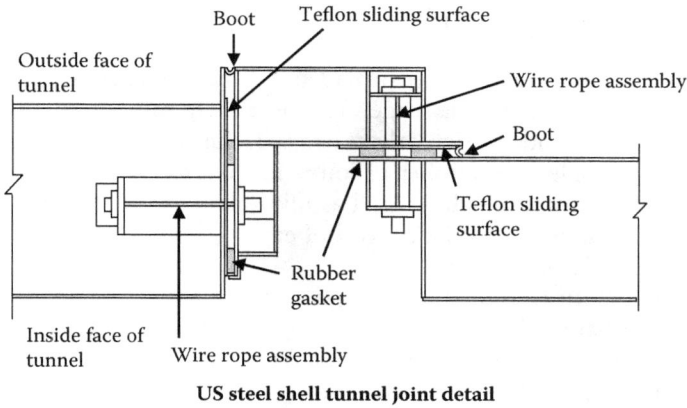

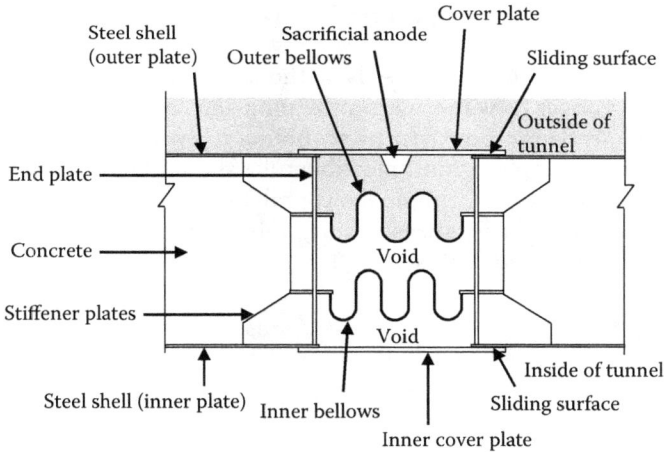

Figure 10.19 Seismic joints in steel tunnels (sections through walls).

The terminal joint was a complex sliding telescopic joint. The joint holds the tunnel vertically and horizontally, but allows longitudinal and rotational movement. Seals are placed around the tunnel to give a watertight seal as the tunnel element moves in and out of the terminal joint. Enough allowance has to be made in the movement capability to allow for some creep that is likely to occur over time with repeated seismic movements. Stops are provided to prevent excessive movement, which would result in the joint opening.

In 1989, the BART tunnel withstood the earthquake that destroyed many of San Francisco's bridges. It was inspected and reopened three hours after the earthquake, and was the only fully operational public transport system in the area.

CONSTRUCTION JOINTS

Wherever construction joints are provided in concrete, particular care must be taken to guard against leakage. This is clearly much more critical to concrete tunnels, and particularly those without an external membrane. As a basic principle, the number of joints in the external perimeter of the tunnel should be minimized. As well as minimizing the points of potential weakness, this assists with the control of early age cracking and shrinkage cracking due to joint restraint.

It is quite practical to plan to have joints only at the base of the walls of the tunnel so that the base is cast first, then the walls and roof together as a second concrete pour. Some tunnels have joints at the top of walls beneath the roof slab also, but this is unnecessary, and should be avoided in the planning. The joints at the bases of walls have a natural compression built in from the weight of curing concrete above, whereas this is less so at the tops of walls.

At construction joints in tunnels without membranes, it is imperative to provide a barrier to water within the joint interface. Although concrete surfaces should be pre-prepared by scabbling to ensure that a good key is achieved, a waterstop or grout injectible tube should be provided. It is also cost-effective to include a simple hydrophilic seal on the joint interface as a back up to these. These measures are relatively cheap compared to the potential costs involved with sealing leakage at a later date.

Chapter 11

Durability

An immersed tube tunnel is buried in a marine environment, and gaining access to the outside surface for inspection or repair is impractical. The costs associated with excavation are likely to be prohibitive, and the ability to perform reliable inspections and remedial works underwater would be questionable. Therefore, it is necessary from the outset to ensure that the tunnel structure is inherently durable and will require little or no maintenance over its life, and that any maintenance work that is required can be carried out from within the tunnel without disruption to its operation.

It is possible to achieve this desired level of durability through simple measures in design and construction. This chapter sets out the measures that are needed and the aspects to be considered in the design to ensure durability is achieved in the completed tunnel. When talking about durability, we are really talking about two things—the durability of the materials used in construction and the watertightness of the tunnel structure. Watertightness is required of the structure, including its joints, and prevents seepage that causes corrosion of reinforcement or steelwork. For steel tunnels, this will relate to defects in the welds of the steel shell; for concrete tunnels, this will relate to the quality of the placed concrete. Watertight concrete is achieved by eliminating through-cracking in the concrete and other defects in the construction and by using good-quality impermeable concrete. The same principles of durability are common to both concrete and steel immersed tubes, although each form will have different priorities.

WATERTIGHTNESS

The fundamental method of ensuring the durability of the tunnel structure is to make it watertight. Watertightness must be achieved both at the tunnel joints, as described in Chapter 10, and throughout the structure itself. A question that is asked on most immersed tunnel projects at some stage is, "What does watertightness actually mean?" Is it simply sufficient to state in the contract documents that the immersed tunnel shall be watertight?

Table 11.1 Watertightness criteria

Case	Description
In situ and prefabricated concrete structures, including immersed tube tunnel	Free of all visible leakage, seepage, and damp patches
Diaphragm walls Secant piles	Leakage shall be restricted to damp patches on the concrete face and at construction joints. Jetting of water will not be acceptable. Prior to construction of an inner lining, total inflow over a given area of structure shall not exceed 0.12 L/m² per day overall and 0.23 L/day on any individual square meter.

Probably not, as this means different things to different people. It is therefore common to try to describe this further. Table 11.1 gives a typical wording that the authors would recommend to define watertightness.

It is appropriate to define more relaxed criteria for the construction of approach structures using diaphragms walls or secant piles. This type of construction carries an inherently higher risk of leakage due to the construction techniques involved. It is usual for an additional inner concrete lining to be constructed to achieve the higher level of watertightness inside the tunnel and to provide appropriate drainage paths.

Acceptable rates of leakage are often quoted for bored tunnel construction in terms of liters/day ingress of water per square meter of tunnel surface area. Typical values used are in the order of 100 L/day/m². For a bored tunnel with a segmental lining with many thousands of meters of sealing gasket, this is appropriate. Leakage is not detrimental to the tunnel segments forming the lining and there simply needs to be a limit that corresponds to the drainage sump and pumping capacity. This type of criteria is meaningless for an immersed tunnel and should not be used. The ratio of joint length to surface area of tunnel will be in the order of 100 times smaller.

The nature of the joints is also quite different between bored and immersed tunnels. A leak through an immersed tunnel joint suggests a significant defect, compared to occasional variations in the effectiveness of the bored tunnel gasket joints that accumulate to the total leakage inflow. Leaks at immersion joints can have a direct impact on the durability because of the cast-in steelwork components of the joint. Leaks at segment joints would suggest incomplete concreting and post-grouting of the groutable waterstops, and the reinforcement local to the joint may, therefore, suffer corrosion. Leaks at welds in steel shells would be similarly detrimental to the durability of the shell and the internal concrete structure. Therefore, given the greater risk to the structure, and the inappropriateness of the bored tunnel leakage criteria, no leaks are tolerated in immersed tunnel structures and watertightness means that the ingress of water is prevented entirely.

Of course, it is accepted that unexpected events occur during construction and all defects cannot be prevented, and leaks are not at all uncommon. However, experience shows that most leakages are usually identified during the construction period and can be rectified by injection grouting. The intention of the construction contract should be to hand over a watertight structure and an experienced immersed tunnel contractor will know that this will entail a degree of remedial work following the main construction of the tunnel elements.

CONCRETE MIX DESIGN

For concrete tunnels, the quality of the concrete is of paramount importance. If the concrete is intended to provide the primary means of watertightness, then clearly, great attention must be paid to achieving a certain quality. Even if the tunnel is to have a membrane applied, there is still a high standard to be achieved in the concrete production as common practice is to allow for the possibility of minor defects or leakage in the membrane over the life of the tunnel.

For steel tunnels, the concrete will be either acting compositely with the steel or provide an internal lining structure. Although in this instance, it does not provide watertightness, nor is it exposed to the marine environment, it is subject to the aggressive environment within the tunnel and a high level of durability is therefore still needed. Concrete mix design is, hence, an important part of design and construction for all forms of immersed tubes.

Objectives

The mix design process should address the aspects of the concrete described in the following subsections.

Permeability

A low level of permeability is essential for durable concrete in an aggressive marine environment. Low permeability will limit the rate of chloride intrusion into the concrete matrix and, hence, the time at which corrosion due to the presence of chlorides will initiate. A target chloride diffusivity coefficient will need to be achieved.

Density

A target density will be set by the designer to ensure that the assumptions made regarding the weight of the structure are fulfilled and the safety factor against the uplift of the structure is achieved.

Strength

The strength of concrete in an immersed tube does not need to be exceptional. In fact, quite low strengths can be used. This is because the wall and slab thicknesses are often slightly oversized to ensure there is sufficient weight in the structure to resist the uplift. Durability and watertightness are of greater concern and concrete strength is not directly influential on this. Nevertheless, the chosen strength will need to be achieved with certainty. There may also be a need for understanding the rate of strength gain in the early days and weeks of curing.

Low heat of hydration and crack control

The heat generated during curing is of major importance and the mix design will generally need to consider measures to limit the heat development. In addition, the shrinkage properties of the concrete will need to be well understood and some control over the degree of shrinkage might be needed. Other early age properties will need to be understood and may influence the mix design.

Flow characteristics

Immersed tubes are frequently constructed using large-volume concrete pours and the mix characteristics in the wet condition need to be considered carefully. High-flow mixes are often used, with associated admixtures to ensure that segregation does not occur. Retarders are also often used to help manage large concrete pours. Because watertightness is the overall objective, the ability of the mix to flow into the geometric features at immersion joints, around waterbars, around tie bars, and at shear keys is highly important to ensure there is no risk of poor compaction.

Other criteria

Other criteria to be met in the concrete mix design include the following:

- The water/cement ratio is typically specified as a maximum of 0.4.
- A minimum cement content is often specified at around 300 kg/m³.
- To avoid alkali silica reaction, a maximum alkali content is specified for an equivalent Na_2O of 3.0 kg/m³ for a concrete with a mortar content (concrete less coarse aggregate) of 60%.
- A maximum chloride content is often specified as 0.10% of total powder content to minimize the initial chloride in the concrete.

National standards may be more onerous or expansive than the above and should be used if that is the case. Otherwise, the values given offer a good starting point for a concrete mix design.

Extensive preconstruction and production stage testing and validation of the concrete mix should be carried out to verify the properties described. In addition, testing of the properties required for the early age analysis of the concrete should be undertaken. The properties in question are described in the section Early Age Crack Control: Simulation software.

Concrete density and permeability

The density properties of concrete are more significant for floating structures than for any other form of construction. The immersed tunnel is no exception and it is very important to understand the density values used in design and that these are achieved in construction. A starting point in the design process is to take a simple range of published values. Some suggested initial values are described in Chapter 9. As soon as it is possible to undertake the pretesting of the concrete mixes that may be used for construction, this should be done, and the mass concrete properties can start to be refined. Similarly, an initial estimate of reinforcement density should be made to apply a range of reinforcement densities to the range of concrete densities. Reinforcement may typically be anywhere between $100-180$ kg/m^3. Occasionally, reinforcement can be a greater density than this, but it usually suggests the section thicknesses are too slender or there are specific loading circumstances that have had to be accommodated. The reinforcement contribution should be added to the mass concrete to derive the reinforced concrete density to use in design. The minimum concrete and minimum reinforcement densities should be combined to give the lower-bound figure, and maximum reinforcement density should be combined with the maximum concrete density to give an upper-bound figure. These upper- and lower-bound figures can then be used in the buoyancy analysis.

The verification testing of the density will need to be carried out during construction and this is usually done using large-scale samples, often one cubic meter, taken at the time of concreting, and by extensive core samples taken from the completed tunnel structure.

The benefits of cement replacement materials for achieving low-permeability concrete mixes for maritime structures are well documented. Pulverized fly ash (PFA) and ground granulated blast furnace slag (GGBFS) are commonly used, along with microsilica. As well as lowering the permeability, PFA and GGBFS bring the benefit of producing low heat of hydration concrete mixes. It is therefore likely that one of these materials will be used for the hydration properties and will provide the low-permeability requirement as a matter of course.

PFA and GGBFS can be used as a replacement for cement in high quantities. PFA typically can replace up to 50% of the cement content, although it is more commonly used at a volume of 20–30%. GGBFS can replace the cement in the concrete mix by up to 90% of the powder content, though more typically it is used at 70% replacement content. Microsilica, also known as silica fume, does not aid the low heat of hydration properties, but can be a valuable replacement material for improving the permeability characteristics. Because it is extremely fine, it produces a much denser concrete matrix than otherwise would be created. Typically, it is only used in small percentages, up to 15% maximum content of the mix. It can bring a strength increase that is not always necessary, and so, typically, a 5–10% content is used.

The permeability of the concrete should be tested and measured against a target value. A reasonable target for the chloride diffusivity of the concrete at a concrete age of 28 maturity days is that it shall not exceed 7×10^{-12} m²/sec. The diffusivity can be determined by the "rapid chloride migration test," as specified in NT Build 492. If the PFA content exceeds 20% by weight of the concrete mix, or GGBS is used as cement replacement in the concrete mix, the target diffusion coefficient needs to be checked after 56 maturity days because of the slower strength gain.

Alternative approaches exist; for example, in the United States, test methods use rapid chloride permeability testing in accordance with AASHTO T 277 and ASTM C 1202. Variations on the test exist, such as Virginia Test Method 112. These are essentially resistivity tests and the target is set in Coulombs (C). There is some correlation required between the resistivity and permeability and this correlation is well developed and understood. For an immersed tunnel, the target should be not to exceed 1000 C.

Concrete strength

As noted, high strength is not a particular requirement for immersed tunnel structures. The need for weight to achieve safety factors against uplift, and the need for thick sections to ensure watertightness means that concrete strengths can remain quite low and the sections can be usually designed to resist load effects quite comfortably. This may not be the case if exceptional spans are needed for multiple lane tubes, but these are relatively uncommon.

A typical strength of concrete for an immersed tube is in the order of 40 to 45 N/mm² cylinder strength. This would normally be the strength to be achieved at 28 days. However, because mixes are often designed with a high proportion of cement replacement materials, it can be difficult to attain the full strength at this time. The rate of strength development is retarded because the heat of hydration is reduced, hence the chemical reactions within the mix take longer and the rate of strength development is slower. Often, to achieve a specified 28-day strength, a higher strength has to be targeted at 56 days. This leaves the structure with additional strength that is not necessary and may not be taken advantage of in the design.

Often, 28-day strength requirements are stated in project specifications, whereas it could be beneficial to leave the decisions on when this strength is achieved to the contractor who, after all, is responsible for when formwork shutters and falsework supports are released and is usually responsible for the final concrete mix design. The decision on this will depend on the method of procurement. For a traditional engineer's design contract, it is common to specify a 28-day strength to ensure the concrete matures over a short period of time and there will be sufficient strength for striking shutters, accommodating loading from settlements in the casting basin, or applying external loading to the structure. However, some flexibility can still be allowed for to suit the contractor's approach to construction. If the choice is left to the contractor, it is important that any slow rate of strength gain is fully understood, the implications are considered in all of the construction operations that follow, and that it is taken account of in the design for the temporary conditions during construction.

Often, a contractor will have requirements for strength gain that are needed to facilitate the construction process—for example, the Øresund Tunnel had a complex set of construction operations in the early days after the concrete had been poured that imposed load on the structure. It was the responsibility of the contractor to assess and design the mix accordingly. The concrete mix was designed to enable the release of the internal formwork supporting the roof slab in three days and to enable the pushing of each tunnel segment along the skidding beams after five days. As a result, some rapid strength gain was needed that meant the employer's 28-day strength requirement was easily met.

Crack width criteria

Reinforced concrete cracks as it flexes and the opposite sides of a section are placed in tension and compression. The tension in one side of the section will cause cracking and transfer of load into the steel reinforcement within the concrete. This is normal behavior and, provided crack widths are controlled to prevent corrosion, is quite acceptable. This type of flexural cracking is not a threat to the overall watertightness of the structure as cracks extend only a limited depth from the surface.

Crack width criteria should be specified for flexural cracking. Design codes typically stipulate maximum crack widths for different exposure conditions. The most severe exposure conditions should be assumed for an immersed tube. Although these criteria are generally applicable to splash zones in the marine environment, the nature of an immersed tube in terms of it being impossible to access the external perimeter for maintenance or repair means the most severe criteria for design should be used. It is therefore recommended that flexural crack widths of 0.2 mm are used for the design of all external faces of the perimeter walls and slabs of the tunnel cross section. This has proved satisfactory over the years, although the tendency of designers and specifiers is to lower this criteria to 0.15 mm. This is somewhat conservative as 0.2 mm has proved satisfactory. Internal faces are subject to less severe conditions and the criteria could be relaxed. However, it should be noted that the presence of deposits within the tunnel that result from exhaust fumes and road salt brought into the tunnel by traffic create a highly corrosive environment and severe conditions should also be assumed. Specifications vary from project to project, but a good guideline would be as follows:

- Internal faces of perimeter structure: 0.25 mm
- Faces of internal walls: 0.25 mm
- External faces of perimeter structure: 0.2 mm

These flexural crack width limits should be used whether the tunnel relies on watertight concrete or has a membrane applied or even if the concrete is all internal, as with a steel shell tunnel.

EARLY AGE CRACK CONTROL

The watertightness of a concrete tunnel will be compromised if the cracking in the concrete penetrates through the full thickness of the walls or slabs in the external perimeter of the tunnel structure. The risk of this occurring should therefore be minimized such that cracks can be reliably eliminated.

Causes of cracking

If a concrete member is placed fully in tension across its complete thickness, the resulting cracking, known as "through-section cracking," would give rise to leakage. Tension in the structure can be caused by

- Restraint to movement of the completed tunnel
- Longitudinal loading from seismic events
- Loading causing longitudinal bending moments in the tunnel
- Restraint to movement between concrete pours during the construction stage

The first three of these causes are dealt with by structural analysis and design. This section deals with the last of these items—the restraint to movement between concrete pours—which happens in the early age of the structure while the concrete is curing. The others are in-service conditions that must be dealt with in the structural design.

Most tunnels are constructed as a series of concrete pours. Typically, the base slab will be constructed first and the walls and roof as a second-stage pour, or as second and third stage pours. Construction joints are therefore required between the stages of concreting, which gives rise to the most common threat to watertight concrete. As new concrete is placed against old, the new concrete hydrates and begins to cure. The hydration process releases heat and the hardening concrete warms up. As the concrete subsequently cools, it tries to contract, but is prevented from doing so by the reinforcement and the friction interface across the construction joint. The result is that tension in the concrete develops and exceeds the available tensile strength of the concrete, causing it to crack. This behavior is well documented in publications such as CIRIA Guide C660, *Early-age thermal crack control in concrete,* and the typical cracking pattern that arises is illustrated in Figure 11.1.

For land structures, this behavior is simply controlled by limiting the crack widths to a magnitude that is similar to that of the flexural cracks, as this gives the necessary control over durability. For water retaining structures, more stringent limits are placed on the crack widths due to the continual presence of water pressure. For immersed tubes, the cracking needs to be eliminated altogether because of the high water pressures on the external face of the tunnel. Reliance cannot be placed on limiting the crack

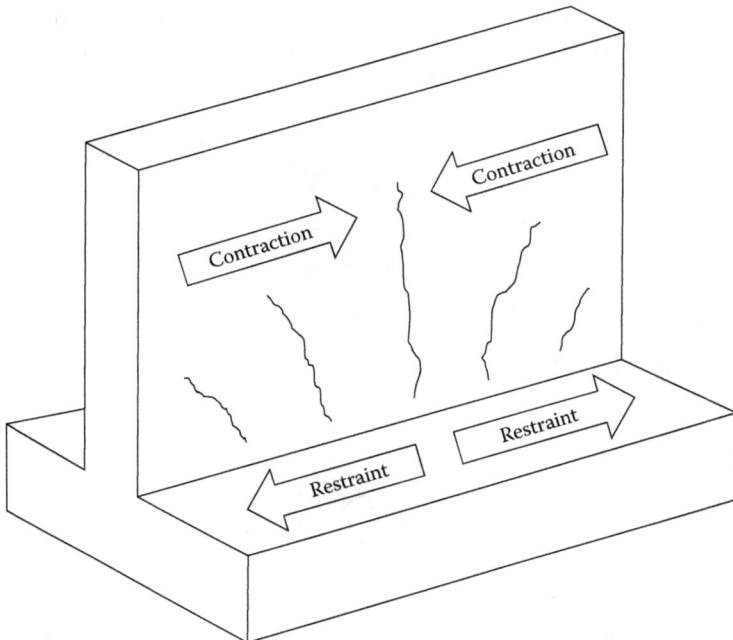

Figure 11.1 Nature of cracking caused by restraint between slab and wall.

width. Although autogenous healing of any cracks will occur to a degree, the water pressures will cause water to penetrate the finest of cracks, with the accompanying maintenance and durability concerns.

The restraint behavior will occur at all construction joints. It is more problematic for thick concrete sections because the temperatures of the concrete will be greater during curing. As a rule of thumb, concrete sections greater than 500 mm thick will need to have some measures applied to eliminate the cracking. As immersed tunnels typically have slab and wall thickness of around 1 m or more, this means that all joints in the external perimeter may give rise to cracking and the curing temperatures and stresses will need to be managed. This applies to joints between base slab and walls, and between the walls and the roof slabs, although more commonly, the walls and roof slabs are cast in one continuous pour to avoid the construction joint. It also applies to any construction joint between the internal walls and the roof slabs. Although the wall is not a watertight component of the tunnel, the restraint it causes may have an effect on the roof slabs that should not be overlooked. If the tunnel elements are monolithic, there will be construction joints between the tunnel sections that run around the full perimeter of the tunnel. This is also a cause of restraint and, potentially, early age cracking, as shown in Figure 11.2.

In addition to restraint caused by external influences, there is also the possibility of internal restraint caused by differential temperatures across the width of a concrete section. If there is a temperature differential through the cross section such that the core is warm but the edges are colder, then the edges will try to contract, but are prevented from doing

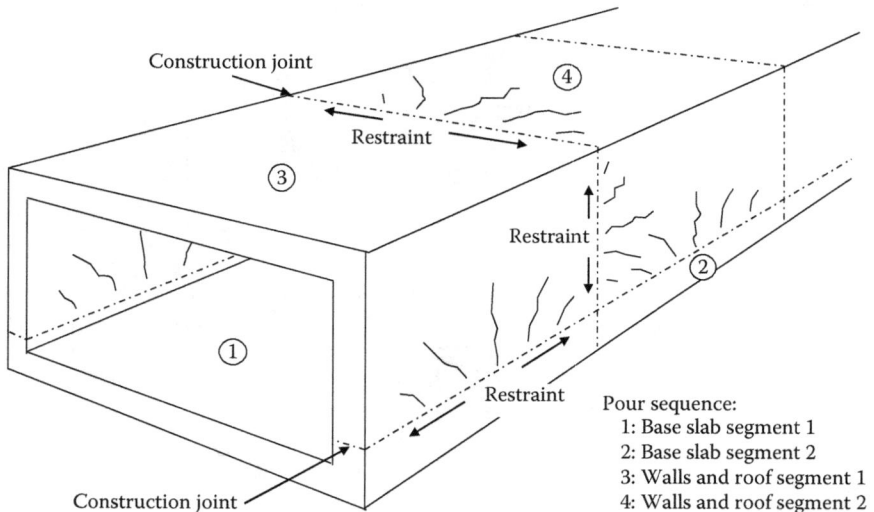

Figure 11.2 Early age crack patterns for monolithic tunnel elements.

so by the core. This could lead to cracking, though it is not necessarily through-section cracking. However, if the situation is reversed and cracking occurs in the core, then there is a risk that the tensile stress causing this cracking can combine with flexural tensile stresses or tensile stresses due to external restraint and culminate in an overall tensile stress level that causes through-section cracking to occur. It is therefore important to control this behavior such that cracking does not occur.

Methods to eliminate cracking

There are a number of measures that a construction planning team can consider to eliminate through-section cracking in a concrete tunnel structure. The first, and most basic and fundamental of these, is in the concrete mix design. The use of replacement materials, such as GGBFS and PFA, will reduce the heat of hydration in the concrete during curing, which will reduce the amount of initial thermal expansion and subsequent contraction that will occur in the first few days of the concrete structure's life. As previously mentioned, the impact of using such replacement materials will be that the concrete has a slower rate of gain in strength, though this may not be a problem, depending on the methods of construction. In addition, designing a mix that has minimal shrinkage in the long term will be of benefit. Selection of aggregates will influence the strength, stiffness, and creep behavior of the concrete and these may need to be considered to optimize the mix design to get the best overall behavior as it cures. Design standards offer some guidance on these properties—for example, Eurocode 2 gives Young's modulus data for concrete with various aggregate types, quoting values for quartzite that must be increased by 20% for basalt and granite aggregates, reduced by 10% for limestone aggregates, and reduced by 30% for sandstone aggregates. This shows the variability that can be obtained and it is recommended that accurate information is obtained from pre-testing of concrete mixes as early as possible in the project cycle.

Concrete cooling

Cooling of concrete during the curing period is a highly effective method of controlling early age cracking. It was developed for immersed tunnels by the Dutch and is now widely used across the world. The method requires that a network of pipes is cast into the reinforced concrete. Shortly after the concrete has been placed, chilled water is pumped through the pipework. As the concrete begins to heat up as a result of the hydration process, the heat is dissipated into the cooling pipes and carried away. As a result, the core temperature of the concrete does not increase to the extent it would otherwise. This limits the thermal expansion that occurs and, in turn, limits subsequent shrinkage as the concrete cools back down.

The tension induced in the structure due to restraint at the construction joints can therefore be limited. With sufficient testing of concrete such that the material properties are fully understood, and analysis of temperature and stress development, the cooling system can be accurately designed to avoid cracking.

The greatest risk of cracking arises at the construction joints between the base slabs and the walls of the concrete tunnel structure. The walls are generally thick enough to generate high temperatures of between 40 and 50°C. Typically, cooling pipes are located in the core of the walls and extend from just above the joint to a point up the wall where temperature gradients have tapered out to a manageable level. They are placed horizontally for the length of the construction pour. For particularly thick walls, a double line of pipes may be required, as shown in Figure 11.3. Pipes may be more closely spaced towards the construction joint where the risk of cracking is greatest. The pipes are connected into a single network and chilled water at around 5°C is pumped into the pipes at the lowest level. As water passes through the pipe network, progressing up the wall, the temperature of the water will increase. It outlets at a high level in the wall, typically at an increased temperature anywhere between 25 and 40°C and is passed back through chiller units to be cooled and passed back into the pipework at the lower level again. A typical cooling pipe layout is shown in Figure 11.3.

Cooling may not be sufficient on its own and insulation mats may be required to control heat radiation from the concrete surface. These can be built into the analysis to determine the insulating properties needed and the durations for which they should be applied. The cooling can be stopped when the temperatures drop sufficiently so that there is no longer a risk of through-cracking. The period for cooling the concrete is typically about

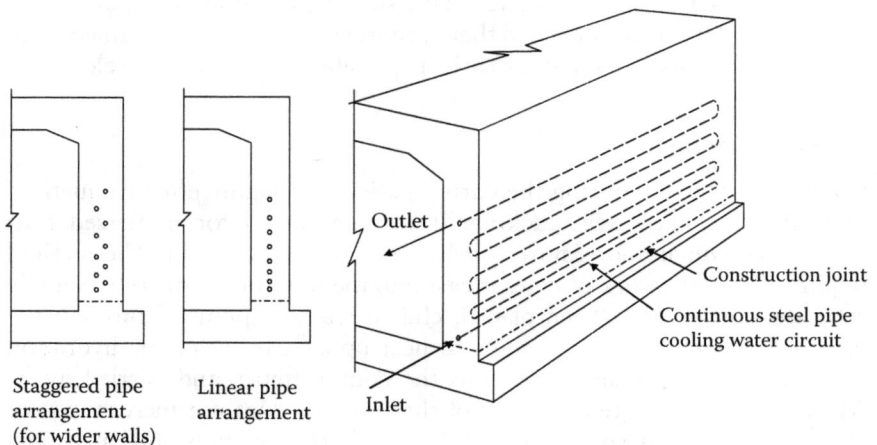

Figure 11.3 Staggered and vertical cooling pipe layout.

three days after the initial placing of the concrete. As an alternative to cooling, it is possible to pre-heat concrete at construction joints to avoid the temperature differential across the joint. However, this technique is less common and infrequently used for immersed tunnels.

Full section casting

A technique that has developed significantly as a result of the Øresund Tunnel is the full section casting technique. This is also described in Chapter 14, and entails pouring the concrete for a single tunnel segment in one continuous operation, starting at the base slab and continuing up the walls and completing the roof slab without any construction joints. This technique has the great merit of avoiding any restraint in the tunnel structure due to construction joints, which are the primary cause of through-section cracking. Cast-in cooling pipes are therefore not required. The method was first developed for small-section utility tunnels, notably in the Netherlands, and later for power station outfall tunnels, such as the ones constructed for the Sizewell power station in the United Kingdom and Pulau Seraya and Tuas Bay cable tunnels in Singapore. For these projects, the small tunnel sections were cast on end, and once the concrete had gained sufficient strength, they were lifted, rotated, and assembled together to form a tunnel element. The short segments were permanently prestressed together.

The Øresund Tunnel was the first large transport tunnel to use this technique. This was not just a scaled-up version of the method used for utility tunnels. It used the same principle for the casting of concrete in one continuous concrete pour, but had to use a different process for assembling tunnel segments into a tunnel element. As the tunnel was some 40 m wide, the weight was such that it was not possible to lift and rotate the tunnel segments. Therefore, they could not be cast on end and had to be cast in the correct orientation and slid together to form a tunnel element. Even with the full section casting method, there needed to be strict control on the ambient temperature conditions, particularly in the Scandinavian winter, when very low air temperatures would cause a problem with thermal gradients through the structural sections. A casting factory was constructed in this instance, with curing tents to protect the tunnel segments from cold temperatures as they were slid out of the factory after casting.

The technique was 100% successful in eliminating early age cracks due to thermal effects and set a new benchmark in the quality of concrete construction that is possible to achieve for immersed tunnels. The technique was subsequently used for the Busan Tunnel in South Korea, although it was adapted there such that the tunnel segments did not need to be slid, as the warmer climate permitted construction in situ with a lighter curing tent. Again, the technique was successful in providing high-quality concrete construction. For long transport tunnels, this method of construction is likely to be the preferred approach, not only for controlling cracking, but also for the many other advantages it offers.

Simulation software

Before sophisticated finite element (FE) analysis techniques were available, the risk of cracking was controlled by limiting temperatures in the concrete. Generally, two criteria were measured:

1. Temperature difference across a construction joint, expressed as D_{ext}. This controls cracking due to restraint arising from the presence of the joint. The limit set for this criterion was commonly 15°.
2. Temperature difference between the core and surface of the concrete, expressed as D_{int}. This controls cracking due to the internal restraint caused by thick sections. The limit set for this criterion was also commonly 15°.

These temperature criteria were quite successful at controlling cracking. However, as it is stresses rather than temperatures that are the characteristic directly affecting cracking, it is becoming more usual to determine the risk based on an evaluation of stress, where the risk of cracking is defined as the applied tensile stress divided by the available tensile strength. This risk of cracking is generally set at 0.5 or 0.7 to ensure there is a margin of safety in the analysis. The Øresund Tunnel utilized both temperature and stress criteria and was a useful exercise in verifying that temperature criteria, while not as precise as stress criteria, were a sound method of controlling the risk of cracking for a simple wall-to-slab interface. However, the temperature difference criteria would not have been sufficient to fully control cracking in the complex full section casting situation.

To accurately predict the behavior of concrete during its early life, it is necessary to undertake some fairly sophisticated analysis of the tunnel structure. This analysis takes account of the precise material properties, the pour sequence, the application of cooling and curing measures, and the ambient conditions, including support to the structure. The ability of this type of analysis to accurately predict the behavior of the concrete and the risks of cracking has improved tremendously in the last 20 years and it is now seen as an essential tool to both the designer and constructor.

Typically, FE software is used, which has the ability to model the different characteristics of concrete in small time increments, from the moment it is placed in the formwork to a point when thermal equilibrium has been reached and shrinkage effects are no longer significant. The development of material properties with time can either be input to the analysis based on precise pretesting, or can be idealized using maturity curves. The properties that are needed to undertake the analysis, and that are generally obtained by testing, are as follows:

- Shrinkage behavior—drying shrinkage and autogenous shrinkage
- Thermal expansion coefficient
- Adiabatic heat development
- Modulus of elasticity
- Tensile strength
- Compressive strength
- Creep coefficient
- Specific heat capacity
- Poisson's ratio (can be taken from design code)
- Thermal conductivity (can be taken from design code)
- Specific heat

Understanding the creep behavior of concrete is very important and the relaxation of young concrete can be taken into account using an aging Maxwell chain with input based on the creep functions of the CEB model code MC 90, ACI codes, and documented relaxation tests that have been carried out on young concrete samples. The temperature dependency of relaxation can be accounted for using an Arrhenius equation.

Other typical input parameters to include in the FE simulations are as follows:

- Support conditions
- Shutter type and thickness
- Cyclical ambient air temperatures
- Insulation mats
- Cooling pipe networks
- Direct solar radiation
- Wind velocity

All of these can be modeled quite precisely and can be varied with time. The analysis model considers a series of time steps that can be as short as one or two hours over critical periods of the curing process, and at each time step, the model will assess the temperature evolution and the consequential effects this has on structure deformation and the stresses within the structure. Stresses need to be calculated in three dimensions as there will be transverse flexural stresses combining with longitudinal stresses arising from restraint due to the construction joints. Graphical output in the form of temperature and stress contour plots and variations of temperature and stress over time at selected points enables the designer or contractor to assess whether the proposed curing regime complies with the relevant specifications. Examples are shown in Figure 11.4.

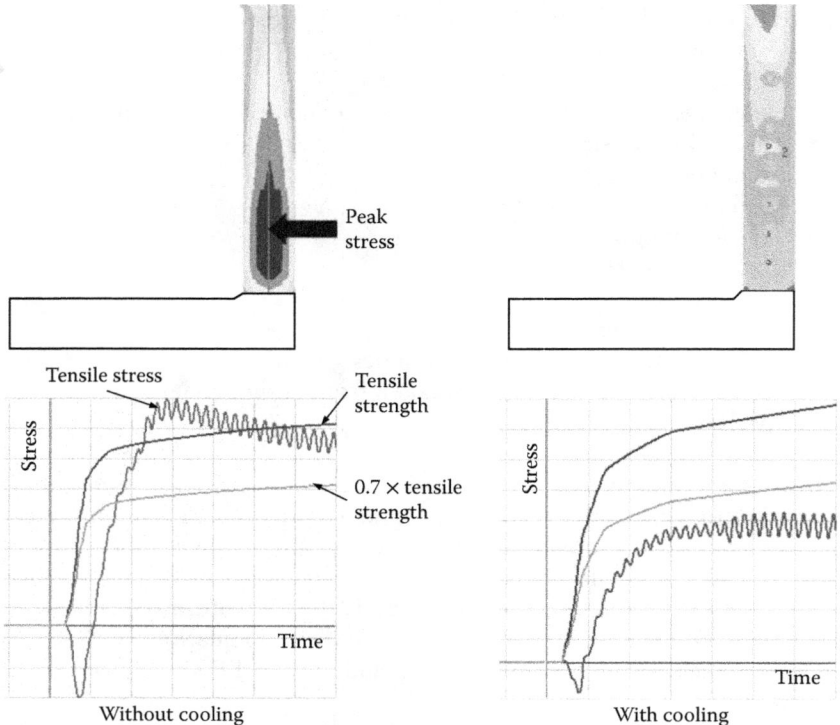

Figure 11.4 Stress contour plot and tensile stress development graphs for case of a tunnel with no cooling compared to case of a tunnel with cooling.

This analysis also provides an accurate measure of the maximum temperatures that may arise. This could be of concern in thick sections where temperatures in excess of 70°C give rise to deleterious effects in concrete, such as delayed etteringite formation, which is an expansive process that can occur in the concrete paste if cured at a high temperature. Cooling and curing can be optimized to avoid any such risks occurring.

MEMBRANES

Waterproofing membranes are often applied to the external structure of immersed tunnel elements to aid watertightness. They come in a variety of materials, but are not generally considered reliable enough to provide the only means of waterproofing. They are applied to concrete tunnels that are formed by tunnel elements constructed as monolithic structures. The risk of concrete cracking for monolithic tunnel elements is higher than for segmental tunnels and, therefore, the membrane provides a backup to the concrete structure, should cracking occur. Segmental concrete tunnels do not require membranes as the concrete can reliably be made watertight through curing and cooling measures.

Membranes are applied to the underside of the tunnel element by casting the concrete onto the membrane. This requires a membrane of steel or hard plastic. A rigid membrane such as this can also be used for the walls. Flexible sheet membranes or spray-applied membranes need to be applied to the roof of the tunnel elements and can be applied to the walls, providing an overlap to the rigid membrane.

Steel membranes are generally in the order of 6–10 mm thick and are provided with shear connections to key them to the concrete. The steel may or may not be considered a structural member, as would be the case with a single shell steel tunnel. Generally, only sufficient shear connections are provided to attach the membrane to the concrete without any composite action resulting. The arrangement of shear connections should ensure that there is no significant separation between the steel and concrete as the concrete shrinks as it cures and ages. There is an advantage to use continuously welded angles as shear connectors as this enables the contact surface between the steel and concrete to be compartmentalized. If there is then a leakage through the membrane, the water is restricted to a defined zone. If shear studs are used, then there are potentially long water paths along the steel/concrete interface. Welding is a critical activity for steel membranes and there should always be a testing regime implemented to check the quality of welding that is achieved. This is particularly important as welding will be carried out under site conditions rather than shop conditions and workmanship may be variable.

To ensure the durability of steel membranes, they can be protected in a number of ways. They can be coated with epoxy paint, typically 500 microns thick. They can also be protected with cathodic protection systems, usually externally mounted sacrificial anodes, or be designed with a thickness that is based on predicted rates of corrosion. It should be recognized that with a sacrificial thickness, it is difficult to accommodate pitting corrosion, which may have a greater rate of corrosion than the general level of surface corrosion. Because of the uncertainty in the rate of pitting corrosion, the use of sacrificial thicknesses is not recommended.

An alternative to the steel membrane is to use a tough plastic membrane such as high density polyethylene (HDPE) sheet. There are a number of products available on the market that can be applied effectively as membranes to the underside and walls of a tunnel element. They are typically in the order of 4–6 mm thick and offer a durable, tough membrane. The plastic sheet is manufactured with a T-shaped rib on one side, which enables a key into the concrete to be achieved and a compartmentalization approach similar to a steel membrane, to limit the spread of water if there are any leaks in the membrane. There are risks with a plastic membrane that need to be dealt with. In particular, there is risk of damage to the membrane from puncture due to handling materials, such as the reinforcement bars, which need to be assembled into a cage on top of the membrane. Although the plastic is relatively robust, it could still be punctured by items being dropped onto it. Therefore, particular thought is needed as to how the reinforcement cage will be supported and whether cages can be pre-assembled to limit the amount of work required above the membrane.

The plastic sheet can be heat welded in situ to produce a continuous membrane for the underside of the tunnel. Pre-formed corner pieces are produced to ensure the correct geometry is achieved. Plastic membranes may need to be sheltered from direct sun in hotter climates as the membrane can distort with large temperature variations, which might pose difficulties in terms of the sheet buckling and reducing or increasing cover depth to the structure reinforcement. This is something that can be managed, and this type of membrane has been used successfully on a number of tunnels, such as the Preveza-Aktio Tunnel in western Greece. The same problem of distortion of the membrane due to temperature variations can also apply to steel membranes.

Membranes for the walls and roof of the tunnel can also take the form of sheet-applied or spray-applied materials. The modern preference is for spray-applied membranes as they are now considered to give better, more reliable adherence to the concrete and better uniformity of the membrane. Whereas, at one time, there was concern about their ability to bridge cracks in the concrete surface, the materials now available have been developed and extensively tested to ensure they are able to do this.

Sheet-type membranes, which are typically bituminous sheets, require a very high standard of workmanship. These are typically applied as a series of overlapping sheets that are torch applied to heat up the bituminous material to glue the sheets to the concrete structure and to seal the overlaps between the sheets. The risk of leakage is potentially greater due to the reliance on the operative's judgement to get an even temperature application to obtain good bond and sealing.

Spray materials come in a variety of materials. These include epoxy and methyl methacrylate resin elastomers, acrylics, water-based products, and polyurethane and polyurea membranes. The elastomer membranes are the most tried and tested on immersed tunnels, but others may be acceptable. Whichever material is selected, the characteristics of the applied membrane that need to be considered are as follows:

- Resistance to water pressures appropriate to the depth of the tunnel
- Ability to bridge cracks greater than the expected flexural crack widths used in the design
- Resistance to abrasion from placing backfill material
- Resistance to damage from minor impacts arising from handling of tunnel elements
- Resistance to aggressive materials, for example, diesel or petrol and chemical attack
- No degradation under UV exposure
- Durability in a sea water environment under exposure to chlorides
- 100% adhesion

Many of the spray-applied materials are also available in a liquid form that can be hand brush applied. Spraying is preferred as a more uniform coating and applied thickness is likely to be achieved. The thickness of the spray membrane is in the order of 2–4 mm, which is necessary to give a degree of robustness. Even so, care needs to be taken to avoid impact damage by plant or materials during construction and inspections are necessary prior to completing the fit out of tunnel elements to pick up any damages so that they can be repaired before immersing the tunnel elements in water. The membranes need to be sufficiently robust to resist abrasion from the backfilling operations, where granular fill will be placed around and above the tunnel elements after they have been immersed. The risk of damage to a membrane during construction of the elements and during transport, immersion, or backfilling is a further reason why membranes are not considered a primary method of providing watertightness, and are applied as a second line of defense to the main concrete structure.

It is difficult to use steel plate or plastic sheet as a waterproofing membrane on the roof. Using it as permanent formwork and placing the concrete under the membrane is difficult, although possible, and it is done during the construction of the internal lining of a steel tunnel element. These are generally circular shells; for a rectangular concrete element, great care is needed to make sure that the large flat roof space is completely filled with concrete and there are no voids caused by air being trapped under the steel plate during the concreting operation. Similarly, it is difficult to fix such a membrane to the roof after it has been poured. For these reasons, the roof of a monolithic concrete tunnel element is normally waterproofed with a flexible membrane applied after the roof has been cast.

Originally, the roof membrane was a fiber-reinforced bituminous membrane. This required careful detailing of the joint with the steel membrane on the walls to make sure the joint between them was watertight. Generally, some form of steel clamping system was provided. This added another detail that required careful attention to avoid any possible water paths through the membrane. In addition, the bituminous membrane was vulnerable to damage both during towing and placing and from the tunnel backfill while that was being placed. Thus, the membrane itself needed protection. This was provided by a layer of concrete protection some 150 mm thick, either cast in situ on the membrane, or formed of precast concrete slabs. Of course, this concrete protection needed to be anchored to the structural roof to ensure that it remained in place, so steel fixing anchors were used. But, these had to penetrate the bituminous membrane to reach the roof slab, so paradoxically, the membrane had to be penetrated again to have its protection attached. Nevertheless, despite these difficulties, many such bituminous membranes were successfully installed on the roofs of immersed tunnels. If a spray-applied membrane is used, then it is relatively easy to overlap it with the steel or plastic membrane. The spray-applied membrane may also need similar concrete protection from the backfilling operations.

CORROSION PREVENTION STRATEGIES

Risks of corrosion exist in many elements of an immersed tunnel structure and a clear corrosion prevention strategy should be developed that is compatible with the chosen design life for the structure. Each component of the structure should be considered in turn and the various risks and potential causes of corrosion considered and mitigated to an appropriate degree. The compatibility of the mitigation methods must also be checked between the different components of the structure—for example, it would be inappropriate to install a cathodic protection system to one item of steelwork and a coating system to another that it is connected to. A holistic approach should be taken.

The risks of corrosion and the types of mitigation measure will vary according to the type of structure and the principle adopted for achieving watertightness. A monolithic concrete tunnel element may have different measures applied, compared to a segmental concrete tunnel element. Both of these will be different from a steel tunnel element. Tables 11.2 through 11.4 provide checklists of items to be considered when developing a comprehensive corrosion strategy for a tunnel. Three tables are presented for three different structural forms, and the various risks and components considered in turn.

Table 11.2 Corrosion strategy checklist for monolithic concrete tunnel elements

| | | Structural form: Monolithic concrete tunnel element | |
| | | Watertightness: Provided by reinforced concrete, external membrane, and joints | |
Structural element	Risk	Mitigation	Criteria and design consideration
Primary concrete structure	Corrosion due to chloride	Cover depth	Design to standard for extreme marine conditions
		w/c ratio	Design to standard, typically <0.4
		Permeability	Design to standard
		Chloride diffusivity	Set target, typically 7×10^{-12} m²/sec
		Flexural crack width	Design to standard, typically <0.25 mm
		Protection to prestress (if used)	Protection caps to anchorage and duct grouting procedures
	Leakage or corrosion due to through-cracking	Cooling or full section casting	Control risk of cracking to eliminate cracks
			Needs simulation, based on risk of cracking <0.7
			Needs accurate concrete properties from testing to undertake a representative simulation
	Carbonation	Coating	Example: silane treatment
		Cover depth	Design to standard, typically 55 mm internal
	ASR	Choice of aggregates	Design to standard
		Concrete mix design	Design to standard
	Delayed ettringite formation (DEF)	Limit maximum temperature during curing	<65°C
			Needs temperature simulation to verify compliance
			Needs accurate concrete properties from testing to undertake representative simulation
	Sulfate attack	Concrete mix	Design to appropriate standard, e.g., BS8500
		Cover depth	Design to standard, typically 75 mm external

(Continued)

Table 11.2 Corrosion strategy checklist for monolithic concrete tunnel elements (Continued)

Structural element	Risk	Mitigation	Criteria and design consideration
Steel membrane	Leakage through welds	Appropriate weld types, specification and testing	Test procedures to recognized standards Undertake 100% testing during fabrication
	Pitting or general corrosion	Coating	Select appropriate coating material and design life Combine with CP
		Sacrificial thickness	Identify minimum thickness and add based on predicted corrosion rates. Not recommended approach for general corrosion.
		Cathodic protection	Initial provision of SACP or ICCP system: Design type and location of anodes Decide on design life Detail for robust connections to structure Include monitoring system Future provision of ICCP system: Specification for cast-in connection plates and electrical continuity of reinforcement Monitoring for initiation of corrosion
	Bacterial corrosion	Coating	Select appropriate material and application coating design life to be as long as possible
		Sacrificial thickness	Determine minimum thickness
	Stray current corrosion	Earthing and bonding	Design to standards
Applied membrane	Leakage at defects	Material selection	Durable, flexible material needed, with good adhesion properties not susceptible to workmanship defects
		Prevent damage during placing	Coating thickness typically >2 mm and material to have high resistance to abrasion and impact and able to bridge cracks
		Thickness testing	For sprayed membranes; to manufacturer's or recognized standards
		Adhesion testing	For sheet membranes; to manufacturer's or recognized standards

Table 11.3 Corrosion strategy checklist for segmental concrete tunnel elements

Structural form: Segmental concrete tunnel element

Watertightness: Provided by reinforced concrete and joints

Structural element	Risk	Mitigation	Criteria and design consideration
Primary concrete structure	Corrosion due to chloride	Cover depth	Design to standard for extreme marine conditions
		w/c ratio	Design to standard, typically <0.4
		Permeability	Design to standard
		Chloride diffusivity	Set target, typically 7×10^{-12} m²/sec
		Flexural crack width	Design to standard, typically <0.25 mm
		Protection to prestress	Protection caps to anchorage and duct grouting procedures
	Leakage or corrosion due to through-cracking	Cooling or full section casting	Control risk of cracking to eliminate cracks
			Needs simulation, based on risk of cracking <0.7
			Needs accurate concrete properties from testing to undertake representative simulation
	Carbonation	Coating	Example: silane treatment
		Cover depth	Design to standard, typically 55 mm internal
	ASR	Choice of aggregates	Design to standard
		Concrete mix design	Design to standard
	DEF	Limit maximum temperature during curing	<65°C
			Needs temperature simulation to verify compliance
			Needs accurate concrete properties from testing to undertake representative simulation
	Sulfate attack	Concrete mix	Design to appropriate standard, e.g., BS8500
		Cover depth	Design to standard, typically 75 mm external

Table 11.4 Corrosion strategy checklist for steel tunnel elements

Structural element	Risk	Mitigation	Structural form: Steel tunnel element
			Watertightness: Provided by steel shell and joints
			Criteria and design consideration
Steel shell	Leakage through welds	Appropriate weld types, specification and testing	Test procedures to recognized standards Undertake 100% testing during fabrication
	Pitting or general corrosion	Coating	Select appropriate coating material and design life Combine with CP
		Sacrificial thickness	Identify minimum thickness and add based on predicted corrosion rates
		Cathodic protection	Initial provision of SACP or ICCP system: Design type and location of anodes Decide on design life Detail for robust connections to structure Include monitoring system Future provision of ICCP system: Specification for cast-in connection plates and electrical continuity of reinforcement Monitoring for initiation of corrosion
	Bacterial corrosion	Coating	Select appropriate material and application coating design life to be as long as possible
		Sacrificial thickness	Determine minimum thickness
		Earthing and bonding	Design to standards
	Stray current corrosion	Sacrificial thickness	Determine minimum thickness

Internal concrete structure	Corrosion due to chloride	Cover depth	Design to standard, typically 55 mm internal
		w/c ratio	Design to standard, typically 0.45
		Permeability	Design to standard
		Chloride diffusivity	Set target, typically 7×10^{-12} m^2/sec
		Flexural crack width	Design to standard, typically 55 mm internal
		Early age and shrinkage cracking	Limit widths in accordance with design codes, typically <0.25 mm
	Carbonation	Coating	Example: silane treatment
		Cover depth	Design to standard, typically 55 mm internal
	ASR	Choice of aggregates	Design to standard
		Concrete mix design	Design to standard
	Sulfate attack	Concrete mix	Design to appropriate standard, e.g., BS8500
		Cover depth	Design to standard, typically 55 mm internal

CATHODIC PROTECTION

Cathodic protection (CP) systems are widely used on immersed tunnels to protect the main reinforcement within concrete tunnels, the steel structure of steel shell tunnels and steel waterproofing membranes, and the steel components associated with immersion joints. Eurocodes cover the general approach to designing cathodic protection systems for structures in seawater and cathodic protection of reinforcement in concrete. The Eurocode for cathodic protection of submarine pipelines is also a useful reference. Additional guidance is available from organizations such as Det Norske Veritas (DNV) and Norsk Sokkels Konkuranseposisjon (NORSOK).

To explain why this type of protection system is needed, it is necessary to understand the causes of corrosion, the likely rates of corrosion, and the limitations of other protection measures. Two different approaches to applying CP are possible and these are described below.

Causes of corrosion

The various causes of corrosion either to the structural reinforcement, steel shells or membranes, or embedded steel components are listed in Tables 11.2 through 11.4. A potential mitigation measure for each of these is the application of a cathodic protection system. This may take a number of forms, but in summary, such a system can help protect against the following types of degradation:

- Corrosion due to ingress of chlorides through the external concrete cover
- Direct attack of exposed steel components fixed in/on the concrete
- Corrosion due to corrosion cells established due to accidental electrical contact between embedded items and structural reinforcement
- Corrosion due to ingress of chlorides (from de-icing salts) of reinforcement on the inside faces of the tunnel
- Carbonation of internal concrete, leading to loss of alkalinity around reinforcement in the internal faces of the tunnel structure
- Atmospheric corrosion of exposed internal steel components

Left unchecked, these processes can lead to the formation of corrosion product, cracking of concrete, loss of material, and associated loss of structural capacity or watertightness.

Rates of corrosion

Where it is important to protect against the loss of material thickness—for example, for watertight steel shells or for immersion joint end frames—it is necessary to understand the potential rate of corrosion so that the consequences to the structure are understood and a protection system can be designed. The corrosion rate of steel in marine environments is remarkably steady. Initially, the corrosion rate of exposed steel in seawater can be rapid, at approximately 0.33 mm per year, but it would then be expected to reduce over a period of months to a steady rate of approximately 0.11 mm per year after a period of around 2 years. These figures are based on various pieces of research and give a good estimate in steady state conditions.

Corrosion rates of carbon steel buried in the seabed mud are lower but more variable than in seawater. Rates of attack can be accelerated compared to seawater due to either sulfate reducing bacteria (SRB) effects or differential oxygenation. SRB are commonly found in harbor mud environments and give rise to aggressive conditions for carbon steel due to the production of hydrogen sulfide, which is mildly corrosive to carbon steel. The rate of corrosion is proportional to the metabolic activity of the SRB which, in turn, is determined by the available nutrients in the local environment. The essential nutrients comprise sulfate ions, available in abundance from seawater and a carbon source, typically contributed by land runoff water, sewage, and industrial effluent.

SRB may give rise to localized shallow pitting corrosion rates in the order of 1 mm per year or up to 2 mm per year in extreme cases. In unpolluted harbors, the SRB activity will be lower, and corrosion rates in the order of 0.5 mm per year would be more typical. It is often the case that the highest risk of SRB activity will be in the top 1–2 m of silt in the harbor, due to pollution, but of course, this layer is removed by dredging over a large area, and so the risks, once the tunnel has been backfilled, are generally quite small. For the design of a cathodic protection system, a suitable average rate of corrosion to assume, for the structure buried in granular backfill over a period of, say, 15–20 years, would be in the order of 0.08 mm per year.

Protection methods

Options for protecting steelwork against corrosion include the following:

Coatings: Unless they can be re-applied, these are only effective for a proportion of the design life of the tunnel. Typically, a high build epoxy system could be relied on for up to 40 years in a buried submerged marine environment. However, accelerated corrosion may occur at defects in the coating system. For buried components with no future maintenance access, it is considered insufficient to rely on a coating system alone, given that the usual design life is around 100 years. Epoxy coated reinforcement is not widely used in immersed tunnels. Worldwide, there is considerable variety of opinion on the cost-effectiveness of its use.

Sacrificial thickness: Although some provision may be made for SRB attacks, generally due to the nature of localized pitting corrosion and crevice corrosion, reliance on the sacrificial thickness of steel is not an appropriate solution. It is also an expensive approach to apply a general increase in thickness when corrosion will take place at very localized spots on the structure.

Corrosion inhibitors: For exposed components, these are only effective if they can be replenished and so may only be suitable for a small number of items in the tunnel. Some chemical additives are used in concrete mixes in parts of the world—for example, calcium nitrate is used in the United States as an inhibitor to prevent chloride-initiated corrosion of reinforcement. This can be expensive and the proportions required can be highly sensitive and make matters worse, if not applied correctly.

Change of materials: Use of higher-grade steels than carbon steel may offer a solution, but this comes at a price. Stainless steel reinforcement would be cost prohibitive, although a reasonable case might be made for immersion joint end frame components. However, this approach is generally only taken for small embedded components.

Improvements to concrete mix: Achieving high-density, low-permeability concrete with the use of cement replacement materials will have a major benefit in slowing the diffusion of chlorides through the concrete and, hence, delaying the onset of corrosion. They will not, however, prevent the onset of corrosion altogether and this does not aid in preventing corrosion that might occur in cracked zones or as the result of defects.

As can be seen, the conventional methods of corrosion protection all have their drawbacks, and this leaves only one real alternative—cathodic protection. Such systems can either be passive or active and can be installed from day one of a tunnel's life for components directly exposed to chloride or, for structural reinforcement, at a point determined in the future by monitoring to assess when corrosion is likely to be initiated.

Sacrificial anode cathodic protection (SACP)

The principle of a sacrificial anode system is that an anode is connected to the item requiring protection and that the relative potential between the two items creates a corrosion cell that corrodes the anode preferentially to the item to be protected. The potential difference is created using dissimilar metals. Therefore, anodes are generally made of zinc or aluminum to generate the corrosion cell. Sacrificial anode systems can be used to protect embedded components, the structural reinforcement within the concrete, or both. Separate systems are preferred for reinforcement and embedded items as the relative current demand will be high for the reinforcement compared to the embedded item. This means that unduly large anodes would be needed to give adequate protection to the embedded item. Ideally, embedded items, such as the immersion joint end frames, should be isolated from the structural reinforcement to avoid this effect. The immersed tunnel should also be isolated from the cut-and-cover tunnels, which are likely to have different surrounding environments, usually less severe, and routes to earth that may act as a current drain for the system protecting the immersed section.

Anodes are mounted on the exterior of the tunnel structure and are connected to the reinforcement within the tunnel, or embedded steel components (Figure 11.5). If protecting the reinforcement, the reinforcement cages must be in good electrical continuity throughout the structure. This may require welding of reinforcing bars at particular intervals along the tunnel and at a number of points around the transverse cross section of the tunnel. This might be coupled with the double tying of the reinforcement bar intersections with steel tie wire to get the required electrical connectivity. If the tunnel is a segmental construction, some continuity cabling will need to be installed across the segment joints inside the tunnel. This will comprise outlet plates embedded in the concrete that are connected to the reinforcement with copper cables with brazed connections. Outlet plates are then connected across the joints by external copper cables secured to the outlet plates with clamps. Depending on the design, electrical continuity cabling may also be required across the immersion joints between the tunnel elements.

Once installed, the anodes will be effective immediately. An initial installation has an advantage over a system fitted in the future, when corrosion is perceived to be initiating. This is because the steel reinforcement embedded in the concrete is already in a passive condition and only a low level of current is needed to maintain this condition. If corrosion has been

Figure 11.5 SACP anodes mounted on the exterior of the Bjørvika Tunnel.

initiated, then the steel is already de-passivated and a greater level of current is required to re-passivate the steel in order to protect it from corrosion.

Impressed current cathodic protection (ICCP)

The principle of an impressed current system is that a DC electric current is passed through a network of anodes that surround the structure that is to be protected. The current travels into the structure via the submerged backfill and returns to the original supply to complete the circuit. The structure, therefore, becomes the cathode of an electrochemical cell and is protected from corrosion.

An impressed current system can be applied to the tunnel structure at any time during its life, but if this is envisaged, then certain provisions should be made from the outset. The same provisions for electrical continuity are necessary as for a galvanic SACP system. A groundbed anode system, usually comprising mixed metal oxide–coated titanium anodes, can be installed into the sea bed around the tunnel and the system is powered by a number of transformer rectifiers placed within the tunnel. The groundbed anodes can be installed at the time of construction or at a later date, but as it is difficult to remove extensive fill from around the tunnel, consideration should be given to their installation at the time of construction. However, the advantage of installing the ground bed anodes at the time when they are needed is that the groundbed anode system can be designed to meet the precise need at the time, taking into account the actual rates of chloride ingress and corrosion that have been monitored, and the remaining design life of the structure. If designed and installed at the outset, it is likely to be designed on a somewhat conservative basis and be an over-sized and more expensive system.

The preparation of tunnel elements for the introduction of ICCP at a future date is a relatively common approach, where the tunnel owner decides the structure will be monitored to assess the risk of corrosion initiation throughout its life. When the risk reaches a certain threshold, usually determined by the extent of chloride diffusion through the concrete cover zone, then the system can be activated. This has a lower initial cost, but requires regular monitoring and interpretation of monitoring results by a specialist, and then a subsequent retro-fitting of the final components of the system to get it up and running.

Monitoring

Whatever cathodic protection system is installed, it is a good idea to carry out monitoring to check the effectiveness of the system. If a SACP system is introduced, it will be designed for the life of the structure, no replacement of anodes over that period, and little or no maintenance. However, it would be prudent to monitor the rate for degradation of the anodes to check the system is behaving as expected. If anodes are being consumed at a faster rate than expected, there may be a need to replace them if possible or supplement the system.

For an ICCP system, monitoring of the condition of the structure is required to check the progression of the corrosion front through the concrete. This will enable a decision to be made as to when to activate the system. Embedded ladder-type probes are available to do this. A number of different probes are required, however, and all monitoring systems require auxiliary and references electrodes, an electrical contact to the rebar, an estimate of the surface area of rebar being polarized, and a corrosion rate monitor.

Monitoring equipment should be installed at a number of representative locations along the length of the tunnel. One set of equipment per tunnel element is desirable to monitor the main tunnel structure, but on longer tunnels, this spacing may increase. For systems applied to specific components, such as immersion joint end frames, a representative number of the frames should be monitored. Ideally, they would all be monitored as it is costly to retrofit monitoring devices, should the CP systems not perform as intended.

Monitoring systems can be remote and fed back to operations and maintenance buildings, or be local within the tunnel and readings taken during planned maintenance activities when access into the tunnel is available. Monitoring frequency is typically quarterly to start with, but when steady state conditions are reached, then this period could be extended to annually.

Interference and interface with earthing and bonding strategy

For any CP system, an assessment of potential electrical interference is needed. This could occur between different isolated parts of the tunnel, for example between the cut-and-cover and immersed sections. It is possible for CP systems to induce stray currents and this needs to be assessed and guarded against. Similarly, a CP system may be interfered with by other CP systems operating in the area.

For rail tunnels and for the mechanical and electrical installations within a tunnel, there will be a number of systems of earthing and bonding of the installed equipment. Any cathodic protection system needs to be compatible with this and separated from earthing rods that may act as current drains to the CP system.

DURABILITY OF WATERTIGHT SEALS

A question often arises as to the durability of the rubber seals used in the immersion joints to provide the watertightness of the tunnel. This has already been touched upon in Chapter 10. The Gina and Omega seals are both formed with SBR rubber, which is a natural rubber product and is highly durable. The material can degrade under UV exposure and, to some degree, due to oxygen presence. The environment of the seals means that neither of these causes are of great concern. Suppliers of the seals can provide extensive test data, including accelerated aging tests, and these are generally found to be satisfactory. The current types of seals have been used for over 50 years and the authors are not aware of any incidences of material degradation. The creep behavior of the rubber materials is well documented and this is allowed for in the design process. Occasionally, production defects occur, but these are usually uncovered during the quality control checking processes.

Other important seals are the internal seals in the tunnel that are installed in the structure joints, the ballast concrete, and the road surfacing. Suppliers will not guarantee long life for such seals, and so, these have to be expected to be replaced during the life of the tunnel. This is not easy beneath the ballast concrete, so the selection of those seals should be made, assuming they will not be replaced, even if their design life is not underwritten by a supplier. An important consideration is the resistance of the materials to corrosive products that may be spilt in the tunnel and find their way to the joint seals. Some high-grade materials, such as NBR rubber, are available that are resistant to acids and alkalis, petrol-based products, and solvents. The cost of these seals is relatively low compared to the remainder of the works, and so, it is false economy to skimp on the specification of these items.

INTERNAL COATINGS

The internal coating of the tunnel performs several functions:

- Protection of the internal tunnel walls. If they are concrete, it protects against deterioration from carbonation and saline ingress due to road salts; if they are steel, then it prevents corrosion due to moisture in the atmosphere and from the aggressive environment.
- Providing a reflective coating to assist the lighting conditions in the tunnel.
- Facilitates cleaning of the tunnel walls.
- Cosmetic, to provide the driver with a suitable driving environment.

The choice of coating may depend on the construction of the tunnel and the options include just painting the interior surfaces, facing the walls with blockwork or tiles, or fixing a secondary cladding system.

In a concrete tunnel, the calcium hydroxide formed during the hydration of cement ensures that the concrete is alkaline with a high pH value. The pH is normally above 12, which is sufficient to inhibit corrosion to the reinforcement. Carbonation is the absorption of carbon dioxide by the concrete. This carbon dioxide reacts with the calcium hydroxide in the concrete and reduces its pH value. This reduction in the alkalinity of the concrete lowers the protection it gives to the reinforcement and can lead to corrosion and consequent spalling and delamination, and a reduction in the strength of the structure. In tunnels, particularly road tunnels, there is a high level of carbon dioxide from the exhaust fumes of the vehicles. This increases the risk of carbonation deterioration in the structure compared with structures in the open air.

A similar reduction in alkalinity is caused by the absorption of salts by the concrete, particularly sodium chloride, which is used as a de-icing agent on roads and is carried into tunnels by the wheels and spray of vehicles. In fact, the penetration of concrete by chloride ions is the main cause of corrosion in concrete highway structures.

One of the best defenses against chloride attack is dense impermeable concrete. The concrete used in immersed tunnels is often relatively impermeable because of the low water–cement ratio and the cement replacement used to reduce the heat of hydration of the setting concrete as part of the measures against early thermal cracking. However, this is not sufficient on its own. Roadside concrete structures are often protected against chloride attack by coatings such as silanes. These coatings are absorbed into the concrete and act as water repellents at an ionic level, preventing water and salts from entering and deteriorating the concrete.

Although such coatings will prevent chloride attacks, they are not so effective at preventing carbonation because although the carbonization process requires the presence of water, there is the possibility of water penetrating the concrete from the outside of the tunnel. The simplest method of protecting the tunnel walls is to paint them, and there are many coatings on the market. The properties required from a concrete paint coating are as follows:

- Good adherence to the substrate
- Protection of the surface against chloride attack and carbonization
- Provide the visual characteristics, reflectance, and color, required as part of the overall tunnel lighting design
- Resistance to the abrasion and chemicals used during the regular cleaning of the tunnel

In some early tunnels, it was thought desirable to both impregnate the surface with a silane type coating and then paint the walls. The silane, however, reduced the adhesion of the paint and the practice was discontinued.

Regular cleaning of the tunnel is an essential aspect of the tunnel maintenance. Tunnels are dusty and dirty places. There is dust from the road, which will contain road salts and other corrosive agents, and dust particulates from vehicle emissions. These build up on the internal surfaces of the tunnel and will eventually cause deterioration of the structure. Before this happens, they will reduce the effectiveness of the lighting system. This is designed on the basis of certain color and reflectance parameters, which will reduce as the tunnel becomes dirtier. Thus, the average road tunnel requires cleaning at intervals of around every three months. The abrasive nature of the mechanical cleaning equipment, such as brushes and high- pressure water jets, requires a robustness from the paint system, which is an important factor in choosing the appropriate system.

The tunnel walls can also be protected by blockwork or tiles, and many early road tunnels were faced with ceramic tiles. These are, however, relatively costly nowadays and the practice has largely been discontinued, although, in some tunnels, they have been retained for decorative purposes. They are still used in tunnels in the United States.

Chapter 12

Foundations

Immersed tunnels are frequently located in poor ground conditions. River beds, estuaries, and marine environments naturally feature sands, silts, alluvium, and muds, often at depth, where ancient watercourses have existed or deposition has taken place over millions of years. It is natural to expect that such conditions may require a significant foundation solution, but this is not the case with an immersed tunnel; in fact, they are an ideal form of construction for this environment as the loading to the soil is relatively light and often the foundation solution can be quite simple. This chapter explores the different foundation solutions that can be used and features that are particular to the immersed tube method.

The foundation of an immersed tunnel can be considered to be made up of two parts:

1. The foundation layer placed on the dredged surface immediately beneath the tunnel structure
2. The deep foundations in the substrata below the level of the dredged trench

The foundation layer beneath the tunnel elements can be formed in a number of ways, but the choice is essentially between a sand foundation, a gravel bed, or a grouted foundation. There are also combinations of gravel and grout that can be considered. Where additional foundations or ground improvement measures are required in the substrata below the foundation, there are many techniques that can be used, including soil mixing, sand compaction piles, stone columns, conventional piles, and material replacement techniques.

FOUNDATION LAYER

The tolerances achievable with dredging equipment means it is not possible to place a tunnel directly onto a dredged surface. Backhoe dredgers are probably the most accurate method of dredging, achieving a tolerance of 200–300 mm. Cutter suction dredgers are less accurate and can deliver tolerances in the region of 500 mm. This depends on the material being dredged, with greater accuracy coming in hard clay, and accuracy decreasing as the particle size of the dredged material increases. The irregularities in the finished surface would not enable the correct alignment of the tunnel to be achieved and may cause undesirable loading conditions. Therefore, a thin layer of bedding material needs to be placed between the underside of the tunnel and the top of the dredged trench. There are two fundamental approaches that may be followed to form this bedding layer:

1. Place the tunnel elements in the trench onto temporary supports, underfill the space between the tunnel elements and the surface of the trench, and release the temporary support.
2. Lay a close tolerance foundation layer at the base of the tunnel trench that the tunnel elements can be placed directly onto.

These can be applied to any form of immersed tunnel, whether steel or concrete, and the choice of method will depend on a number of factors. For example, water depth will influence the choice of method as it may affect what operations are able to be undertaken by divers and, hence, point to a particular solution. Material availability is another factor in the choice and if costs of material import are high, this can dictate the use of locally sourced materials. There are minor differences between the sand and gravel foundation systems in terms of the design, mostly in the settlement behavior, but these are generally not significant enough to drive the choice of foundation. Liquefaction potential is often a driving force behind the decision between foundation types. Sand gradings have limited resistance to liquefaction, so grouted or gravel foundations are more highly favored in seismic areas.

Both forms of foundation require marine plant and equipment that will need to be developed for the project. These methods—and the equipment required—are described further below, with the advantages and drawbacks of each solution identified.

SAND FOUNDATION LAYERS

There are two principal methods of forming a sand foundation: by sand jetting or by sand flow. Sand jetting is carried out from the side of the element by equipment moving along the side of the tunnel. Sand flow is carried out by pumping a sand/water mix through pipework to a number of outlet points in the underside of the tunnel. Both have the objective of completely underfilling the space between the tunnel and the dredged trench bottom. Sand jetting has been used successfully on tunnels in Europe and Asia, but it is less used now as the level of control achieved with sand flow is greater. Nevertheless, both techniques are described as they are both quite valid construction methods. The sequence of placing the elements and the sand foundation is similar, whichever type of method is used to place the sand.

Sand foundations are a very effective solution and can be formed with reliable, proven construction methods and are easily accounted for in design. However, there are some particular issues to consider, especially if weighing the advantages and disadvantages of a sand foundation compared to a gravel bed foundation.

Sand jetting method

With this technique, the tunnel element is temporarily supported above the bed of the trench. This can be achieved in a variety of ways, but the most typical is to support the primary end of the element on the previous element, with the secondary end being supported some 600–1000 mm above the bed on two hydraulic rams, one on each side of the element. The space left for the sand foundation is governed by the space needed to operate the sand jetting equipment under the element. The sand is then injected into the space under the element. Original sand jetting equipment is shown diagrammatically in Figure 12.1. It was developed in the 1930s by Christiani and Nielsen, the Danish contractor that was instrumental in the development of concrete immersed tunnels.

The sand jetting equipment is mounted on the tunnel roof and can travel longitudinally along the tunnel. It is fitted with a delivery pipe that runs down the outside of the tunnel wall. At the bottom of the wall is

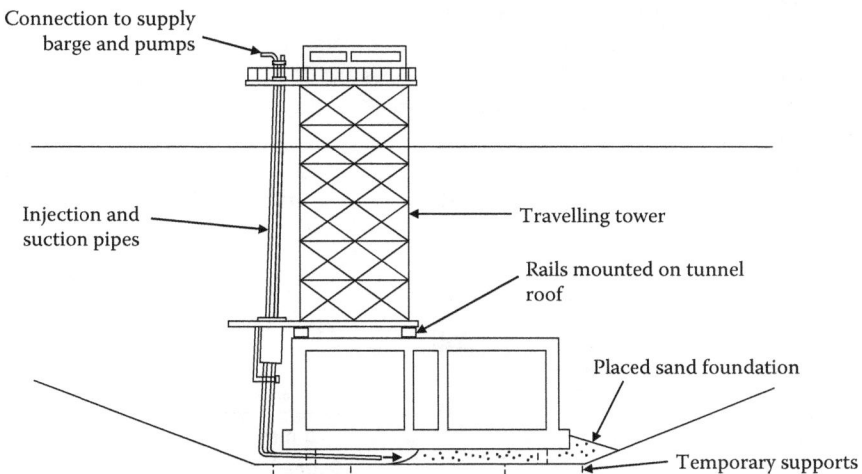

Figure 12.1 Sand jetting equipment.

an elbow that turns the pipe through 90° to run under the tunnel. The elbow swivels so that the end of the pipe can sweep a series of arcs under the tunnel. The delivery pipe actually incorporates three pipes: a central delivery pipe and two suction pipes. The sand/water mix is pumped into the space under the tunnel through the delivery pipe. The concentration of sand in the delivery mixture is normally in the range of 10–20% by volume. The suction pipes remove an amount of water equal to the water in the sand/water mixture. With this arrangement, a regular flow pattern is established. The sand does not swirl around, nor is it washed away, but is deposited as a solid wall of sand filling the space under the tunnel element.

The return water is monitored for its sand content. When the sand in the return water is similar to that of the mixture in the delivery pipe, the space is full and the pipe can be moved to its next position. Injection may be completed from either side of the tunnel or from a single side, depending on the width of the tunnel structure. As a strip is completed, the gantry moves on and the injection pipes are moved back to their start position. Monitoring the jacks at the secondary end also indicates how the space underneath the element is being filled as the load on the jacks reduces. Typically, around 2000 m³ per day of sand foundation can be placed with this method. The jetting pattern can be adjusted with this method if the grain size, or gap height, is different from expectations.

This method produces a uniform sand foundation under the tunnel element. A relatively coarse sand is used with a uniform grading with a uniformity coefficient between 2 and 4. It should be clean and free from silt or clay. While crushed sands have been used, natural river or marine sands give better results. The mean grain size (d50) is between 500 μm and 2 mm. The sand/water mix is approximately 10% water, although this can vary up to 20% if required. From experience, these properties result in a mixture that flows well and results in a uniform layer of sand to support the tunnel. A typical grading for the jetted sand is described in Table 12.1.

The placed sand can be described as low to medium relative density (0.3–0.6). The porosity of the placed sand is around 40%, which corresponds to a void ratio (volume of voids divided by volume of solids) of 65–70%.

Table 12.1 Sand grading for jetting

Percentage passing	Sieve size
d_{90}	<5 mm
d_{50}	500 μm–2 mm
d_{10}	>150 μm

When the foundation is complete and the temporary jacks are released, the element settles on to its sand foundation. The sand has a low relative density and a low initial stress. Field measurements (Romhild and Rasmussen) have shown larger deformations than in laboratory tests. As a general rule of thumb, the settlement in the sand foundation during this operation, is of the order of 10–20 mm, and this is allowed for by setting the element correspondingly high on its temporary jacks.

The sand jetting apparatus can also be modified to extract any silt that might build up under the element before the foundation is placed. By injecting water, rather than a sand/water mixture, the silt is disturbed and put into suspension, which is then extracted by the return pipes. Although this procedure is possible, it is better not to rely on it and to clean the trench before the element is immersed. Nevertheless, it is a possibility, should something untoward occur.

Sand flow method

With the advent of the tunnel building program in the Netherlands in the 1970s, it became apparent that a different technique was required. The depths of the tunnels were increasing and the waterways were busier than those of previous immersed tunnels. The sand jetting equipment riding on the tunnel roof and traversing the river presented an unacceptable obstacle to navigation. Additionally, the coarse sand needed was relatively expensive. Thus, a new method was developed in which the sand/water mixture was pumped directly into the space under the tunnel through holes in the tunnel floor. This was called the sand flow method and has, subsequently, been widely adopted as the usual method of placing tunnel foundations.

Initially, the idea was to fit a revolving nozzle under the tunnel that would distribute the sand/water mixture. However, early tests quickly demonstrated that the revolving nozzle was unnecessary and the jetting operation itself resulted in the sand filling the space under the element. The deposition mechanism is that when the mixture is first pumped, the sand settles out when the velocity of the pumped water can no longer transport the sand. So initially, a circular-shaped wall of sand is formed around the discharge point. As the pumping continues, the mixture breaks through this wall at its weakest point and deposits more sand on its outside slope. Eventually, this space also fills with sand and the frictional resistance to the flow builds up, resulting in the flow taking an easier path and breaking through an adjacent point on the circular wall. This revolving process repeats, the diameter of the sand deposit increases, and a pancake is formed, filling the space under the element. As the pancake nears completion, the flow rate is reduced so that the canal carrying the mixture is gradually filled up. By positioning discharge points in the tunnel floor at appropriate spacings, the whole area under the element can be filled, as shown in Figure 12.2.

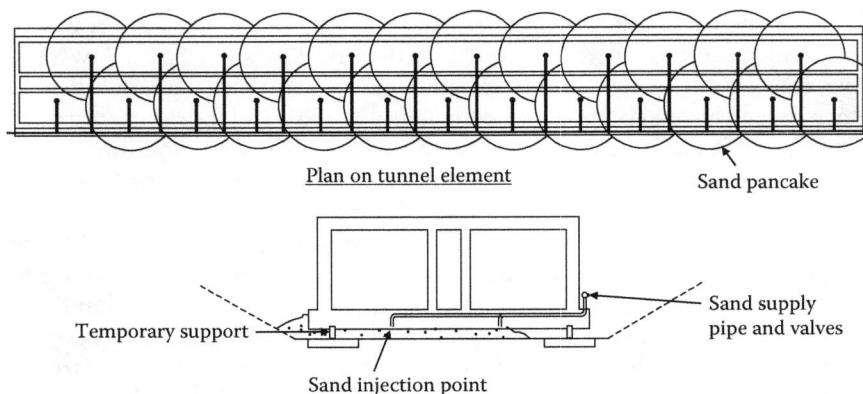

Figure 12.2 Sand flow arrangement.

Initially, the sand was pumped from inside the tunnel through holes in the tunnel floor. However, this required valves through the base slab, which must be effectively sealed afterwards so that the tunnel is watertight. Valve pipework was capped with a welded steel plate and encased in the ballast concrete, but still carried a risk of leakage or even flooding during the temporary condition, if some valves were to fail. To avoid this potential weakness, the delivery pipes are now cast into the tunnel floor and joined to external pipes carried on toes outside the walls. In this way, there is no direct connection between the inside and outside of the tunnel that could result in leakage. The external delivery pipes are routed along the toes to a pumping plant near the shore so there is no additional floating plant in the river, but the method does introduce an additional risk because divers are needed to couple the external pipes and adjust the valves. However, this is not normally a very hazardous operation. Pipework can be seen in Figure 12.3.

The typical diameter of a completed pancake at an injection point is around 12 m. The diameter depends on the size of the discharge hole and the pressure of the discharge. There is a limit to this process, as during the discharge, an upward pressure is applied to the underside of the tunnel, effectively trying to lift it. This requires more temporary ballast inside the element to hold it down. A balance has to be struck between the upward force, which is normally limited to about 3000 kN, and the number of discharge ports. As a result, the spacings of the discharge ports are typically between 10–15 m. The number of lines of injection points along a tunnel element and the interval for injection along each line need to be set to achieve full coverage of the base area. If the sand grading and proposed spacing of the discharge points have not been proved previously, it is common for tank tests to be undertaken to verify that the pattern of the discharge points results in a satisfactory foundation with the sand proposed.

Figure 12.3 Pipework for sand flow. (Courtesy of NRA/DirectRoute.)

Table 12.2 Sand grading for sandflow

Percentage passing	Sieve size
d_{90}	<900 μm
d_{50}	150–500 μm
d_{10}	>100 μm

As well as reducing interference with river traffic, the sand flow method can also be used with finer sand than the sand jetting method. A coarse sand would result in a smaller-diameter pancake and, therefore, require more discharge points. In principle, the method can be used with any type of sand, but a typical grading for use with the sand flow method is shown in Table 12.2.

The concentrations of the sand–water mixture are similar to those for the sand jetting method and the properties of the placed sand layer are similar in both methods. Thus, similar settlements are experienced as the tunnel element is lowered on to the sand. Sand flow is a slightly faster operation, with placing rates of 3000 m³ per day compared to 2000 m³ for the sand jetting method.

Sequence of construction

The sequence of construction for a sand foundation is as follows:

1. Dredge trench.
2. Dredge pockets in the trench base for temporary foundation pads for all or a number of tunnel elements.

3. Place temporary support foundation pads in trench.
4. Immerse tunnel element no. 1.
5. Mobilize sand mixing/pumping barge (if not pumping from shore).
6. Progress sand foundation along the length of the newly placed element, stopping short of the immersion joint.
7. Place tunnel element no. 2.
8. Progress sand foundation from end of tunnel element no. 1 and along tunnel element no. 2.
9. Release temporary supports from tunnel element no. 1.
10. Place locking fill to sides of tunnel element no. 1.
11. Repeat process from 7 through 10 for remaining elements until tunnel is complete.

Divers can monitor when the sand flows beyond the edge of the tunnel base slab to ensure that the complete underfilling of the tunnel element is achieved. The locking fill must be placed soon afterwards if there is any risk of erosion of the sand layer due to scour. As mentioned in Chapter 9, it is necessary to monitor the support reaction during underfilling as this exerts an upward force on the element and reduces the reaction. With a nose/chin arrangement, a jack or load cell can be built into the support system to monitor the reaction.

Risk of sedimentation

A prerequisite of the sand flow and sand jetting methods is a clean unobstructed space between the tunnel underside and the top surface of the dredged trench. One potential danger of this method is that if the watercourse is carrying a high level of sediment, it can be deposited and settle in the base of the trench beneath the tunnel elements. This, in itself, will not prevent the sand flow operation from being undertaken, but layers of uncompacted loose silt within the sand foundation may cause excessive, uneven, and unpredicted settlement of the tunnel elements when they are released from their temporary supports and the weight is transferred onto the foundation layer.

In a watercourse carrying a high sediment load, it is desirable to immerse the tunnel elements as soon after completion of dredging and trench cleaning as possible, and to perform the sand flow operation as quickly as possible after the elements have been immersed. Sand foundations can be installed within a few days, but there will be a period after a tunnel element has been placed before the sand foundation can be installed when there is a risk of sedimentation occurring in the dredged trench, which could result in fine materials finding their way into the space beneath the tunnel element. Equally, debris being transported along the sea or river bed could find its way into this space. Sedimentation causes problems at this stage of construction as it is difficult to remove.

The sand flow/jetting process will not necessarily displace sediment and there are a number of case histories to back this up, which show that the subsequent settlement behavior has been exaggerated. It is therefore very important in planning the marine activities for forming the foundations to assess the risk of sedimentation and plan for monitoring and cleaning procedures, should it occur. If there is a risk of sedimentation, then exposure time between placing an element and underfilling it should be minimized. If necessary, it is possible to install a curtain around the bottom of the element to keep the sediment out. Compared to a gravel bed foundation, a sand foundation offers a higher-risk solution in high-sedimentation areas. If undertaking cost comparisons, the ability to inspect and clean out sediments should be assessed. The cost of removing a tunnel element, re-cleaning the trench, and re-immersing it will add significantly to the construction cost as well as causing delay. Steps should be taken to minimize this risk from the outset of any project.

Water depth/currents

The sand jetting and sand flow methods will both require divers to assist the work. Sand flow is the most demanding as it requires divers to attach the pumping hoses to inlet valves along the tunnel, one by one, working from one end of the tunnel to the other. Or, if the pipework is continuous, they have to activate the valve systems to allow progressive underfilling. As water depth increases and divers' working time reduces, the use of divers becomes less desirable, and more automated methods of forming the foundation should be preferred. Sand foundations could be more susceptible to currents than gravel foundations because of the risk of fine material being eroded during construction. For these reasons, sand foundations have been used more commonly in shallower inland waters, where the environment is more protected and divers can work more easily. However, tunnels with sand foundations have been constructed in fast-flowing canals and rivers, particularly in the Netherlands, where the sand flow technique was first developed.

Construction tolerances

To assess the required material quantities and rates of production, and to prepare realistic designs, it is necessary to understand the construction tolerances that can affect the thickness of the sand foundation. The tolerance on the accuracy of the dredged surface is the key parameter to consider. This will depend on the dredging plant employed, along with the water depth and the accuracy of monitoring and measurement of the dredging operation that is expected. Typically, a tolerance of ± 250 mm may be achieved, and this needs to be applied on top of the minimum thickness of sand foundation. For example, if the minimum thickness of sand is set at 300 mm, and the dredging tolerance is ±250 mm, the variation in sand bed thickness should be 300–800 mm.

In addition to this, there will be a placing accuracy for the tunnel elements when set on their temporary support jacks; this will be much smaller, at around ±25 mm or less, but should nevertheless be added to the minimum sand thickness. In the example given, this would mean a thickness of 300–850 mm. Therefore, the average thickness will be 575 mm. The variation about the mean thickness will most likely be random and have no effect on settlement behavior. Local variations are typically smoothed because the stiffness of the structure will even out the settlement, but designs cannot easily cope with any significant step changes in thickness that may occur as a result of stopping and starting dredging work or a change in the equipment set-up between shifts. Dredging procedures should aim to avoid this.

Settlement

If layer thicknesses are known and the density/void ratio of the sand after placement can be reasonably predicted, then it is possible to accurately assess the degree of settlement that can be expected. As discussed above, it is necessary to assess the impact of tolerances on construction and to ensure there are measures in place to avoid soft materials within the foundation layer arising from sedimentation. Settlement will be mostly immediate and occur as the weight of the tunnel element is transferred from its temporary supports onto the sand layer. At that time, the element will only have a notional overweight due to ballasting. Further settlement will arise as the backfill is placed around the tunnel and ballast exchange is carried out within the tunnel to give the structure its permanent condition safety factor against uplift. It is common to set the tunnel elements high to account for initial settlement in the sand layer. This is generally of the order of 10–20 mm.

The effect of incomplete underfilling should be considered in design. Some project specifications have required a "loss of support" condition, where a stipulated percentage of the support area is considered to be not effective. This is based on experience in the Netherlands, where investigations were undertaken to assess the completeness of the foundation layer as a result of unexpected settlement behavior. This is quite a conservative approach to design but, perhaps, is appropriate if designing in isolation from the contractor. In a design and build environment, it would be more appropriate to assess the risks based on the techniques to be employed, the methods of monitoring planned during the underfilling, the assessed reliability of the operation, and the results of any trialing undertaken on site prior to the main works. This is also discussed in Chapter 9.

Loss of support may occur for other reasons than just the ineffectiveness of the sand flow or sand jetting operation. The substrata beneath the sand layer must also be assessed as to whether there is a risk of loss of sand materials into it. For example, the Øresund Tunnel was constructed in a trench dredged in limestone. There was a significant risk of sand material being lost in fissures in the limestone, which was a driving factor in the choice of a gravel bed solution for that project.

It is important that the adequacy of the foundation is verified during construction. Because the space beneath the tunnel is difficult to access, it is difficult to carry out testing to verify that the underfilling operation has been successful. Monitoring of the volume of materials used as the operation progresses is a reasonably reliable method of monitoring. To account for the actual dredged depth and the variation in the level of the dredged surface, a bathymetric survey of the trench should be carried out. This will enable a reasonably accurate calculation of the required material volume to be made. However, this can only be implemented as a means of control over a reasonably long length of the tunnel as there will still be some loss of material and there will be variations in the mix proportions and individual pancake geometry, which means it is difficult to monitor the volumes accurately on the micro scale, or pancake by pancake. Because of the way the pancake is formed, the sand should always completely fill the depth below the element so that only the lateral extent of the filling needs to be checked. It is important that divers monitor the injection points to ensure that sand being injected reaches the next injection location to ensure there is overlap of the pancakes. This is the most reliable method of monitoring. It could be possible to test through the base of the tunnel using ultrasonic methods or similar methods, but this is not straightforward and is generally not done unless a settlement problem arises and the cause needs to be determined. Remedial injection, usually of cementitious grout, through the base slab can be carried out if settlement problems arise and this has been undertaken on a number of tunnels.

Behavior in seismic conditions

Sand foundations can perform satisfactorily under seismic load, provided some thought is given to the risks at the design stage. The issue to guard against is preventing liquefaction. This can be achieved by the selection of an appropriate grading or through stabilization. Experience has shown that a normal sand foundation will cope with a Richter magnitude of 4 without any special treatment. For more severe events, it is possible to mix the sand with up to 5% of a coarse-grained cement or clinker to act as a binder. The grain-size distribution of the cement is adjusted to the size distribution of the sand so that segregation is avoided. If possible, the particle size distribution of the cement should match that of the sand. The foundation will then have a uniformly distributed content of hydraulic binder, which, when it hydrates, will cement the sand grains and give a degree of cohesion, which will decrease the sensitivity of the sand foundation to liquefaction in marginal cases. In highly active zones, a gravel or grouted gravel solution is more likely to give an appropriate solution.

GRAVEL FOUNDATION LAYERS

Gravel bed foundations exhibit different characteristics than sand foundations. Immediate settlement is more predictable and smaller, but the major advantage is in the construction sequence. The gravel bed is placed before the element; it can be cleaned and, if necessary, repaired before the element is placed, and the time between forming the foundation and placing the tunnel element can be kept to an absolute minimum to avoid contamination or disturbance to the gravel. This was one of the driving factors behind its choice on the Øresund Tunnel where, due to risks of seaweed accumulation, it was considered beneficial to use gravel. The time to mobilize a sand placing barge, although short, may be slightly longer and, therefore, has a slightly higher risk of debris accumulating in the void beneath the tunnel element.

Gravel beds offer improved performance of foundations in seismic conditions. Although it is important to guard against shakedown effects, which may cause the gravel layer to compact, the gravel will not suffer liquefaction and can be used to relieve the buildup of excess water pressures in the substrata and reduce the risk of liquefaction.

Gravel is generally more expensive than sand, but overall, the cost of a gravel bed is broadly similar to a sand foundation. No delivery pipework is needed in the tunnel element, there are no temporary support jacks with their large concrete pad foundations, and possible siltation problems are avoided. Gravel beds are also a more common choice of foundation for deep water and offshore environments. The method of installation lends itself to these circumstances and the materials are less sensitive to strong currents. They are formed by placing gravel by fall pipe, grab, or dumping from barges and then screeding to achieve a uniform top surface. There are a number of techniques used for this and the choice will depend on the level of accuracy required on the top surface of the gravel.

Gravel materials can be used from a number of sources although the durability of the stone is important to ensure there is no degradation with time that could lead to long-term settlement. Grading is usually quite uniform to avoid the need for compaction, and an internal angle of friction of 35–40° is desirable, particularly if forming a discontinuous bed with individual berms where the stability of the berms is important. Stone size can vary within a wide range; the nominal stone size may be typically in the region of 50–75 mm, but could be coarser.

Many tunnels throughout the history of immersed tunnels have used gravel bed foundations; notably, the early steel tunnels in the United States that were founded on gravel because steel tunnels, which were often quite narrow, were able to tolerate the achievable surface tolerances well. They continue to be used today for steel and concrete tunnels and the methods of forming a simple gravel bed with an underwater screeding frame are little changed from the first techniques used. The original screeding frame method has been joined by the Scrading™ method developed in the late 1990s and subsequently used on several tunnels in Europe. Currently, both techniques are in common use.

Underwater screeding frames

Screeding frames are relatively simple pieces of equipment. A steel frame spans across the width of the tunnel foundation and supports a beam, or rail, that travels in the direction of the tunnel to push the gravel into a uniform thickness layer similar to a bulldozer on land grading material to a level.

Gravel should be placed to a slightly greater thickness than the intended final thickness of the bed such that the screeding frame is just required to remove a slight excess layer of gravel. The operation must be planned so that too much gravel does not build up in front of the screeding beam as this could cause the beam and frame to ride up over the gravel, thus not creating an even layer, or preventing the screeding beam from operating correctly. Solutions to this problem have been developed, including shaping of the screed rail to ensure that gravel is displaced sideways to the edges of the trench. Alternatively, short breaks in the gravel bed can be created to allow the excess gravel to be pushed into a void and the screed rail to pass over it.

The tolerance that can be achieved with a screed rail will depend on the ability to accurately level the four corners of the frame before the screeding takes place. This can normally be achieved to an accuracy of ±2 to 4 cm. But, this will depend on the water depth, condition of the dredged trench, survey methods, and the degree of control over the operation that can be exerted. Deep water, offshore conditions are therefore likely to achieve a coarser surface level than shallow water inland conditions.

Screed frames can be designed to be operated from the surface, as shown in Figure 12.4, can be operated by divers, or can be operated entirely remotely. An example where a remotely operated frame was used is the Bosphorus Tunnel in Istanbul, where the elements are up to 55 m

Figure 12.4 Screeding frame used on the Fort McHenry Tunnel. (Photo courtesy of W. Grantz.)

under water. The finished level created was to a coarse tolerance and, in this instance, the tunnel elements were temporarily supported and the gap between the tunnel and the gravel was filled with grout.

Scrading™

A different approach to forming a gravel bed was developed for the Øresund Tunnel project and has subsequently been used on a number of immersed tube schemes. The technique achieves a close tolerance finish on a gravel bed. The Øresund Tunnel structure was originally designed for a sand foundation and the design therefore did not take account of any significant imposed deformations due to surface tolerance in the foundation layer. The foundation solution was changed to gravel and this then became an important design consideration as the gravel was overlying limestone and any variations in gravel thickness would impose deformations on the tunnel structure. The gravel-laying procedure had to achieve a certain tolerance so that the design of the structure would not need to be changed. This resulted in the development by Boskalis of the multipurpose pontoon (MPP), a floating barge that could span the width of the tunnel trench, held in position using spuds on either side of the tunnel trench, as seen in Figure 12.5. The barge was not fully fixed to the spuds and could

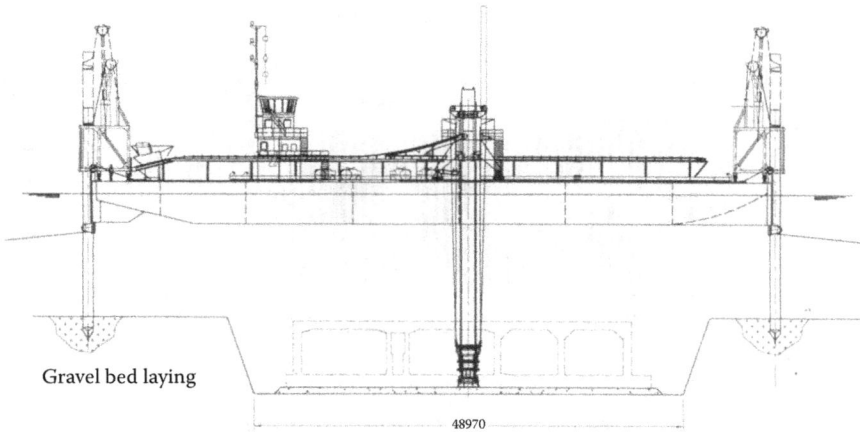

Gravel bed laying

48970

Figure 12.5 Arrangement of multipurpose scrading pontoon. (Courtesy of Øresundsbron/Boskalis.)

move vertically with the sea swell. In addition, the barge could move along the direction of the tunnel relative to the spuds. A fall pipe was mounted on one edge of the barge that could traverse the width of the barge to deliver gravel to the dredged trench over its full width. The fall pipe was fitted with a screeding shoe that would screed the gravel as it traversed the width of the trench, so the gravel was deposited and leveled in the same operation. Real-time adjustment of the screeding shoe enabled an accurate level to be maintained, taking account of any movement of the barge due to sea swell.

The MPP was equipped with a conveyor system to unload gravel from a barge and feed it to the fall pipe. The fall pipe was 1.2 m in diameter and created a strip of gravel across the trench, with a bearing width of 1.65 m. When one strip was completed, the barge would move along the trench and create the next strip of gravel. The process continues until strips have been laid over a sufficient length to support a tunnel element, creating the geometry shown in Figure 12.6. While this does not offer continuous support, it is a sufficient bearing area for the tunnel such that there are no significant settlements. The intermittent nature of the gravel bed offers some advantage in that any unevenness in the surface will be flattened out by the tunnel element when it is placed on the bed, as excess gravel can be pushed into the voids. It also allows the water to escape as the tunnel element is placed on the gravel.

The accuracy of this method of construction is quite high and achieves a tolerance of ±25 mm, taking into account the accuracy of the survey. The fall pipe is also fitted with a sonar survey system and can be used to measure the levels that have been achieved on the gravel bed surface.

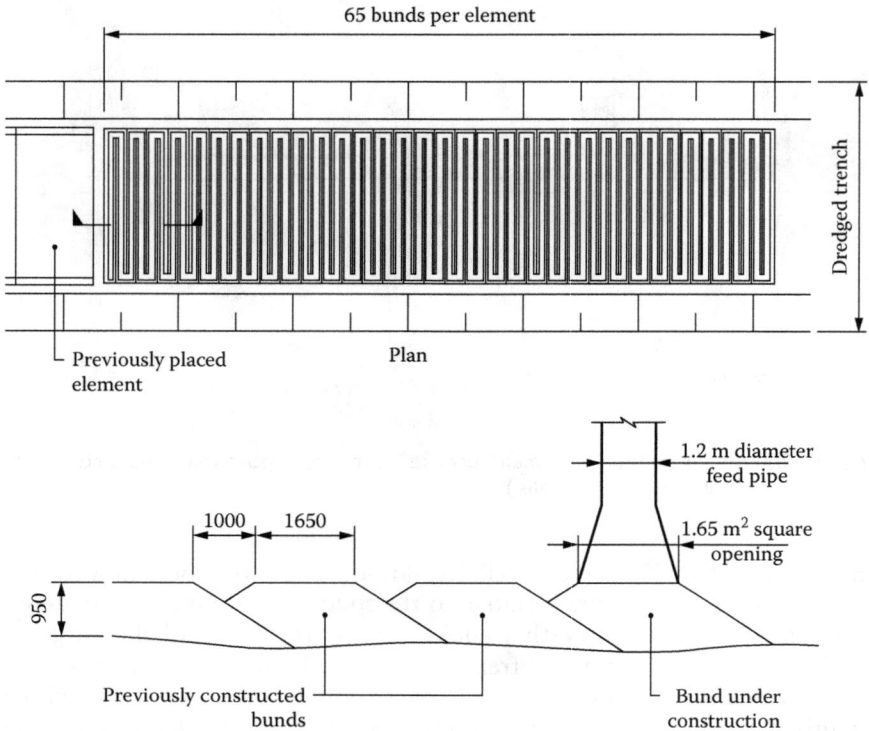

Figure 12.6 Geometry of a gravel bed.

This is done before the barge moves to its next position so that remedial works can be carried out immediately.

This equipment has now been used on the second Benelux and Caland tunnels in the Netherlands, and the Bjørvika Tunnel beneath Oslo Harbor. The same principles of creating gravel berms were used on the A73 Roer Tunnel in the Netherlands, but with different equipment, as this was constructed beneath a small inland waterway within a narrow cofferdam.

Settlement

Settlement within a gravel layer will be very limited. It comprises two principle elements: initial bedding in of the tunnel element when it is placed as it takes up any unevenness in the surface of the gravel and immediate elastic settlement, according to the stiffness of the layer and applied load. There should not be any long-term settlement within the layer.

Settlement will depend on the thickness of the gravel layer and this will, in turn, depend on the tolerance of the dredged surface, the screeding tolerance that can be achieved, and the minimum thickness required to ensure

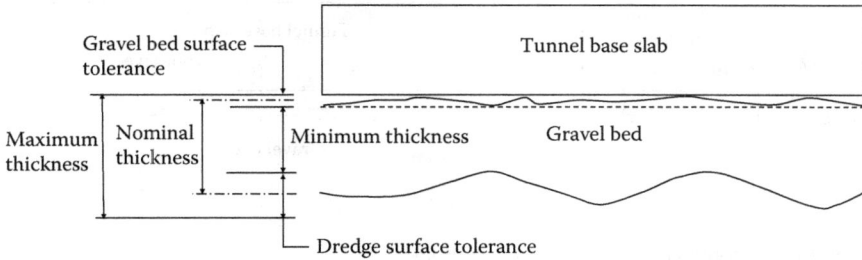

Figure 12.7 Average thickness of a gravel bed, taking into account tolerances.

that the gravel layer is continuous and that it suits the placing methods. By this, it is meant that a certain minimum thickness of gravel will need to be placed as it would be difficult to construct a thin layer of gravel using fall pipes, grabs, or dumping methods to place the material. The impact of construction tolerances on the layer thickness is illustrated in Figure 12.7. Typically, gravel beds will be in the order of 0.5–1.0 m thick and the total amount of settlement expected would be in the region of 1–2 cm.

The quality of the dredged surface may influence the settlement behavior. If there is substantial soft or disturbed material on the surface of the trench, which is quite common, there may be a tendency for the gravel to bed into the underlying materials when loaded. Some projects have chosen to place geotextile materials along the bottom of the trench to prevent this. This was done on the Bjørvika Tunnel project, for example, to prevent the gravel penetrating into the clay below.

GROUTED FOUNDATIONS

Grouted foundations can be used in circumstances where it is difficult to form conventional sand or gravel foundations. These circumstances are usually due to deep water or strong current conditions, or in seismic regions, where this method is used to create a stable foundation layer in the event of an earthquake. For example, at great water depth, it can be difficult to achieve a close tolerance finish on the top of a gravel bed, so it may be more appropriate to screed the gravel bed to a "rough" tolerance, set the tunnel elements on temporary supports, and fill the space between gravel and tunnel with a pumped cementitious grout. This type of solution has been used for the Aktio-Preveza Tunnel in western Greece, the Bosphorus Tunnel in Istanbul, and on tunnel projects in Japan.

The grouted solution tends to be more expensive as equipment for placing and screeding a gravel bed is still required, although it can be simpler than the plant required to form a close tolerance gravel bed. In addition, however, there is the additional cost of the equipment and materials for the

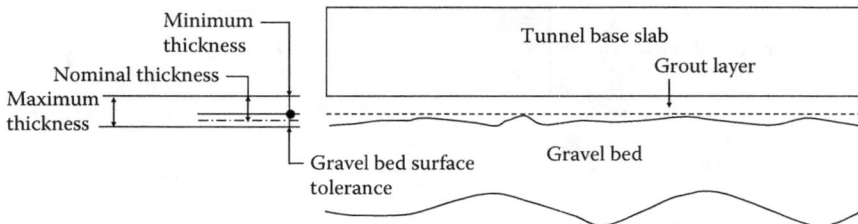

Figure 12.8 Nominal thickness of the grout layer.

grouting operation. The grout layer can be quite thin and only needs to be in the order of 100–150 mm thick. In the same way that a sand flow foundation thickness is dictated by dredging tolerances and flow characteristics, the grouted layer thickness is dictated by the tolerance on the surface of the gravel bed and the flow characteristics of the grout beneath the tunnel. The tolerance on the surface of the gravel bed may mean the nominal thickness of the grout layer is closer to 300 mm, as shown in Figure 12.8.

Because the cost of cementitious grout is high, relative to other backfill materials, and a large surface area is required to be treated, the costs soon become significant in terms of the project budget. It is desirable to keep the thickness of the grouted layer to a minimum and to ensure that materials are not wasted either by grout escaping to the sides of the tunnel or by permeating extensively into the underlying gravel layer. It is uncommon to use a grout solution on its own and, generally, it is used in combination with gravel to keep material costs under control.

Injection of the grout is carried out through pipes set into the structural base slab of the tunnel. Pipes will be spaced at regular centers along the tunnel and may be in a number of parallel or staggered lines, depending on the width of the tunnel structure and the expected flow of grout from an individual injection pipe. Pipes are fitted with valves and waterbars around their perimeter to protect the tunnel from ingress of water due to the high external water pressure. Although the pipes must prevent water ingress, it is necessary to have control valves that permit fluid to enter the tunnel so that the grouting operation can be monitored. Only by opening a valve slightly and allowing water or grout to enter the tunnel can the extent of the grout be monitored. As soon as grout is detected in the water, grouting is stopped at the previous valve. Injection can then commence at that point and the next pipe is opened to check when grout reaches it.

Grout pipes should be sealed after use with a plug of watertight material to ensure there is no risk of long-term seepage of water through the pipe. They will be left filled with grout material, but this may not be sufficient on its own, and a protective plug is desirable. This may be a simple concrete plug placed in situ. If the injection pipe is finished within a recess at the top of the structural base slab, the in situ concrete plug can simply fill this recess. Reliance should not be placed on the ballast concrete within the tunnel to do this job as there is usually a discontinuity between the structural slab and the ballast concrete.

Grout materials must be chosen so that they will flow readily. However, they must remain thixotropic and not dilute with the water present outside the tunnel. Some penetration into underlying gravel should be expected, but the grout should be sufficiently viscous so that this is not extensive. It may be very important not to clog the gravel material with the grout if, for example, the tunnel is in a seismic zone and the gravel is intended to act as a water path to relieve excess pore water pressures that could build up in a seismic event. Even with these requirements, it is necessary to be able to mix the grout with simple equipment locally to the injection points, with good control over quality such that grout with the same consistency and properties is produced.

Typically, the components of a grout will include additives to control the flow and viscosity of the mixed grout. The strength of the grout is not a primary concern that drives the mix design. It needs to be no more than the underlying gravel or ground as the layer is there to transfer uniform load, not to act as a structural component. To prevent loss of material to the sides of the tunnel, locking backfill can be placed down each side of the tunnel element before the grouting operation is started. If this is not possible or practicable, then simple curtains or a skirt can be attached to the bottom side edges of the tunnel elements to provide containment. This form of foundation has been used on the Bosphorus Tunnel in Istanbul, as shown in Figure 12.9. This figure illustrates the difficulties in carrying out this operation among the other equipment present in the immersed tunnel, such as ballast tanks.

Figure 12.9 Underbase grouting on the Bosphorus Tunnel. (Photo courtesy of W. Grantz.)

Grout bags

Grout bags or mattresses can be used to form a grouted foundation in the manner described above, but they offer some simple advantages. The grout bag naturally confines the material and, therefore, the grout can be provided in a desired area of support and the need to fully underfill the tunnel with grout can be avoided. This would save on grout materials, but the grout bags come with some additional cost, which needs to be balanced with any material saving. They are also normally grouted from pipework that is external to the structure, and hence, no internal operations are needed and pipework and valves through the base slab of the tunnel are avoided. They offer a reliable solution and can be specifically designed to suit a project circumstance and to suit the tunnel specific geometry.

An example of grout bag use is the Fort Point Channel Tunnel in Boston, Massachusetts. There, the tunnel elements were supported on piles, and grout bags were used to fill the space between the pile head and the underside of the tunnel. Grout bags are attached to the tunnel element while it is floating. The tunnel element is set on temporary supports and injection can be carried out from the quayside or a work barge above the tunnel that carries materials, pumping equipment, and mixing equipment. Grout bags are initially compressed and, as the grout is injected, they expand to fill the void between the tunnel and the underlying ground or foundation. The bags have a designed shape that allows them to expand to fully fill the void without uneven expansion or the risk of snagging on features of the tunnel.

The grout will usually be a neat non-shrink grout that has good flow characteristics and a retarded set so that there is no risk of the grout hardening before the grout bag is fully expanded and filled. This method of construction is a little unusual and the grout bags tend to be bespoke designed. Therefore, they should be prototype tested to verify the design of the bag and method of filling functions correctly.

PILED FOUNDATIONS

If there is a need to control settlement, piling is an effective way to do this. Most forms of piling can be carried out underwater, though simple driven solutions would be favored from the point of view of construction simplicity. Piling solutions can be used in transition zones to smooth out settlement profiles at the junctions between the immersed tunnel and the approach structures. Piles can either be connected to the underside of the tunnel or be "floating," where load transfer from the tunnel to the pile is achieved through the soils, backfill, or geotextile layers between the tunnel structure and the tops of the piles. Piles are generally located under the walls of the tunnel, though a floating pile solution would need a pile group under the whole tunnel.

The selection of pile type is no different for an immersed tunnel than for any other marine structure and will largely be determined by ground conditions, the required load capacity, the water depth, and plant required and possible to use at the site. The usual balance of cost and efficiency will need to be assessed by those undertaking the construction works along with consideration of all the site constraints that may dictate or preclude certain solutions.

The most complex area of design and construction of a piled foundation is to make a connection between the pile heads and the tunnel elements. This is quite unusual and best avoided, and thus has only been done a few times. A number of methods have been developed to make a positive connection, such as

- Grouting
- Adjustable pile heads
- Geotechnical solutions
- In situ underwater concreting

Grouting offers an effective way of forming a connection between the underside of the tunnel and the tops of piles, provided there is an effective way of containing the grout. This can be done either with curtains or by using grout bags that inflate as the grout is pumped in to expand and fill the available gap.

The potential movement of the tunnel must be considered. Thermal expansion or seismic-induced movement could cause sliding at the interface, or if the level of friction is high enough, deflection of the pile head. The pile head will need to be designed with this degree of loading, and measures taken to ensure that either the sliding interface is able to move or, if the pile head deflects, measures to accommodate the rotation and possible crushing of the edges of the grout layer may need to be considered. Rubber pads could be used effectively to deal with this behavior. The solution used on the Fort Point Channel Tunnel in Boston is shown in Figure 12.10, where grout was injected through tubes cast into the structure, and the grout was contained between the underside of the tunnel and the pile head by a flexible collar attached to the top of the pile.

Adjustable pile heads have been used on a number of tunnels in the Netherlands. These are rather complicated details, as can be seen in Figure 12.11, and are by no means a common approach. The concrete pile solution has a top part that can be raised by the injection of grout so that it comes into contact with the underside of the tunnel element. The steel pile solution features a sleeve within the base slab of the tunnel element that is pushed downwards to make contact with the top of the pile as it is injected with grout. Interlocking teeth provide a degree of horizontal restraint at the interface. Tunnel elements are supported on temporary supports in both

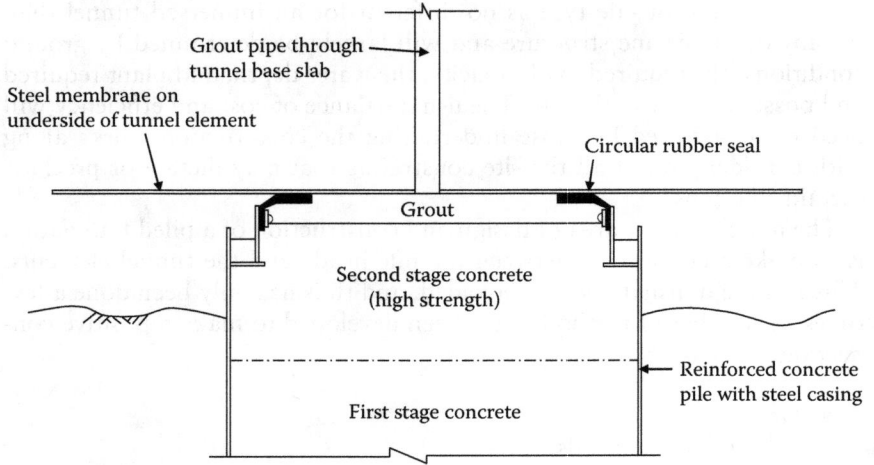

Figure 12.10 Fort Point Channel pile head detail.

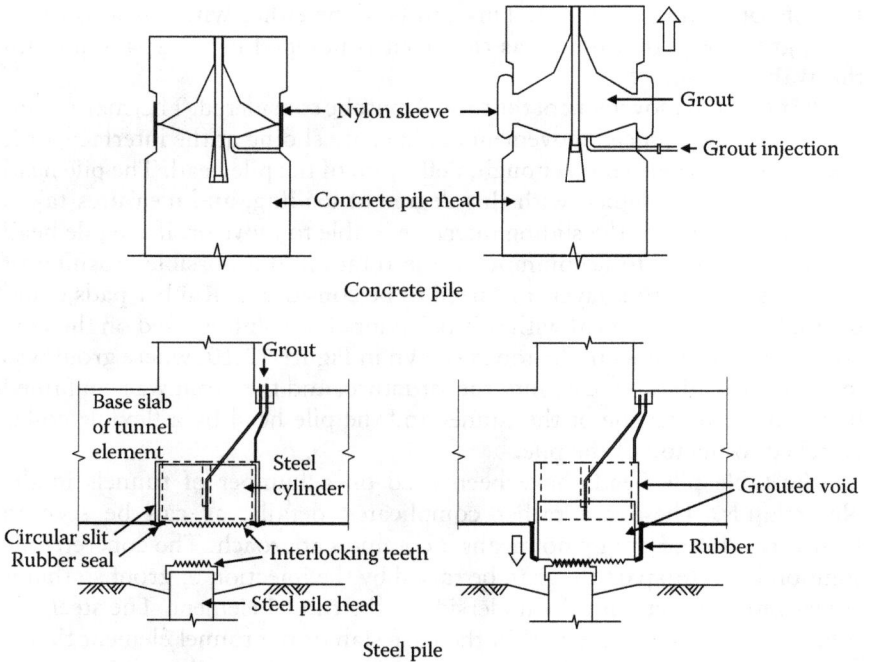

Figure 12.11 Adjustable pile head detail used in the Netherlands.

instances until the permanent connection between the piles and the tunnel element in achieved. Simpler methods of achieving load transfer should be found, if possible.

In situ underwater concreting is possible in principle, but the practicalities of working beneath the tunnel elements make this highly undesirable. While divers are used to working in such environments in the offshore industry, the quality of construction that can be achieved with in situ works has to be questioned. There are sensible alternatives available, so this type of work should be avoided from a health and safety viewpoint, diving time should always be minimized, and certainly, divers working beneath tunnel elements should be avoided.

Floating pile foundations

The principle of a floating pile foundation is that load from the tunnel is transferred through the soil to a group of piles that are not in direct contact with the structure above. The piles pick up load through friction or bearing and act as a pile group to improve the bearing capacity of the soil beneath the tunnel and reduce settlement. This type of foundation can be difficult to justify in design terms, and care is needed to ensure that the desired load transfer can actually be achieved.

The question arises as to how effective load distribution is and whether the piles can effectively work as a group. At depth, it is reasonable to assume the piles act as a group, but locally at the surface, they are likely to act individually as a series of independent point supports beneath the tunnel. A clear philosophy is needed by the designer to develop the backfill and foundation layer details and to analyze settlements and the structure correctly.

For effective load transfer to the pile heads, geotextiles may be needed to reinforce the tunnel bedding layer so that it can transmit load. This is relatively costly to provide and is not always practical if deep water or high currents are present. However, if it can be achieved, then there may be greater confidence and less construction risk compared to trying to make a physical connection between the piles and the underside of the tunnel.

Design methods exist to take account of the global settlement of a floating pile group, such as the equivalent footing method in the American Association of State Highway and Transportation Officials (AASHTO) design codes. The pile group is assumed to act at a depth of two-thirds of the pile length that extends into firm materials, also described by Duncan and Buchigani in 1976, and seen in Figure 12.12. The question still remains as to how the load transfers uniformly into all of the piles in the group at a high level, to make the design method valid.

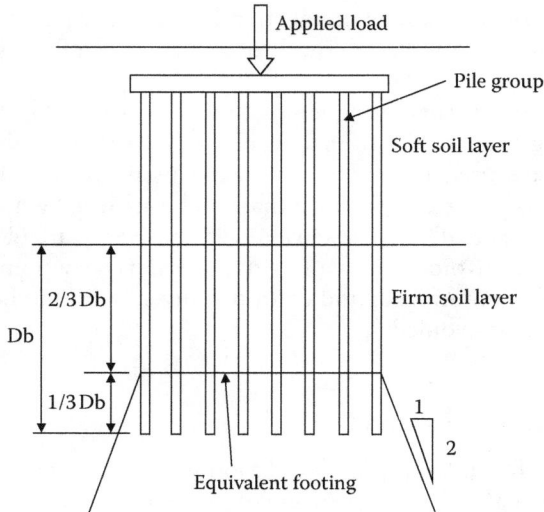

Figure 12.12 Floating pile equivalent footing design assumption.

This type of solution has been used on the Bjørvika Tunnel in Oslo's harbor, where floating piles were used to create a transition in the foundation between the immersed tunnel and the approach tunnel at one end. In this instance, piles were driven into the over-consolidated clay once the trench had been dredged, and geotextile was placed in the bottom of the trench. This served a dual purpose: to ensure that the gravel bed did not migrate into the underlying clay, and to assist in confining the gravel layer such that load transfer could take place between the tunnel and the piles.

GROUND IMPROVEMENT

Ground improvement may be required for a variety of reasons. Immersed tunnels are often built in the relatively poor fluvial or glacial deposits of rivers and estuaries, with sands and gravels often interspersed with layers of silt and clay. These soft layers may lead to unacceptable differential settlements along the length of the tunnel. Bearing pressures under immersed tunnels are usually relatively low, generally of the order of 20–30 MPa, often less than was exerted on it by the soil that was there before the tunnel was constructed. Therefore, it is usually differential rather than absolute values of settlement that govern design. Excessive settlement can affect the alignment of the road or railway within the tunnel and may cause problems for the tunnel structure in terms of causing concentrated load effects that could lead to cracking or damage to the tunnel structure or loss of water-tightness in the structure or the joints. Hence, ground improvement may be necessary beneath the footprint of the tunnel.

Bearing pressure under the tunnel also increases as the tunnel hits the shoreline. Although the tunnel alignment is rising, the river banks and the shallows adjacent to the bank will put greater loads on the roof of the tunnel. If soft subsoil layers are present here, then either absolute or differential settlement can be larger than can be accommodated by the structure, leading to a need to improve the ground.

Another common reason for ground improvement is in response to the seismic loadings that may have to be accommodated. The main issue here is the liquefaction of the ground around the tunnel during an earthquake. Because liquefaction occurs in saturated soils, the types of loose materials susceptible to liquefaction are frequently found in the beds of waterways. The tunnel alignment cannot usually be altered to avoid the material and the structure itself cannot be made liquefaction resistant, so the only viable option is to improve the soil parameters to reduce its susceptibility to liquefaction. This can be done by improving the strength, density, and/or the drainage characteristics of the materials.

The assessment of the potential liquefaction hazard is a specialist subject, especially if the soils are layered and variable, with different fines contents and cohesiveness. The state of knowledge on liquefaction assessment also changes with every earthquake as the data set increases. For example, the 1999 Kocacli earthquake in Turkey, the 2000 Duzce earthquake in Turkey, and the 1999 Chi-Chi Taiwan earthquake indicated that material with high fines contents may be more liable to liquefaction than previously assumed.

Ground improvement may also improve the stability of soils to minimize dredging volumes. This might be advantageous in very soft alluviums and silts that have a small angle of repose. A simple cost balance will indicate whether this is worthwhile; that is, is the cost of the improvement less than that of the dredging that would be saved? This might also be necessary if the natural dredged trench profile undermines existing infrastructure and so needs to be steepened to avoid the impact.

If the soils need to be improved, there are a number of recognized techniques that may be chosen:

- Granular replacement
- Stone columns
- Sand compaction piles (SCPs)
- Soil mixing

With these techniques, it is possible to improve bearing capacity, the shear modulus of the soil, resistance to liquefaction, and the stability of underwater slopes.

Ground replacement

The simplest method of ground improvement is to excavate any poor materials and replace them with better-quality engineering material. Soft sediments, organic material, peat, or any other low-strength or compressible layers can be removed by normal dredging techniques and replaced by granular material that is graded to need no compaction and can be placed by grab, fall pipe, or barge dumping, as required.

This method of ground improvement is generally used adjacent to the shoreline, where limited areas of softer material can be removed and replaced relatively easily. It is often used at the transitions between the immersed tunnel and the approach tunnels to smooth settlement behavior at the interface. There is commonly a firm foundation beneath the approach tunnel and a soft foundation beneath the immersed tube. This could lead to differential settlement, but by introducing a triangular wedge of granular material that is thickest at the approach tunnel and reduces in thickness as you move away from the approach tunnel, the differential settlement can be controlled. An increase in fill on the tunnel roof, combined with a lens of very soft material, is often most easily dealt with by replacing the soft material with better granular material.

An advantage of using this method of ground improvement is that it uses plant that is already required for dredging the trench for the immersed tube. It is also a very simple approach that can be readily monitored by normal survey techniques. The extent to which it can be done may be limited by the dredging equipment and the depth to which it can operate. It may also be limited according to the volumes of material that need to be disposed of and imported, as these can soon become substantial and the cost and time for the additional dredging and backfilling works need to be considered. This will be exacerbated if the side slopes to the dredged trench are shallow, as any additional depth of dredging will require a large total volume, due to the additional materials to be removed from the side slopes.

Getting to the material that needs to be replaced may involve removing material above it, and if this is in the waterway, then considerable volumes of material may have to be removed. This raises environmental issues because of the increased dredging operation as well as having to find a disposal site for the removed material. On the Conwy Tunnel in the United Kingdom, there were considerable depths of glacial-lake clays under one of the approach cut-and-cover tunnels. The excavation for these cut-and cover-tunnels required the lake clays to be dewatered so that suitable side slopes could be constructed. The low permeability of these clays meant that it would not be possible to dewater these within the timescales required by the contract. So, the clay was excavated and replaced with granular material before the excavation for the permanent cut-and-cover tunnel commenced.

A further example of this method is the Limerick Tunnel in Ireland, completed in 2009, which had areas of peat and poor ground immediately below the tunnel; these were dredged and replaced with granular backfill. The cost of this solution was compared to other simple improvement techniques, including piling, soil mixing, and stone columns, but because the treatment was relatively shallow, a simple replacement approach was taken.

Stone columns

Stone columns are an effective method of improving bearing capacity and reducing both the total and differential settlement. They can also be used for improving the behavior of soils in a seismic event by improving the shear modulus of the ground and controlling the ground water to prevent liquefaction. They provide vertical drainage pathways that can serve two functions. This can increase the rate of consolidation settlement, which may be useful in limiting differential settlements. They also provide pathways for water to escape during a seismic event. By providing a pathway for excess pore water pressures to be dissipated, the soils are prevented from reaching a liquefied state. Stone columns were used for this purpose on the Aktio-Preveza Tunnel project in western Greece. This tunnel has suffered many earthquakes since construction and the soils have behaved as expected and no disturbance to the tunnel from liquefaction has been evident. The two methods of installing stone columns are described in the following subsections.

Replacement method

In this method, a sleeve is lowered from a crane barge into an augered hole. The auger removes spoil and the enables the sleeve to penetrate. Once the required depth is achieved, the sleeve is filled with stone and withdrawn, leaving the stone column in place. There will be some relaxation of the stone into the surrounding soil as the sleeve is withdrawn and the sleeve will be topped up as it is withdrawn to ensure that a complete column is formed.

Displacement method

This technique removes no soil from the ground; therefore, it will densify the ground, and so would be more effective. A tremie pipe is pushed or vibrated into the soil to the required depth. The tremie has a special driving shoe that also enables stone to be released. As the pipe is then withdrawn, gravel is released through the bottom of the pipe. There will still be relaxation of the stone into the soil as the tremie pipe is withdrawn, and the stone will be topped up, as necessary.

The choice of replacement or displacement method may depend on the soil type and the degree to which the soil needs improving. The plant required for each includes a work barge, craneage for the sleeve or tremie pipe, craneage for a hopper for loading stone into the pipe and for the replacement method, a piling rig for the auger, and plant to remove the spoil. In addition, facilities to unload the stone from barges and stockpile and handle it into the hopper are required. This can result in quite a large operation and a need for a large work barge. To minimize the cost of the plant and equipment, and the handling of materials, the displacement method is generally preferred, provided the soils are suitable for this.

The limitations on this method relate to depth. The craneage is dictated by the length of sleeve or tremie pipe needed as this needs to be lifted clear of the water. It must also be possible to lift the hopper above the tremie pipe to load it with stone. The method has been used to install stone columns up to 30 m below the sea bed level.

It should be noted that the quantities of stone may be difficult to assess precisely as they will depend on the relaxation of the stone column into the soil as the tremie pipe is withdrawn. This must not be forgotten when estimating costs for this solution and comparing them to other options that may be possible.

Diameters of sleeves and tremie pipes are typically 600 mm to 1.0 m. The stone columns will be typically spaced on a grid at approximately 2 m centers, either on a square or diamond grid. The choice of spacing and arrangement will depend on the design conditions that are to be achieved for bearing capacity, density, or for the relief of pore water pressures.

The grading of the stone tends to be very narrow to ensure self-compaction of the stone within the column and to maintain a water path through the column. There will inevitably be a transfer of sediments into the stone as the stone relaxes into the soil. The diameters of the columns are generally quite large, and so it is considered that a water path will be preserved, even though some clogging from silt will occur.

If the stone columns are used to prevent liquefaction, it is important to provide a water path to the surface of the sea or river bed and there will need to be a connection at the heads of the stone columns to a permeable layer that connects through the backfill around the tunnel to the bed level. This may require a gravel bed foundation layer immediately beneath the immersed tunnel structure and coarse backfill to the sides of the tunnel.

Soil mixing

Soil mixing, also known as compaction grouting, is a technique that has been widely used in the Far East and offers the deepest ground improvement possibilities. The technology and plant were developed in Japan. The technique uses a set of augers to penetrate the ground and as they are withdrawn, cement is injected and mixed with the soil. Hydration occurs due to the presence of ground water and a column of strengthened material is formed.

Typically, the plant required for this will have a set of augers mounted in a line or pair of lines that all operate together and form a wall or panel of treated material. Anywhere between two and eight augers has been used. The sequence of treatment usually forms a square pattern of these walls in the ground. Typically, an eight-auger arrangement would create a panel of approximately 2–3.5 m.

This method of ground treatment can improve bearing capacity and the stiffness of the soil mass, and can be used in soils with high liquefaction potential to ensure that the soil does not collapse if liquefaction occurs in the surrounding ground. The treatment can be carried out to depths of up to 70 m below the sea bed level and the method is suitable for deep water and offshore conditions. It was used on both the Bosphorus Tunnel and the Busan Tunnel to improve the soft cohesive material around the tunnel, which contained sufficient fine material to be susceptible to liquefaction.

Sand compaction piles

Sand compaction piles (SCP) are a method of densifying the ground using displacement techniques. They have been used for enhancing bearing capacity, stabilizing the ground, and reducing settlement for a variety of structural types. They are suitable for marine installation and have been used successfully to densify loose deposits to increase their safety factor against liquefaction.

Similar to deep soil mixing, the techniques were developed in the offshore industry in Japan and they are well documented in the publications of the Japanese Port and Harbor Research Institute and the Japanese Geotechnical Society. Two methods of installation are possible: either with a vibro composer method or the SSP method. Both require a sleeve to be vibrated into the ground, which is filled with sand that is released as the sleeve is withdrawn. The difference between the two methods is that the vibro composer releases a certain column height of sand and is then used to compact it before releasing the next column of sand above. The SSP method features a vibrating sleeve that releases the sand in stages as it withdraws vertically.

SCPs are commonly laid out on a triangular or rectilinear grid. For offshore applications, 1.5–2 m diameter SCPs are commonly used and depths of soil up to 70 m can be treated.

The design methodology for SCP is well documented in the literature (JGS 1998). Two design procedures can be used. The first uses design charts based on case histories, which is appropriate for sandy soils with a fines content of less than 20%. The second approach calculates soil improvement in terms of voids ratio, relative density, and fines content.

Validation of ground improvement

Measurement of the ground improvement is conventionally assessed by carrying out standard penetration tests (SPT) between SCP locations and computing an average N value. By comparing the post-installation N value with the original N values for the site, the degree of compaction of the soil can be determined. For design, a target N value must be determined based on the available site investigation information, either CPT tests or SPT tests.

It has to be recognized that there is a good deal of empiricism in the design of foundation solutions such as sand compaction piles. Furthermore, there is wide scope for problems to be encountered during construction. Consequently, a robust design approach must be taken to ensure that the foundation is adequate.

Chapter 13

Earthworks

There are always many challenges in completing the earthworks for an immersed tunnel. There is the dredging and backfilling of the underwater trench, the installation of the foundations for the tunnel elements, the excavation associated with cut-and-cover approach tunnels, and, possibly, the construction of a large earthworks casting basin in which to build the tunnel elements. Each of these requires a high level of geotechnical expertise to develop an appropriate solution.

DREDGING

Dredging is a fundamental and important part of any immersed tunnel construction. There has to be a trench at the bottom of the waterway in which to place the immersed tunnel elements. The bed material will often be soft alluvial or fluvial deposits that can be readily removed by modern dredging plant. Reasonably stiff clays and some limestones can also be dredged successfully. Harder rock material can be removed if encountered as isolated intrusions into the trench, and this can be done by blasting or specialized plants. If a lot of hard rock is present, then the immersed tunnel may not be the right solution. Dredging appears to be a relatively straightforward activity, but in reality, it can present some complex problems. Dredgers are expensive to operate, and to be economic, they need to maintain high utilization and production rates. There are many different types of dredgers, each with its own advantages and disadvantages. The final choice of dredger for any particular scheme depends on its suitability for the soils, cost, and availability. Unless there are particular reasons for either using or not using a particular type of dredger, the choice of type and size is usually best left to the dredging contractor. It is therefore important to seek specialist advice on the dredging operation when planning an immersed tunnel.

The impact of the dredging has to be considered at the outset. It will interfere with the use of the waterway and can have environmental implications. At the feasibility stage, on some crossings, the immersed tunnel has been ruled out at an early stage because of fears over the dredging operations. These fears are often overstated and modern equipment and procedures greatly reduce the dredging impact.

There are several types of dredgers and the appropriate one to use depends on the volume and type of material to be removed, the time available, the water depth, and where the material is to be taken. The main types of dredgers that are generally used on immersed tunnel projects are as follows:

- Backhoe
- Clamshell
- Cutter suction
- Bucket ladder

A backhoe dredger is essentially a hydraulic excavator mounted on a pontoon. Sizes of the excavation buckets vary considerably, ranging from small buckets of 0.5 m³ to buckets of 10 m³ or more. The larger machines are easily capable of dredging to depths of 20–25 m that would cover the requirements of most estuarial or river immersed tunnels. Backhoes are also suitable for removing stiffer material because the hydraulic arms can exert considerable horizontal as well as vertical forces. Production rates vary considerably, depending on the material and depth to be excavated, but a large modern machine should be capable of removing 500 m³ per hour at depths of 20 m.

Backhoe dredgers are popular in Europe, whereas clamshell dredgers are more common in the United States and Asia. A clamshell dredger consists of a pontoon-mounted revolving crane that is fitted with a grab or clamshell (Figure 13.1). The grab is lowered to the bed on the hoisting wire and opened and closed either by wires or a hydraulic system. In theory, a clamshell dredger can operate at any depth and is only limited by the length of wire on the winches. For some mineral extraction sites, depths of 100 m have been reached; however, accuracy decreases with depth. But, clamshell dredgers can easily cope with the depth requirements of immersed tunnels. The size of the clamshell used depends on the material to be excavated—larger grabs are used for soft soils, whereas in more cohesive soils, a smaller heavy grab may be preferable. Grab sizes vary considerably, the largest being a 200 m³ machine, but that is a very specialist piece of equipment and would not be used on a tunnel scheme. Grabs in the range 5–15 m³ are more normally used in tunnel schemes.

Backhoe and clamshell dredgers are particularly useful in confined areas close to the approaches. They can dredge with great accuracy around the edges of cofferdams. With either type, the excavated material is placed in barges for transportation to a storage or disposal site. A key factor in using either type of dredger is the environmental requirement for the prevention of

Figure 13.1 Clamshell dredging on the Marmaray Tunnel (27 m³ grab). (Photo courtesy of W. Grantz.)

contamination. These might simply be limits on the percentage of excavated fine material that can be spilled into the water during the excavation. In busy historic waterways, the bed material may be contaminated and no spillage at all allowed during the operation. In these circumstances, special shrouded buckets are available to prevent material from being spilled as the grab raises the material up through the water. Silt curtains are also often used around the dredging operation to contain any spillage and stop it from being released into the waterway.

Because the grabs also contain water as well as the excavated material when they reach the surface, there is a tendency to allow the water to drain off the disposal barge, so that it does not just transport a lot of water to the disposal site, and to make sure that the barge contains the maximum volume of solid material. Again, environmental restrictions may prevent this practice. Such restrictions result in the barge transport being less efficient, but that is a price that has to be paid. In some sensitive locations with very contaminated material, it has been necessary to cover the full barges so that no contaminated material is lost overboard during the voyage to the disposal site. These requirements would all have to be agreed upon with the harbor and environmental authorities during the preparatory stages of the project so that the appropriate operating regime can be defined.

Bucket ladder dredgers used to be the most common form of dredger. They consist of a continuous chain of buckets on a ladder that cut into the bed at the bottom of the ladder and empty their contents into a chute at the top of the ladder. Their high maintenance costs and noise of operation has resulted in their being almost entirely replaced by backhoe and suction dredgers.

Cutter suction dredgers are often used on immersed tunnels as there are many of them that dredge at depths of 20 m or more. A cutter suction dredger has a revolving cutter head that loosens the bed material, which is then sucked up a pipe and pumped either into barges or ashore through a pipeline. The type of material that they can dredge is more restricted than a backhoe but, in general, they can cope with the soft material encountered in a lot of immersed tunnel trenches, including soft rock. Whether they are suitable depends on the quantity and the method of disposal. They are particularly suitable if the dredged material can be pumped directly from the dredger through a pipeline to a disposal site. This can be 1–2 km away and the pumping is carried out by using intermediate pumping stations along the pipeline. Pipelines, which are about 1 m in diameter and float on the water surface, are generally preferred to discharge into barges, but they can interfere with navigation and are subject to currents and wave action. These issues can be overcome by laying vulnerable sections of the pipeline on the river bed.

Figure 13.2 shows a cutter suction dredger working close to the river bank on an immersed tunnel scheme. Note the floating disposal pipe taking

Figure 13.2 Cutter suction dredger on Limerick Tunnel. (Photo courtesy of NRA/ DirectRoute.)

the dredged material straight to the sedimentation ponds on shore, adjacent to the site.

The disposal site for pumped disposal has to be adjacent to the river and be large enough to be able to store the material while the water that it has been mixed with settles out and is drained back into the river. It may be necessary to have a series of settling lagoons to enable the sediment to settle out, so that when the water is returned to the river, it is clean enough to meet any environmental standards. Once the material has been drained sufficiently, it can either be left in place and used as part of the landscape or, if it is suitable, re-excavated and reused, possibly in the tunnel project itself. Here, it can be used to backfill the trench or form, for example, part of the approach works to the tunnel.

If barge disposal is used, then a suitable disposal site has to be identified. This is becoming increasingly difficult as regulatory regimes become stricter and approvals and licenses are required. The traditional method for disposing of dredged material is simply to dump it at sea by bottom dumping from the barge. This may still be the best option, but other possibilities must be considered. Land-based disposal, via a pump ashore facility, is an alternative. The presence of contaminated material may complicate the issues as the material may require special treatment before a particular site can be used and, indeed, may require the use of a special disposal site. These can all be costly processes that also affect the techniques and production rates of the dredging operation.

Immersed tunnel trenches are large excavations. The bottom is typically 5 m wider than the tunnel section. The underwater side slopes vary according to the material. In the soft surface layers, slopes of 1:5 or flatter can be expected, even up to 1:8 in very soft conditions. Clay and silt can be excavated at steeper slopes, say between 1:3 and 1:5, with sand excavations at around 1:3 and stiff clays and weak rock at 1:1 to 1:2. This often results in a trench in excess of 100 m wide at the bed level and the quantity of material excavated is frequently in excess of 500,000 m³ for even a 1 km long tunnel. With such a large volume, every effort should be made to minimize the environmental impact, either by reusing the material on the tunnel as backfill or using it on some other nearby scheme or in a reclamation project. The reuse of this material can have a significant impact on the cost as well as the environment, so it has to be considered at the planning stage. Suitable sites have to be identified and any restrictions on their use made clear. Whatever method of disposal is considered will involve a regulatory regime that has to be complied with.

A generally smooth bottom to the excavation can be achieved with backhoe and cutter suction dredgers and modern instrumentation allows them to dredge the profiles and side slopes accurately. Clamshell dredgers are not so accurate, as the wire-based system produces an irregular bottom profile. This disadvantage can be mitigated by increasing the dredging tolerance so that any peaks do not interfere with the tunnel foundation. The tolerance on the final level in the bottom of the trench varies according to the material being excavated and the equipment being used, but will be in the order of ±250 mm. The dredging contractors should be consulted to confirm what they can achieve in the particular circumstances as this information is needed to design the foundation layer of either sand or gravel.

The final cleaning of the trench is often done by a specialist dustpan-type dredger. This is really like a large vacuum cleaner, where the material to be removed is simply sucked up a pipe. Some dustpan dredgers also incorporate high-velocity water jets to loosen the material before it is removed.

Water injection dredging (WID) may also be used for the cleaning and maintenance of the dredged profile. This method uses water jets to put accumulated sediment back into suspension in the water column such that it is carried away by the currents. This method can be used as an effective means of cleaning while never exceeding the natural levels of sediment that would normally be expected to arise in the watercourse.

The dredging operations are carried out in stages, depending on the needs of the project. For example, a typical sequence might be to first remove contaminated material close to the surface. This is followed by the bulk dredging of the trench, trimming of the foundation, and finally, cleaning the trench immediately before the tunnel element is placed. Different equipment might be used for these stages and the sequence will need to be carefully planned because of the mobilization costs associated with the dredging plant. It has a high capital cost and needs to be used efficiently; it should not be idle on a site waiting to carry out its operation.

The effects on navigation have to be considered. At some stage, the dredger will have to be in the navigation channel. In busy waterways, such as Hong Kong Harbor, the navigation fairway had to be diverted temporarily, first to the north, then to the south. In less busy waterways, communication may be sufficient for the dredger to pull aside when a ship wants to pass. These arrangements have to be discussed and agreed upon with the navigation authorities well in advance as it takes time to set up and issue the necessary notice to mariners.

BACKFILL

Locking fill

Having dredged the trench and placed and founded the tunnel elements, the remaining trench around the tunnel must be backfilled. Locking fill is the name given to the fill that is placed on each side of the tunnel up to about half the height of the tunnel (Figure 13.3). Its purpose, as its name suggests, is to give initial stability to the elements, locking it into position

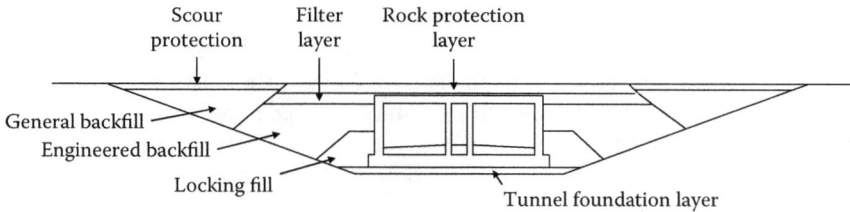

Figure 13.3 Typical tunnel backfill.

and preventing any lateral movement. The material is placed after the element has been lowered on to its foundations. Because the tunnel element is susceptible to movement while it is just resting lightly on the bed, the locking fill must be placed carefully and evenly to avoid applying any horizontal loads to the element, which might move it sideways. It should be placed in layers that are usually no more than 600 mm thick, and are placed evenly on each side of the tunnel to minimize the out-of-balance horizontal force and eliminate the risk of movement. If a sloping profile is adopted for the backfill, then the horizontal shoulder of the locking fill should extend at least 2 m from the wall of the tunnel before the downslope starts. The top slope of the locking fill depends somewhat on the properties of the material, but should not generally be steeper than 1:2.

The placing operation has to be controlled to prevent shock loading from moving the tunnel element. For this reason, bottom dumping of the locking fill is prohibited and it is either placed through a fall pipe or by grabs. The multipurpose pontoon that has been used for gravel bed laying can also be used to place the locking fill, which makes best use of the equipment. The locking fill must be a clean, sound, hard, durable granular material that is free draining and will compact naturally underwater. The grading should be such that it will not be susceptible to liquefaction under the design seismic conditions.

General backfill

The remainder of the trench above the locking fill is filled with general backfill. The requirements for this are not as specific as for the locking fill. In general, the only requirements are that it can be placed satisfactorily and is free from either chemical or organic contamination. However, there are usually additional restrictions imposed by the river or environmental authorities. There is often a requirement that the backfill is non-cohesive as the loss of fine material during the placing of cohesive material would be unacceptable in the watercourse. It is also good practice to place good-quality engineering fill adjacent to the tunnel so that its settlement characteristics are predictable and there is no long-term consolidation. This also leads to the adoption of non-cohesive material. Outside this, a lower-quality material can be used (Figure 13.3).

The placing method is also less restrictive than for the locking fill. As long as the backfill does not segregate, bottom dumping is allowed, although care is needed close to the tunnel walls if there is a waterproofing membrane. Placing by grab or fall pipe may be needed adjacent to the walls to prevent damage to any such membrane. The difference in height of the fill on either side of the tunnel can also be greater as the locking fill is giving the element lateral stability. It is often possible to re-use material that was dredged from the trench if a suitable storage site can be found between operations. If the general backfill is of good quality, then it may also be suitable as locking fill and the only difference between the two layers would then be in the method of placing.

Tunnel cover

The trench is normally backfilled to the existing bed level so that, after construction, there is no disturbance to the flow in the waterway, which returns to what it was before construction started. However, the tunnel is a valuable asset and will be in place for a long time, so precautions must be taken to prevent the tunnel backfill being removed by scour or other changes in the bed of the waterway. These may not be envisaged at the time of construction, but 20–50 years later, there may be dramatic new requirements for, or changes in, the waterway. To achieve this, a scour protection layer is placed over the backfill. It could extend some 15 m each side of the tunnel so that it will deform into the depression made by any scour and maintain its protection to the tunnel backfill.

The scour protection is a rock layer placed on top of the backfill. The size of the rock depends on the stone weight required to prevent it from being washed away by the currents in the waterway. For a typical waterway with only small navigation, a 750 mm thick layer of nominally 150 mm single sized stones should be sufficient, and for a larger waterway, a 1.5–2.0 m thick layer of stone with D_{50} of 500 mm is typical, but calculation will always be needed to verify this. The stone should be hard, durable, and inert.

Placing the rock blanket over the tunnel requires care so that the tunnel roof, and in particular, any waterproofing layer, is not damaged in the operation. This may require placing by grab directly over the tunnel. The top level of the rock protection is normally placed at the level of the existing seabed, although this does not have to be the case. Provided that the navigation and hydraulic requirements are satisfied, there is no reason why the top of the tunnel should not be above the bed level. The immersed tunnel is then contained within an underwater embankment. In such a configuration, it is even more vital that the scour protection performs its function and maintains the fill around the tunnel and protects it from impacts.

Equally, there are situations where it is not necessary or required to backfill the tunnel trench back to the existing bed level. The protective layer can be placed and the trench left as a depression in the bed. Normally, such a trench would then be backfilled over time by natural sedimentation. Hydraulic studies would be needed to verify that the trench does not have a tendency to migrate upstream or downstream before it is backfilled.

Some deep immersed tunnels have been planned, where the trench is not backfilled to the original bed level, in order to limit the loading on the tunnel roof. This requires a strict monitoring regime to be imposed to check that the level of material on the tunnel roof is within the design parameters and to trigger the maintenance dredging if it approaches the limiting levels. In such cases, it may also be possible for the depression above the tunnel to migrate along the bed. So, it is not enough just to check the depth of the trench above the tunnel; its location relative to the tunnel is also needed.

Release of dragging anchors

The rock protection also performs another function. It protects the tunnel from damage caused by dragging anchors. In navigable waterways used by large ships, anchor sizes can be considerable and, if one caught on the top corner of a tunnel, it could lead to damage or displacement of the tunnel element. Originally, concrete immersed tunnels were built with chamfered top corners to prevent an anchor catching on the corner, but this has now largely been superseded by designing the rock protection such that it releases the anchor from the sea bed.

A ship's anchor achieves its function by digging into the sea bed. Larger anchors may penetrate soft sediments by several meters when they are dropped. Smaller anchors will gradually penetrate down into the sediment as they are dragged. Therefore, the anchor being dragged will travel through the sediment just beneath the bed level, at the height of the tunnel roof. When the anchor encounters the rock protection, the larger rock size chokes the anchor and forces the anchor out of the bed. It then bounces across the top of the rock protection until it re-engages with the bed on the other side of the tunnel.

There are various options available for the arrangement of the anchor release rock. It can either be placed over the whole width of the tunnel and extend at least 15 m either side of the tunnel or it can be placed as release bands at the extremities of the normal rock protection, as shown in Figure 13.4.

Filter criteria

With various layers of different material around and above the tunnel, it is important to ensure that each layer stays in place and is not susceptible to the migration of material through adjacent layers. Normal Terzaghi filter criteria are used to establish compatibility between the various layers.

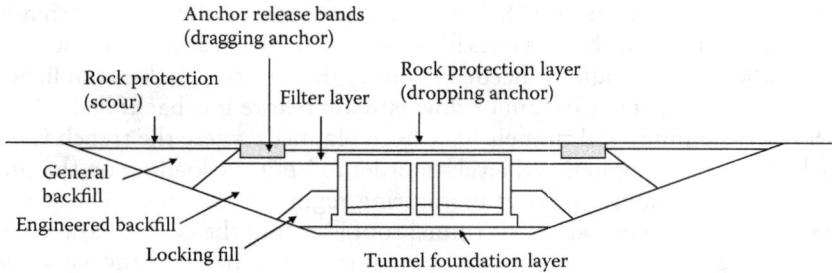

Figure 13.4 Anchor release bands.

Occasionally, however, it is necessary to introduce geotextile materials into the backfill to act as a separating layer. The main difficulty with this is during placing, when the geotextile has to be weighted down. This is often achieved by fixing lengths of reinforcing bar to the geotextile so that it sinks to the bed as it is unrolled.

CUT-AND-COVER TUNNEL EARTHWORKS

We do not propose to set out in detail the design and construction of cut-and-cover tunnels as there are many other books on the subject. Here, we shall concentrate on those aspects that are relevant in the construction of immersed tunnels.

Perhaps the simplest form of cut-and-cover tunnel is constructed in an open earthwork excavation. This requires a large dewatered excavation adjacent to the river bank or shoreline, which often extends into the waterway. When considering this type of construction, one of the first things to establish is if any temporary intrusion will be allowed into the waterway and, if so, to what extent. This may require hydraulic modeling to determine any changes to the flow regime, both in terms of current velocities for navigation and accretion, and scour patterns in order to demonstrate its acceptability. Figure 13.5 shows a typical large open earthworks excavation for a cut-and-cover tunnel.

The earthwork bund is often constructed by end tipping, starting at the bank. Given that the excavation for the cut-and-cover tunnel will be quite deep, this enclosing bund often extends some way into the channel. At an early stage, the material present in the area of the excavation must be considered carefully to assess the likely side slopes required to establish a safe, stable excavation.

The excavation for the cut-and-cover tunnel not only has to be stable, it also has to be watertight to enable construction to take place safely within it. Water can get into the excavation in three ways. Firstly, it can overtop the

Figure 13.5 Øresund Tunnel cut-and-cover tunnel construction. (Photo courtesy of Øresundsbron.)

surrounding bunds. For this to happen, in practice, would be unusual as it is relatively easy to design for, but it does point out the need for the bunds to be high enough to withstand high tides and surges as well as overtopping from wave action. The design return period for such events does not have to be as long as for the permanent works, as the excavation will only be in use for a limited time, in the order of one to two years.

Secondly, the surrounding bund also has to prevent water permeating through it and flooding the excavation. This is usually achieved by installing some form of watertight barrier in the center of the surrounding bund. The simplest approach is to use a sheet pile wall, which can be easily removed when the excavation is backfilled. For this to be successful, the toe of the sheet piles must key into an impermeable layer in order that water does not simply run under the bottom of them. If no such impermeable layer is present, then they must extend deep enough so that the hydraulic gradient under the wall limits the amount of water flowing under it to an amount that can be controlled within the excavation. Other forms of cutoff wall are available, such as cement bentonite walls, and these are also commonly used. These are easily removed when the bund they are constructed in is removed.

Thirdly, water must be prevented from coming up through the base of the excavation. If the base of the excavation is in an impermeable layer, then the cutoff wall will be sufficient and the only issue is the stability of the base against heaving as the excavation proceeds. This can be prevented by a series of dewatering wells that lower the water table sufficiently under the excavation to prevent heave occurring. Often, two or three rings of

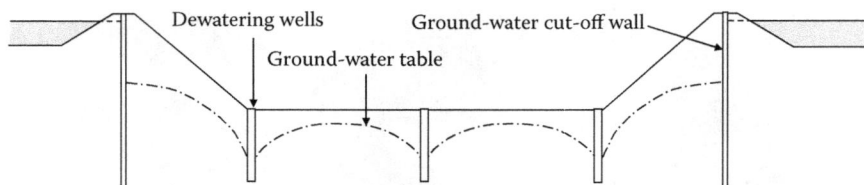

Figure 13.6 Dewatering cut-and-cover excavation.

dewatering wells are required around the excavation to control the water table during construction, as shown in Figure 13.6.

Once the excavation is stable, secure, and watertight, the cut-and-cover tunnel is constructed as a conventional reinforced concrete box at the bottom of the excavation. When the cut-and-cover tunnel has been completed, the seaward end is sealed with a watertight bulkhead, and has a Gina gasket sealing plate fitted, similar to those used at the ends of the immersed tunnel elements. Alternatively, part of the closure joint may be installed at the end of the cut-and-cover tunnel. The end of the cut-and-cover tunnel is then ready to receive and be joined to the end of the immersed tunnel element. But, at this stage, the end of the cut-and-cover tunnel is still in the dry at the bottom of the excavation.

To expose the end of the cut-and-cover tunnel to the water, a second inner bund can be constructed over the completed cut-and-cover tunnel on the landward side of the initial enclosing bund. This is usually conveniently situated along the permanent shore line so that it reinstates the existing shoreline. The remaining excavation behind this bund can then be backfilled while the seaward end of the original bund is removed, allowing the water to flow in around the exposed end of the cut-and-cover tunnel. The dredging of the trench can then be completed up to the exposed end and the immersed tunnel placed against it (Figure 13.7).

An open earthwork excavation is not the only way of constructing a cut-and-cover tunnel. If it is not possible to construct temporary enclosing bunds out into the waterway, then the cut-and-cover tunnel can be built behind the existing shoreline. Construction can be in a temporary dewatered excavation similar to that described above, but behind the existing bank, or can be constructed in a structural slot using a diaphragm wall or secant pile techniques. These diaphragm walls or secant piles can either be simply temporary or they can form the permanent walls of the cut-and-cover tunnel. The various forms of construction are described in more detail in Chapter 8, which covers tunnel approaches. Another technique widely used is to construct a temporary excavation within steel pile walls. Because of the depth of the excavation, a combi-pile wall is often used, which consists of discrete tubular piles with sheet piles linking between them. Such a wall was used

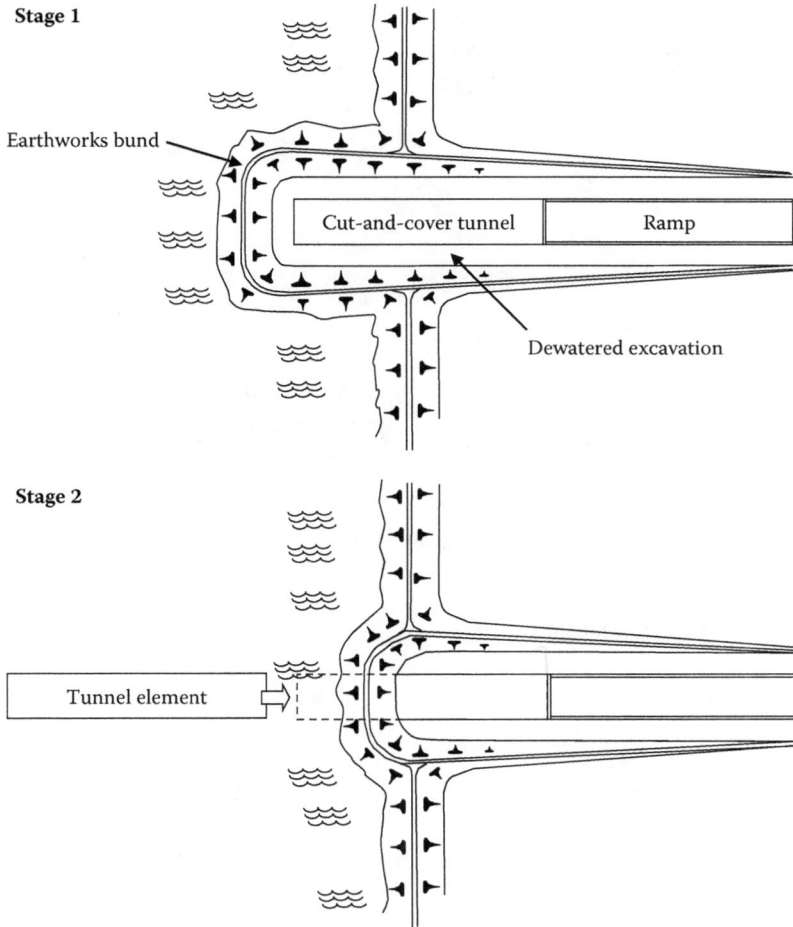

Stage 1

Earthworks bund

Cut-and-cover tunnel

Ramp

Dewatered excavation

Stage 2

Tunnel element

Figure 13.7 Exposing the end of the cut-and-cover tunnel.

in the construction of the cut-and-cover section of the Limerick Tunnel in Ireland, as shown in Figure 13.8.

Whatever method of construction is chosen, the seaward end of the tunnel needs be fitted with a bulkhead and a gasket sealing plate. So, even with a diaphragm wall or secant pile construction, it will still usually be necessary to construct a short length of in situ concrete box so that the bulkhead and sealing plate can be installed on the end to receive the immersed tunnel. Then the existing river or sea wall will be removed to allow underwater access to the exposed end.

Figure 13.8 Limerick Tunnel cut-and-cover construction. (Photo courtesy of NRA/DirectRoute.)

CASTING BASIN

Every immersed tunnel scheme needs somewhere to construct the tunnel elements. Many of the early immersed tunnels, particularly the concrete tunnels built in the Netherlands, were constructed in large open earthwork excavations. These were often large enough to enable all the tunnel elements to be built simultaneously.

The earthworks issues with building a casting basin are similar to those of building the cut-and-cover tunnel section. You have to construct a large open dewatered excavation adjacent to a waterway. It does not necessarily have to be adjacent to the tunnel site, although that has often been the case. It is becoming more difficult to find sites for earthworks casting basins because of the increasing pressure on land availability and more stringent environmental requirements. If, for example, we consider a basin that is to accommodate six 120 m long elements, each 25 m wide, then the floor area of the bottom of the basin will be approximately 260 m × 150 m. The basin floor will be approximately 9–10 m below the level of the water in the adjacent waterway to enable the elements to be floated away, and the top of the enclosing bund should be some 2–3 m above high water level to prevent flooding and overtopping. So, the crest-to-floor height of the excavation is some 12–15 m that, at a slope of 1 in 2, adds some 60 m to the floor dimensions. The overall footprint of the basin is, therefore, about 300 × 200 m, and it will still need a working area around it. Finding an area this size adjacent to the waterway can be difficult. Such sites are rare in built-up areas, and away from the towns, river banks are important to wildlife and the general environment, and gaining approval is not an easy matter. However, such sites can be found and large open casting basins have been built. Often, open discussions can dispel many of the perceived objections to such a basin.

The key issues for construction are as follows:

- Water exclusion
- Slope stability
- Foundation conditions
- Working platform
- Backfilling and reinstatement

Water exclusion and slope stability have been mentioned earlier in connection with cut-and-cover tunnels. The issue with the casting basin is one of scale. The volume of material to be excavated and stored or disposed of is considerable and the volume of groundwater that may have to be pumped is large. The groundwater pumping could be required for a period of 12–18 months. This could have a considerable effect on the groundwater regime in the area. It could deplete the groundwater aquifer to the detriment of local or regional water supplies. The settlement caused could damage adjacent structures. And discharging that volume of water into the waterway could, in itself, be an issue if it is significant enough to alter the local characteristics of the water in the waterway.

Detailed discussions with groundwater authorities will be required to establish the extent to which the groundwater can be extracted. Groundwater modeling might be needed to assess the effects of the dewatering on the aquifer and a monitoring regime set up to verify that the predicted effects are not being exceeded. Baseline monitoring is also required to establish the groundwater regime before construction starts. This monitoring also continues after the dewatering is stopped in order to monitor how the groundwater regime returns to normal.

As the site is next to a waterway, there is possible conflict between saline water in the river and potable water in the aquifer. The pumping operation can lower the water table over a considerable distance. If this is a potable aquifer, then care must be taken in decommissioning the pumping regime. If it is simply turned off, it is likely that saline water from the river will quickly penetrate into, and pollute, the aquifer. The saline water is closer and there is a plentiful supply. To prevent this, a phased close down of the operation is required so that saline intrusion is prevented and the natural groundwater regime is allowed to return to its original condition.

The settlement in the surrounding area that is caused by the dewatering must be assessed. Generally, settlement contour plots are developed, which enable predictions of which adjacent structures are likely to be affected. Whether a particular building is likely to be affected by settlement depends on the foundation construction of the structure. Often, old buildings adjacent to waterways are founded on shallow rafts or on timber piles, both of which are susceptible to movement as the soil is dewatered. Timber piles are also likely to deteriorate if they dry out. Condition surveys will be required of any structure that could be affected by the works. This is as much to safeguard the tunnel constructor from spurious claims that the construction affected a particular building as it is to monitor the effects of construction.

If the effects of dewatering are too great, then it may be necessary to install recharge wells. These pump water into the ground around the site, and therefore limit the effects of the extraction operation. The design of such a groundwater recharge scheme should be carried out by specialists.

Once the excavation has been established, the construction of the immersed tunnel elements is straightforward. However, there is one important geotechnical aspect that affects construction: the floor of the casting basin. A working platform is needed on which to construct the elements. Generally, this is provided by a 500 mm thick layer of granular material, with the tunnel elements being constructed directly on it. The granular layer can be accurately screeded to form the correct bottom profile for the element. Although they may often appear straight, immersed tunnel elements are usually curved in the vertical direction and often horizontally as well.

The granular layer also serves another purpose. The elements are finely balanced between floating and sinking and are designed to have only a small freeboard. When the element is being floated up in the basin, there is a possibility that it will stick to the floor. If that happens, there is a risk that the element may suddenly free itself and rise uncontrollably to the surface. The granular layer under the element enables the water to flow under the element when the basin is flooded, making sure that the uplift distributes evenly over the underside of the element. Often, as an additional safeguard, porous pipes are laid in the gravel under the element. If the element looks like it is sticking, then air can be pumped through the pipes and the airflow from the pipes will be enough to loosen the gravel and enable the element to float.

Other treatments for the basin floor are possible. No-fines concrete has been used, but that is expensive. Sand has also been used, contained between a grillage of concrete beams that are cast to the correct geometry for the underside of the element. Either the steel bottom plate or plywood formwork can then be placed directly on to the sand. If plywood formwork is used, it is provided with fixings so that it does not come off and flap about during the immersion process. If that happened, the placing of the sand foundation could be severely impeded.

When the elements have been constructed, the basin is flooded and the elements towed out. If the basin is to be reused, then the entrance channel is resealed and the basin pumped out, the dewatering re-established, and the next batch of elements constructed. The number of times a basin is to be re-used will determine whether to construct a floating dock gate that can be removed and later replaced, or to simply dredge through the casting basin earthworks wall to gain access.

If the basin is no longer required, it may have to be backfilled. If it cannot be used as some form of marine amenity, it represents a deep water basin, which can be dangerous, so it should be backfilled. There are two ways of backfilling the basin: either in the wet or in the dry. Both options are expensive, so if another use can be found for the basin, that would be preferable to backfilling it. If it is backfilled in the dry, it has to be resealed and pumped out. If it is backfilled in the wet, a free-draining, self-compacting material is required. Backfilling that depth with a cohesive material would probably leave a very soft, boggy area, which would be a danger to the public. Either way, a lot of material is required.

Figure 13.9 shows the Lee Tunnel casting basin, which was incorporated into the southern approaches to the tunnel. After floating out the final element, the basin was resealed, pumped out, and backfilled as a land-based operation. The approaches were then constructed through the backfilled basin.

Figure 13.9 Jack Lynch Tunnel casting basin. (Photo courtesy of Cork City Council/ Robert Bateman.)

ARTIFICIAL ISLANDS

A feature of some immersed tunnel schemes, and particularly those associated with long crossings, is artificial islands. Long crossings are often combinations of bridges and tunnels and these artificial islands form the transition between them, as illustrated in Figure 13.10, showing the Hampton Roads Tunnel in Virginia. The island will change the flow patterns in the waterway. Sometimes, this is very important, such as at the Øresund Tunnel, where the flow in and out of the Baltic Sea was a major consideration, and the final shape was the result of considerable hydraulic modeling. In Chesapeake Bay, there was no such constraint and a simple rectangular shape was used. Artificial islands may also be needed for intermediate ventilation purposes, although these should be avoided, if possible, because of the additional cost and complexity they bring to the scheme.

Construction of the islands is often tied in with the dredging of the tunnel trench as the excavated material can be used directly. It is often placed hydraulically within protective bunds direct from the dredger. As islands can be located some distance offshore, the design and construction is often complicated. The water depth away from the shorelines may require a depth of fill that can be considerable. This can lead to a large degree of settlement within the placed material; sufficient time in the program needs to be allowed for this to occur, and compaction methods and drains may be needed to accelerate this. A stable platform will be needed in which to construct the cut-and-cover tunnel. The weight of the fill material may

Figure 13.10 Transition island at the Hampton Roads Tunnel. (Photo courtesy of W. Grantz.)

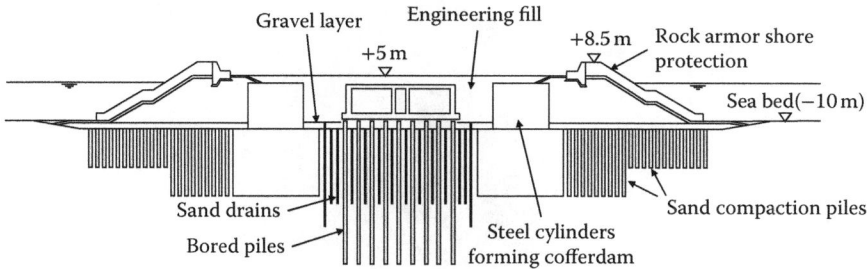

Figure 13.11 Section through an artificial island, based on the Hong Kong Zhuhai Macao Bridge Project (HZMB).

also induce a large amount of settlement in the soils beneath the reclamation. Drains may therefore need to extend to depth, or alternatively, ground improvement may be desirable beneath the footprint of the island to control settlement.

Figure 13.11 shows how complex the earthworks and foundation solutions can become. This arrangement is based on early details for two islands proposed for the Hong Kong Zhuhai Macao link. The islands are formed using many different types of construction in order to deal with the deep water and underlying soft ground.

Access for the construction of the islands will require temporary jetty or quay facilities to transfer plant, labor, and materials to the island worksite. The logistics planning requires much thought, particularly whether the island needs to be enlarged for storage of materials. It is not uncommon for a batching plant to be set up on an artificial island and require the necessary power, water supplies, and associated minor infrastructure to service the construction works satisfactorily.

Chapter 14

Construction and placing of tunnel elements

This chapter gives a general overview of the stages of construction of the immersed section of the tunnel, showing the various methods and techniques that can be used. It covers the construction of the elements and the marine activities of float up, towing, immersion, and ballasting so that the tunnel elements are ready for the internal finishing works. Joints, foundations, earthworks, and finishing works have been covered elsewhere.

CONCRETE TUNNELS

Casting facility

The casting facility needed depends on the type of tunnel being constructed. Several possibilities exist, such as a casting basin, shipyard, dry dock, ship lift, or floating dock. Traditionally, a large open casting basin has been used for a concrete tunnel and a shipyard for a steel shell tunnel.

Casting basin

For a conventional concrete tunnel casting basin site, there are several basic requirements. Access to water—preferably deep—has to be made available so that the elements can be floated away. The geology has to be suitable for dewatering a large excavation with cut off walls and pumping wells. If possible, the material excavated should be reused in the project. Access and services must be available. The power requirements of a casting basin are several megawatts, which is sometimes not available in more remote areas and has to installed at the site. The material at the bottom of the excavation should be suitable for constructing the elements, preferably without the need for ground improvement.

With shorter tunnels, it is usually possible to cast all the tunnel elements in one use of the casting basin. With a longer tunnel, multiple uses may be required. This will require the basin to be flooded to enable the first batch to be floated out. Then it has to be re-sealed, dewatered, and cleaned prior to the construction of the next batch. With a single use, the casting basin perimeter can be breached to gain access by dredging away the surrounding bund, but if it has to be resealed and used again, then a resealable gate might be preferable. Conventional dock gates or floating gates can be used.

Sometimes, it is possible to use an existing facility to build the elements. The Bjørvika Tunnel elements were built in a dock that had been built to construct North Sea oil platforms. The Shek O quarry on the south side of Hong Kong Island was used to build the Western Harbor Crossing and the Airport Railway tunnel. In the Netherlands, the Barrendrecht casting basin was built in the early 1960s for the Heinenoord Tunnel and has subsequently been used for many other tunnels. It is, however, relatively rare to find an existing facility that can be used, unless a previous builder or scheme promoter has had some foresight.

When selecting and preparing to use a casting basin, some careful thought must be made as to the spatial planning. Room for vehicle access between tunnel elements would ideally be provided to enable cranes, concrete trucks and pumps, and general site plant to move freely through the site. If a small number of elements are to be built, it may be possible to site cranes at a high level outside the basin, particularly if retaining walls are used to form the sides of the basin. In this instance, less space is needed between and to the sides of the tunnel elements and they can be squeezed into docks with only 2–3 m between the elements and to the sides of the dock, just sufficient for erecting and striking the wall formwork. Space must be left at the ends of the tunnel elements to enable the equipment for ballasting and the ballast tank components to be taken into the tunnel element. Space at the ends is also needed for lifting in and assembling the bulkheads, for mounting the Gina gasket on the end of the tunnel, and for erecting scaffolding to carry out any prestressing operations. Typically, a 5–10 m space is needed as a minimum at each end of each tunnel element for this work. Generally, little storage space is provided within the basin as it is more expensive to provide the space at the lower level, compared to a higher level around the perimeter of the basin. Once tunnel elements are being constructed, the roof of the completed tunnel elements often provides the most convenient storage location.

Dry docks and slipways

The use of shipyard dry docks is often discussed in the feasibility stages of a scheme. If one can be found, it is ideal as construction of the elements could start very quickly with the minimum amount of site establishment work to set up the dock facility. However, there are often drawbacks. Old dry docks were built for ship construction, so they are often the wrong shape for tunnel elements as they narrow towards the bottom. In assessing whether a dry dock has sufficient depth, allowance has to be made for a foundation layer beneath the elements to create the required profile to the underside of the tunnel, to protect the dock floor, and to permit water percolation under the element for float up. Docks that are available are often disused and in need of refurbishment. This is particularly true of the dock gate. The water depth over the cill must be checked to ensure it is sufficient to float out the tunnel element as well as the water depth in the channel outside the dock. Substantial rehabilitation and modification may therefore be required. In some cases, planning consent may be needed. Nevertheless, large flat-bottomed dry docks do exist, and if one can be found, it is likely to have access to services and be surrounded by working areas for plant and materials. There are a number of recent examples where docks have successfully been used, including the New Tyne Crossing in the United Kingdom, the Marmaray Tunnel in Istanbul, and the Second Midtown Tunnel in Virginia.

Large concrete tunnel elements have not so far been built on slipways. The elements are very heavy and it would be difficult to control the movement as they are launched. Launching of an element would need to be carried out in a very controlled manner to avoid overstressing the structure when it is part in and part out of the water and subject to dynamic loads. The tolerances of the sliding surface for launching are also likely to be very tight, to avoid imposing any unacceptable distortions into the concrete structure. Consideration has been given to using water skids or multiwheel self-leveling platforms to place the element in the water, but this has never been done on large tunnel elements although such techniques have been used on smaller utility tunnel elements.

Construction of concrete tunnel elements

Construction of a monolithic concrete tunnel element is typically carried out in 12 m bays. This length stemmed originally from the 12 m length of reinforcement bars that were available. This constraint no longer applies as longer bars are available, but 12 m length pours are still widely used. The section is built up from a number of individual pours—first the base, then the external and internal walls, and finally, the roof. Typically, the concreting proceeds sequentially from one end of the element to the other, although in some circumstances, other sequences may be used to reduce the effects of induced deformations. On many tunnels, the outer walls and roof slab are combined into a single concrete pour to reduce the number of construction joints in the outer perimeter of the structure. The practicality of this must always be considered according to the size of the concrete pour that this creates. The joints between the pours are normal reinforced concrete construction joints with continuity of reinforcement across the joint and cast-in waterstops in any joints that are in the watertight perimeter of the tunnel section.

The control of the dimensions of the element during construction is very important. If the section thickness varies, either due to inaccuracies in setting out or deformations of the formwork, then the element will either be heavier or lighter than designed. This will, in turn, affect the floatation characteristics and the amount of ballast required to provide the permanent factor of safety against uplift. Continuous monitoring of the as-built dimensions is needed so that, if necessary, corrections can be made as construction progresses, and so that ballasting can be fine-tuned.

As discussed in Chapter 9, the overall setting out of the element must also be considered in relation to the geotechnical characteristics of the ground on which it is being built. The reinforced concrete boxes are heavy and settlement will occur as they are cast. In order to maintain the overall setting out of the element, this settlement must be allowed for.

Associated with this settlement during construction are the stresses that are induced by it. These are locked in during the construction and must be taken into account by the designer in the overall stress analysis of the element. These locked-in stresses may well act as a constraint to the contractor on how the elements are built, so the basis on which the element has been designed must be conveyed to the contractor during the tender period. This aspect is one where the design-and-build form of contract has an advantage as the contractor and designer will collaborate from the outset on the construction sequence and the element is designed accordingly.

As important as the dimensions is the density of the concrete. This, again, must be monitored to ensure that the element complies with the weight and volume assumptions made in the design. Density monitoring is usually accomplished by casting larger than normal cubes, say 800–1000 mm cubes, as these are more representative of the density of the cast concrete than the usual 150 mm cubes. Concrete cores can also be taken from the element to confirm the in situ concrete density.

Segmental tunnel element construction follows the same principles as monolithic, except that match-cast joints that are non-structurally continuous are introduced at 20–25 m intervals. This allows better control of early age cracking during curing and also allows some articulation in the permanent structure, which reduces the longitudinal bending moments. Prestressing is used temporarily to provide structural continuity so that the element can be floated and immersed into position.

The practicalities of large-scale concrete work are the same for both monolithic and segmental concrete element construction. Indeed, they are not dissimilar to any other large-scale piece of infrastructure. The main items to consider in construction planning are as follows:

- Good concrete mix design is required for the ease of placing concrete into tall shutters, so good flow characteristics and compaction methods are needed. This is, perhaps, more important for immersed tunnels than some structures, as they are subject to high hydrostatic pressures, and remedial works for leakage or corrosion are harder to carry out once the tunnel elements are in position.
- Planning of a sufficient supply of concrete for large-size pours is necessary, along with back-up supply and systems to enable continuity of working, should a batching plant suffer a breakdown or interruption. This might be particularly important if the casting basin is in a remote location.
- Care must be taken around details in the structure to ensure that good concrete compaction is achieved. Immersed tunnel structures tend to have many cast-in items for the temporary works and also for permanent installations in the tunnel. There are some particularly sensitive features, such as waterstops at match-cast segment joints that need specific procedures for concreting.
- The accuracy of placement of cast-in items is essential. This is particularly the case for the immersion joint steel end frame.
- Curing procedures must be well planned and implemented. To produce watertight structures, measures over and above what would be used for conventional retaining structures or cut-and-cover tunnels are necessary. The types of measures are described in Chapter 11.

Øresund Tunnel fabrication facility

Occasionally, unusual casting facilities and concreting methods may be required. It was in order to meet a tight construction program that an innovative casting method was developed for the Øresund Tunnel. This is worth describing in detail as the process was quite different from anything undertaken before. The tunnel was 4 km long and the immersed section consisted of twenty 175 m long concrete elements. An exhaustive search for a suitable casting facility did not lead to any obvious candidates. An area of land was available within the Copenhagen dockyard that was large enough to accommodate a limited size casting basin, and this was developed into a factory-like facility. The elements were built using a combination of immersed tunnel and incremental bridge launching techniques. Each element was made of eight 22 m long segments. The elements were built on land above seawater level, then the area was flooded, and the elements floated out through a dock gate.

The layout of the facility is shown in Figure 14.1. The elements were constructed on what was an assembly line. The reinforcement cages were assembled in a large building at the western end of the site. The shed was of double height so that sections of prefabricated reinforcement cages could be lifted over each other into position. The completed segment cages were then slid into the concrete casting bay, which was also within the shed. On the casting bed, hydraulically controlled formwork was placed in and around the reinforcement cage. The concrete for each segment was then placed in one continuous pour. After some initial curing and striking of formwork, the completed segment was then jacked forward to enable the next segment to be match cast against it. The segments were moved on a series of six skidding beams, one under each of the tunnel element walls. Specially made low-profile hydraulic jacks supported the segment on the beams. These had automatic reaction control to prevent deformations in the young concrete and avoid cracking. Once three segments were concreted, the first had completed its curing cycle, and as further segments were concreted, they began to emerge from the shed. The segment casting process was repeated until a complete element of eight segments had been completed. The complete element was then jacked forward on the beams so that it was fully out of the shed in the shallow basin area. Because of the sensitivity of the concrete sections to possible deformations during the sliding process, the design of the beams, jacks, and concrete segments were integrated to minimize the overall deflections during the slide. At this stage, a lot of the temporary installations, such as bulkheads and ballast tanks, were fitted to the elements.

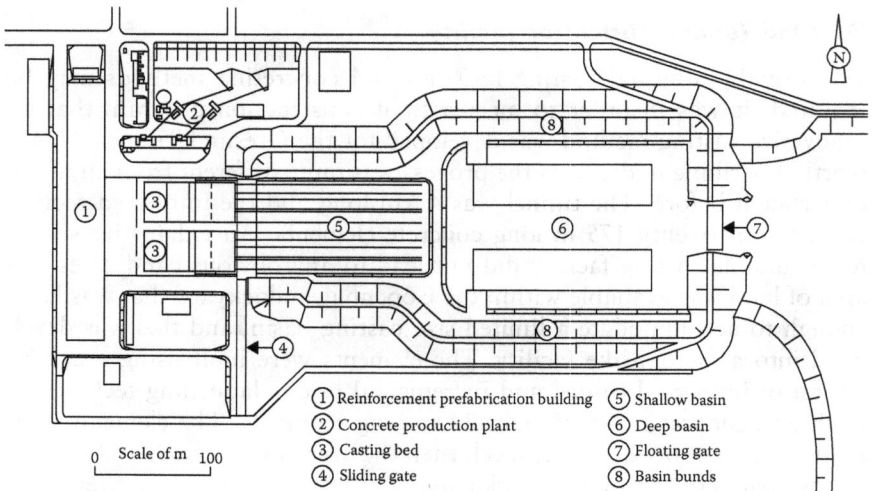

① Reinforcement prefabrication building   ⑤ Shallow basin
② Concrete production plant   ⑥ Deep basin
③ Casting bed   ⑦ Floating gate
④ Sliding gate   ⑧ Basin bunds

0   Scale of m   100

Figure 14.1 Øresund Tunnel casting facility plan.

The shallow basin was surrounded by enclosing the embankment. This was sealed by closing a sliding steel gate behind the element, between it and the concreting shed. Water was then pumped into the basin and the element floated off the beams and moved into the deeper end of the basin. The water was then drained out of the shallow basin and the sliding gate opened to enable the production of the next element, and leaving the tunnel element floating in the deep basin at sea level. After some initial fitting out within the deep basin, the element was warped out of the basin to a temporary mooring station, where it was outfitted with all the necessary equipment for immersion before it was towed to the tunnel site.

With this method, the construction of the elements was achieved to a high standard as all the construction work was carried out indoors. It enabled construction to be continuous and not suffer any delay due to harsh winter weather, and the contractor was able to meet the tight five-year program for the tunnel. Figure 14.2 shows the facility in full production. There were twin production lines and the fastest cycle time for a pair of 175 m long tunnel elements was 44 days with an average cycle time of 55 days.

Parts of this production concept were utilized for the Busan Geoje Tunnel in South Korea. Here, the warmer climate enabled construction to be outdoors in an open casting basin. Full section casting was carried out in a similar manner, but with travelling formwork and curing tents. Again, this method proved successful for the large-scale production of

Figure 14.2 Øresund Tunnel casting facility. (Photo courtesy of Øresundsbron.)

concrete elements. A similar facility to that used at Øresund is currently being used for the 6 km long immersed tunnel on the Hong Kong Macao Bridge Project in China and a larger-scale similar facility is anticipated for the Fehmarnbelt Tunnel between Denmark and Germany.

Facilities like these can be developed for any project, but generally, for most shorter concrete immersed tunnels, an open casting basin, an existing quarry adjacent to the sea, or the areas of the approach structures are the most common and cost-effective choices for construction. This type of facility will only be economic for long tunnels.

STEEL TUNNELS

For steel tunnels, an existing shipyard would be favored for the construction of tunnel elements as all the expertise and material are already available at the site. But, just as a dry dock owner may not want to disrupt his normal business to allow the construction of concrete tunnel elements in his dock, so a shipyard owner may not want to disrupt his normal business to allow the fabrication of the steel shells. Although that is where steel shell tunnel construction started, the construction techniques do differ from ship construction and the tonnages involved are relatively small for a successful yard. Also, the construction of tunnel elements would only utilize a certain set of the skills that a shipyard will have on offer, so it does not make best use of their available resources.

The fact that steel shells can be transported long distances is an advantage in finding a suitable and willing shipyard. The search can extend some way from the tunnel site. In the United States, element transport distances of 2000 miles have been successfully accomplished using semi-submersible barges, and such distances enable a large number of yards to be brought into consideration. In other parts of the world, such large distances would involve transfer between different countries which, although possible in practice, may introduce political difficulties. Major infrastructure projects like tunnels are usually government-funded and the governments may not want to see a significant proportion of their tunnel expenditure go elsewhere.

Where it is not possible to find a suitable yard, it is possible to set up a fabrication site. The space required is not as large as for a concrete tunnel casting basin, only about two-thirds of the size and, provided transportation links are good, the labor and materials can be brought in. An early example of this was the first Hong Kong Cross Harbor Tunnel built in the early 1970s. The steel plate was shipped in from the United Kingdom and the steel shells were assembled on slipways and launched sideways into the water. Figure 14.3 shows the fabrication of the Fort McHenry Tunnel shells on a small site adjacent to the river.

Figure 14.3 Fort McHenry Tunnel construction. (Photo courtesy of W. Grantz.)

The shell is assembled from a series of subassemblies in a modular fashion. First, modules that are typically four to five meters long are fabricated (Figure 14.4). For circular shells, the steel plate is laid flat and the internal longitudinal stiffeners welded to it. Then, these modules are rolled to form a short complete length of the required cross section and any internal transverse stiffeners welded into position. These completed modules are then welded together into larger subassemblies, each about 25 m long.

These subassemblies are then joined together to form the complete element. This final assembly can be carried out directly on the launching slipway. As in the case of a concrete tunnel with an external steel waterproofing membrane, the steelwork has to be watertight and all the welds must be tested for watertightness. The end bulkheads can then be fixed to make the shell watertight. Before the bulkheads are fixed, the reinforcement and other equipment are placed inside the shell as access is very limited once the bulkheads are in position. In this state, the steel shells are quite light and have a shallow draught, so the fabrication facility does not have to be adjacent to deep water. A small amount of concrete may be placed inside, on the floor of the element, to act as ballast or keel and improve the stability of the element.

Figure 14.4 Fabrication of the Fort McHenry Tunnel elements. (Photo courtesy of W. Grantz.)

The completed element is then launched similar to a ship—either sideways or end on. Launching sideways places less stress on the element, and so is the preferred method (Figure 14.5).

The steel shell is then towed to an outfitting facility where the internal concrete lining is placed while the element is afloat. The facility does not have to be extensive, but simply a jetty where the shell can be moored with an adjacent site and the concrete can be batched, as shown in Figure 14.6. This figure shows the outfitting yard for the Fort McHenry Tunnel and tunnel elements can be seen floating with varying drafts, according to the amount of internal concrete that has been placed.

The concreting operation is not straightforward as access to the shell, while it is floating, is very limited. It all has to be transported into the floating shell through holes in the shell roof. The internal concreting operation is carried out as a series of floor, wall, and roof pours. These impose stresses and deformations on the steel shell that have to be taken into account during the design stage. As the element is afloat, the analysis is quite complicated—not only as the incremental concrete loads are applied, but also as the end bulkheads put a hogging moment into the whole element.

Figure 14.5 Launching a Fort McHenry Tunnel element. (Photo courtesy of W. Grantz.)

Figure 14.6 Concreting Fort McHenry Tunnel elements. (Photo courtesy of W. Grantz.)

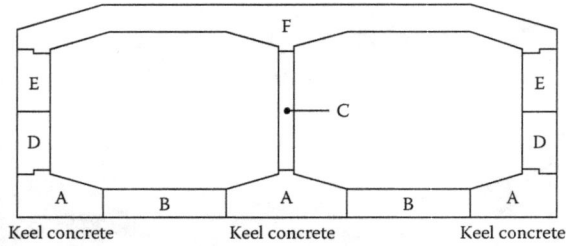

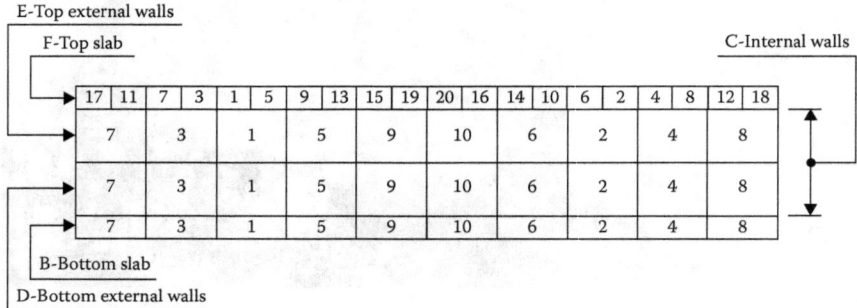

Figure 14.7 Sequence of concreting for single shell rectangular tunnel element.

A detailed concrete placing sequence has to be developed and analyzed to check that the shell is not overstressed during this process. The stages of concreting are illustrated in Figures 14.7 and 14.8 for rectangular single shell and circular double shell tunnel elements. When concreting has been completed, the elements are then taken to the tunnel site and placed, similar to concrete elements.

UTILITY TUNNELS

Because of their smaller size, utility tunnels allow a high degree of flexibility in their construction method. Initially, construction followed conventional methods, but the low weight of the small-size tunnels enabled factory precasting methods to be developed to improve the speed and quality of construction. The construction of tunnel segments using full section casting techniques was therefore developed. The segments were generally cast on-end and then lifted and rotated so that they could be assembled into tunnel elements. The light weight of tunnel elements also permitted the consideration of different methods of handling and moving of elements on land, and facilities such as slipways and ship lifts have been used in addition to conventional craneage for transferring the elements into the water. The

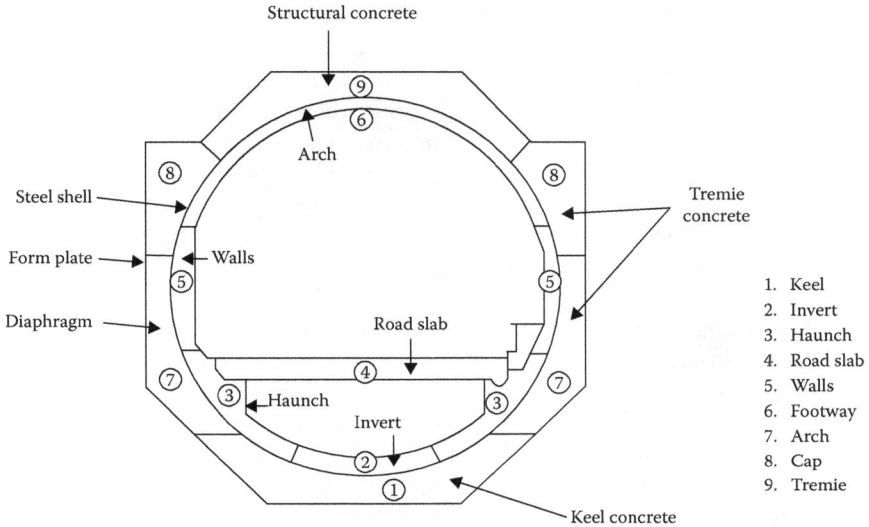

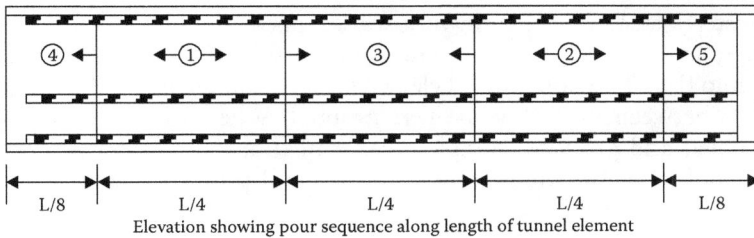

Figure 14.8 Sequence of concreting for circular double shell tunnel element.

weight of elements also allows both buoyant and non-buoyant elements to be considered as the non-buoyant elements may well have weights that lie within the lifting capacities of marine cranes.

The first application of full section casting construction methods was for circular section water tunnels. The first completed was a siphon in Egypt under the Nile at Cairo in 1964. This featured 5 m long segments permanently prestressed together into tunnel elements. This was followed shortly afterwards by a circular intake tunnel in New Zealand, formed by 2.4 m long segments epoxy glued into elements. At the same time the Rotterdam Metro tunnel beneath the Nieuwe Maas river was under construction. This used the full section casting technique on a larger scale for 10 m wide by 6 m high tunnel segments that were 15.5 m long. The segments were assembled into elements of 75 m and 90 m long. The tunnel was completed in

1966 and the metro opened in 1968. Small diameter circular utility tunnels continued to be built in the Netherlands through the 1970s and a rectangular section tunnel was built using very similar methods. Three-meter-long segments were cast on end in a single pour, then transported and lowered into a dock, where they were glued together with epoxy resin under temporary prestressing, and then the final tunnel element was permanently prestressed—a construction method reminiscent of a precast bridge deck construction. The elements were positively buoyant and ballasted externally to immerse them, using floating cranes for support.

Similar tunnels were then built elsewhere in Europe, in France, Germany, and Yugoslavia. In Asia, the first tunnel to be constructed in this manner was the Pulau Seraya Utility Tunnel in Singapore. This comprised twenty-six 100 m long elements, with a cross section that measured 6.5 m × 3.7 m. The elements were built from 3 m long segments with grouted joints that were prestressed together. This project was later to be followed by the Tuas Bay Cable Tunnel, which was built using the same method. The cross section of this tunnel was larger, at 11.8 m × 4.4 m, as it was twin bore and catered for electrically powered vehicles as wells as carrying high voltage cables and an associated cooling system. The 4 m long tunnel segments were constructed on-end in a single pour to produce concrete that was free of thermal and shrinkage cracks and therefore watertight. Once the formwork was stripped, a frame was used to rotate the segments and they were then lifted onto a ship lift, where they were assembled into the 100 m long tunnel elements, as seen in Figure 14.9.

The joints between the segments were grouted. When all the segments were assembled and grouting complete, the full tunnel element was prestressed. Detailing at the joints was important and protective rings were

Figure 14.9 Tunnel element assembly on ship lift in Tuas Bay.

used around the holes for the prestress cables to ensure they would not be in-filled accidentally by the joint grout. No cast-in waterstop was used in these joints as the grout layer was deemed to be sufficiently watertight. Prestress is permanent, so the joints will not open in service. Again, a factory-style production process was implemented that saw rapid production and curing of segments of a high quality. The assembly of segments and the subsequent fit-out process on the ship lift had to be fast and efficient to produce all 22 tunnel elements in a reasonable period of time. Both the Pulau Sereya and Tuas Bay tunnels were negatively buoyant when floated. They utilized buoyancy tanks to assist with transportation and immersion and were placed using marine cranes.

In the United Kingdom, two interesting examples of this construction method were seen for power station cooling water intakes and outfalls. The first was at Sizewell B power station, followed by the South Humber Bank power station at Immingham. The Sizewell outfall was a relatively small octagonal section tunnel formed of 100 m long elements. They were cast alongside a dock at ground level and an innovative sliding system enabled them to be moved and lowered into the dock. The Immingham tunnel was made up of 74 tunnel elements, just under 50 m in length, slightly larger at 3.6 m × 5.9 m. They had a simple immersion joint arrangement with a single rubber gasket. The avoidance of thermal cracking was not as critical for these structures as they would always be wet inside. They were therefore constructed in a conventional two-stage pour—base slab followed by roof and walls together—and the tunnel elements were constructed as monolithic structures. Once the elements were complete, they were jacked up so that a low level multiwheeled self-leveling transporter trailer could be slid beneath the element. The element was then transferred by tractor to a barge, which carried the element to its position over the dredged trench. Shear leg cranes then lifted the element from the barge and immersed it under its negative buoyancy.

While utility tunnels of this type are relatively unusual, the particular construction methods that can be applied allow high-quality rapid construction. Some of the techniques have already been transferred to larger transportation tunnels and this will probably continue as contractors strive to develop more efficient construction methods.

Floating docks

Floating dry docks have been considered for concrete element construction as some of these are large enough to accommodate complete elements. But, as far as the authors are aware, none has so far been used. Logistical problems with getting all the manpower and materials to the dock are one issue that needs to be overcome.

TEMPORARY WORKS EQUIPMENT

To enable a tunnel element to be floated into position over the tunnel trench and then lowered into position, many items of marine plant are used and a lot of temporary equipment has to be fitted to the tunnel element. The design of these temporary works is usually left to the contractor as he knows what equipment he has available to him and what methods he prefers. This is one of the areas that makes design-and-build a suitable form of contract for immersed tunnels, as the permanent and temporary works details can be fully integrated.

Bulkheads

To make the elements watertight so that they can float, the ends have to be sealed with watertight bulkheads. Originally, timber bulkheads were used but these have given way to steel or concrete construction. They are critical to the safety of the construction so should be designed with a high safety factor in mind. The end of an element has many features, such as shear keys and prestress anchor recesses, which give an irregular perimeter for the bulkhead to seal against—a key point to remember when designing the bulkhead supports and sealing details.

Steel bulkheads are generally built of 12 mm thick steel plate, backed by vertical steel columns. The columns are fixed to the tunnel roof with steel brackets and the bottoms of the columns bear against a transverse concrete beam across the floor of the tunnel. The steel plate is usually welded along its perimeter to a cast in steel angle to give full watertightness. Concrete bulkheads are similarly constructed, but with a vertical reinforced concrete wall supported by the vertical steel columns. The bulkheads have to be demolished and removed after immersion to gain access to the tunnel. Removing the steel plate is tidier than breaking up the concrete wall, but both systems are practical.

In a long tunnel, there is an advantage in making the bulkheads demountable so that they can be taken out from the completed elements and reused in later elements. This concept was used on the Øresund Tunnel where, instead of using flat steel plate, a modular system was installed to facilitate the easy removal of the bulkhead. These can be seen in Figure 14.10. Demountable bulkheads such as these require a large number of sealing gaskets between the panels and around the perimeter. This can result in a degree of leakage that has to be managed. Divers can seal the largest leaks externally with any residual leakage being collected inside the element. In principle, a demountable concrete bulkhead could also be used, which avoids the messy demolition, but it is not so practical for multiple use because of its weight.

To enable people to access the joint area between the bulkheads, watertight hatches are provided at floor level. These are the same as the watertight

Figure 14.10 Demountable steel bulkheads. (Photo courtesy of Øresundsbron.)

steel doors used in ships' bulkheads. Other openings in the bulkheads are needed to enable the pipework to be attached for pumping water into and out of the internal ballast tanks, and also to attach air ventilation ducts to enable safe working within the element. The detailing of the watertightness of these openings is important as the tunnel elements may be underwater for some time before the openings are used.

Access shafts

An access hatch is provided in the roof of the tunnel element. This gives access for people to the tunnel element while it is floating, and after it has been placed, and before permanent access is available from the adjacent element. Access while floating is needed for internal inspections, which are particularly necessary if the elements are stored for some time in a flooded casting basin or on temporary moorings. The roof hatch is reached through a long steel shaft bolted to the tunnel roof, as seen in Figure 14.11. The shaft is usually about 1300 mm in diameter and has to be long enough for the top to be always above water level. If long tows are involved, or the elements are stored in a flooded casting basin for some time, then shorter access shafts are used, which are replaced by longer ones for the placing operations.

These long shafts are vulnerable structures and are potentially subject to damage by errant shipping. Any impact could damage the shaft and result in flooding of the element, so the opening at the tunnel roof level is provided with a watertight hatch that is kept closed unless it is actually in use. The shaft fixings are also designed to fail before damaging the tunnel element. With very deep tunnels, it becomes impractical to provide such access towers during placing. They either have to be attached after placing or omitted altogether and reliance placed on gaining access from within

Figure 14.11 Typical tunnel access shaft. (Photo courtesy of NRA/DirectRoute.)

the tunnel. An interesting example of the shaft being attached afterwards was on the Marmaray project in Istanbul. Because of the deep water, the shaft was attached in sections after immersing the first tunnel element. This then served as the principle access into the immersed tunnel while it was constructed, for plant, labor, and materials, as there was no access from the ends of the tunnel until the connecting bored tunnel was completed some time later. The shaft was therefore a very robust and substantial structure in its own right and was relatively large, providing space for a staircase and hoist.

In or close to shipping channels, physical protection to the access shafts may be needed to prevent accidental impacts. During critical operations, it may be necessary to set up a ship monitoring center to monitor and coordinate marine activities. It may also be advisable to have fast guard boats stationed close to the work area to intercept vessels that are inadvertently heading for the works. The monitoring center can also send alarms to the working vessels to warn the crew of an approaching ship.

When access is fully available from within the tunnel, the shaft can unbolted and removed, and the roof hatch permanently sealed. The reinstatement can be carried out by either using pumped concrete from inside the element, or by delivering the concrete down the access shaft before it is finally removed. A typical detail through the tunnel roof slab can be seen in Figure 14.12, along with how this opening is sealed after the shaft is removed.

A steel tunnel element will require a similar access tower during immersion. It will also temporarily have larger access hatches in the roof to permit the internal concrete to be placed. These are sealed after the concreting

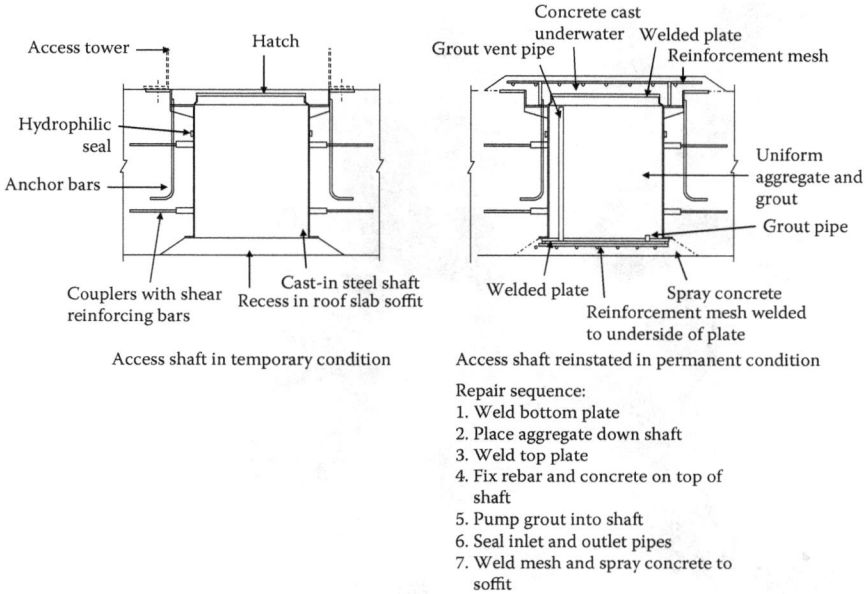

Access tower — Hatch

Concrete cast underwater Welded plate
Grout vent pipe Reinforcement mesh

Hydrophilic seal

Anchor bars

Uniform aggregate and grout

Grout pipe

Couplers with shear reinforcing bars Cast-in steel shaft Recess in roof slab soffit

Welded plate Spray concrete
Reinforcement mesh welded to underside of plate

Access shaft in temporary condition

Access shaft reinstated in permanent condition

Repair sequence:
1. Weld bottom plate
2. Place aggregate down shaft
3. Weld top plate
4. Fix rebar and concrete on top of shaft
5. Pump grout into shaft
6. Seal inlet and outlet pipes
7. Weld mesh and spray concrete to soffit

Figure 14.12 Section through typical access shaft at roof level.

operations have been completed, but this is a simple operation as the tops of the elements are accessible when the element is still floating.

Access towers do not have to be located on the roof of the element. They can be attached to the bulkhead in a snorkel-type configuration, as shown in Figure 14.13. Often used on steel tunnels because of the narrow area available due to the shape of the roof, they also have the advantage of not requiring a hole through the roof, which is a potential long-term weakness in the watertightness of the structure. Their disadvantages are having a 90° bend in the shaft and then needing to be demounted before the next tunnel element is placed.

Bollards

To control the element during float-up, towing, and immersion, or when the element is temporarily moored, bollards are attached to the tunnel roof at the corners and sides of the element. The base plate fixings for the bollards are cast into the tunnel roof and the bollard itself attached to these base plates for towing. After use, the bollards are detached from the element. The bollard loadings will depend on the loads imposed on the element during towing and placing and, indeed, the number of lines used, but generally, the working loads on the bollards are in the 250–500 kN range, although on larger tunnels this can go up to 1000 kN.

Figure 14.13 Snorkel access used on the Bosphorus Tunnel. (Photo courtesy of W. Grantz.)

Other pieces of equipment may be needed for maneuvering of the tunnel element, depending on the complexity and configuration of the winch cables, such as sheaves, fairleads, and winch stations.

Lifting lugs

To place the element, it has to be lowered from whatever sinking rig is being used. This will require steel lowering lugs to be fitted to the tunnel roof. Generally, four such lugs are used and the loads depend on the element being placed, but would typically be in the order of 1000 kN. The lugs are usually fixed to the tunnel roof, as shown in Figure 14.14, but sometimes, because of lack of space or anchorage capacity in the roof slab, they are fixed to the sides of the element close to the top of the wall. For either arrangement, the anchors for the lugs will extend some distance into the structure due to the high loads that need to be transferred. The fixing method is similar to the bollards, with the lugs being demountable from the cast-in base plate after use.

Figure 14.14 Typical suspension lug with lowering winch cable attached. (Photo courtesy of Strukton.)

Temporary supports

Some concrete tunnels and most steel tunnels are placed directly onto screeded gravel beds. Many concrete tunnels, however, are supported temporarily about 0.5 m above the bed to allow a sand foundation to be pumped into the space underneath the element. With this method, the primary end of the element is usually supported on the secondary end of the previous element, or the approach structure, if it is the first element to be placed. The secondary end is supported by a pair of hydraulic rams that project out of the underside of the tunnel near the secondary end of the element and rest on support pads, which are typically 1–1.5 m thick and 5 m square. These rams are installed in the element while it is in the casting basin, before the bulkheads are fitted. An alternative is to support the element on four rams—two at each end of the element. This is not always a critical load condition, but it nevertheless minimizes the longitudinal bending moments that the structure or the temporary prestressing within the structure has to accommodate. If a nose/chin support is used, the unsupported span length will be greater and the bending moments proportionally greater.

There are several methods of supporting the primary end of the element on the previous element. One of the most popular is a nose and chin arrangement, whereby a concrete or steel nose built onto the central wall of the primary end rests on a similar, but inverted, chin on the secondary end of the previous element. Such an arrangement is shown in Figure 14.15, which shows a primary end concrete nose that will rest on the chin, in the space between the bulkheads. Note also the steel columns in the element that provide support for the bulkhead. A small hydraulic ram is placed between the nose and chin to control the vertical positioning of the element.

The second method of supporting the primary end is to use a support beam on the roof of the tunnel. This is attached to the roof of the primary end and cantilevers out over the end of the element; it can be seen in Figure 14.16. As the element is placed, the beam comes to bear on a steel support plate on the roof of the previous element. The beams often catch in a guide, which ensures the correct horizontal alignment of the elements. Some contractors prefer to separate the guide and support functions and have a non-load-bearing male/female type guide. These support systems can also be combined with the jacks that pull the elements together.

When the element is supported in this way, it is held on a three- or four-point support system. The rams at the secondary end are hydraulically coupled so that the loads are balanced between them and the element is not subject to any torsional loading caused by uneven settlement of the secondary end rams.

If a nose/chin or supporting beam is used, the weight of the immersed element will bear either onto the newly placed sand foundation under the previous tunnel element, if the temporary supports have been released, or onto the temporary supports themselves. Either the temporary supports or the sand foundation under the previous element have to be designed to carry this load.

Figure 14.15 Nose and chin end support. (Photo courtesy of Kent County Council/BAM Nuttall/Carillion/Philip Lane.)

Figure 14.16 Primary support beams at sides of tunnel with pulling jack, center. (Photo courtesy of NRA/DirectRoute.)

Given the low density of the sand foundation, it is generally better to keep the temporary supports active until the sand foundation is placed under the subsequent tunnel element. This minimizes the risk of differential settlement occurring due to short-term increased loading at one end of the element.

The ram and jack arrangement is shown in Figure 14.17. The ram is housed within a watertight sleeve, which is cast into the concrete. The components are carefully machined to ensure a tight fit and the sleeve/ram interface features O-ring seals designed to ensure watertightness under high water pressure. The sleeve is anchored back into the concrete to resist the acting water pressures.

The temporary support assemblies should ideally be located as close to the walls of the tunnel element as possible to enable loads to be transferred directly into the walls of the structure. A number of tunnels have achieved this by housing the assembly within a recess in the external walls (Figure 14.18). If this is not possible, the jack can be housed within a frame that is tied into the base slab of the tunnel. In this instance, the design of the tunnel base slab needs to account for punching shear effects. Consideration has been given to placing them in toes on the outside of the element. This simplifies the internal construction, but their operation is more complicated as it requires activation by divers, and this option has so far not been pursued.

The assembly normally features a packing piece between the ram and the jack to enable sufficient extension of the ram beneath the base slab to achieve the required height above the trench in which the foundation layer will be placed. Once the temporary supports are released, the jack and

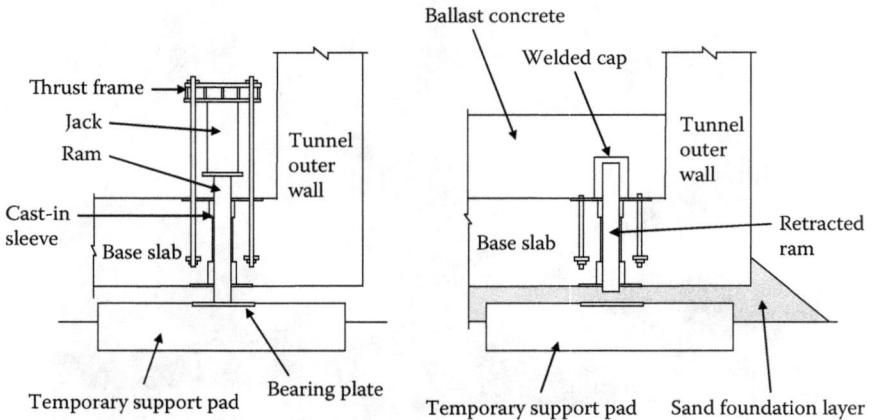

Figure 14.17 Temporary support jack and ram details.

Figure 14.18 Support ram arrangement. (Photo courtesy of NRA/DirectRoute.)

packing piece can be removed, but the ram is sacrificed and left in place in a raised position, with the top of the sleeve welded for watertightness, and enclosed in concrete within the ballast concrete depth, or within structural concrete if set within a recess in the tunnel wall.

The rams cannot bear directly onto the floor of the trench as they would punch straight into it and not provide any support at all for the element. Normally, precast concrete foundation pads, topped with a steel plate, are placed in pockets excavated at the bottom on the trench. The pockets are blinded with gravel to provide a level surface and then the concrete pads are placed. The rams, projecting from the bottom of the tunnel element as it is being placed, land on the steel plates on top of the pads. The steel plate enables the concrete pad to deal with the high bearing stresses imposed by the rams. It also enables the rams to move horizontally across the pad during the final positioning of the element. Given the frequent poor ground conditions that immersed tubes are built on, the size of the slabs can be quite large to ensure that the bearing pressures are adequate and excessive settlement does not occur. A degree of settlement can be accommodated, however, provided it is within the working range of the jack.

Pulling jack

The other equipment required to place the elements is a system to draw the tunnel element being immersed up to the secondary end of the previously placed element and provide the initial compression of the Gina gasket. There are several ways of doing this. It can be done with the winch cables used to position and control the movement of the tunnel element, but more commonly it is done with jacks and couplers. The simplest method is to place a hydraulic jack on the tunnel roof, as shown in Figure 14.16. Then, when the element is placed, a steel cable is attached from the jack to a bollard or bracket on the previous element and the two elements are pulled together. This arrangement originally required the use of divers, but as tunnels have become deeper and technology has improved, remote underwater vehicles can be used to undertake these operations.

Ballast tanks

The last major piece of equipment required to place the element is the internal ballasting system. It is a common misconception that the water is just let freely into the element to give it negative buoyancy. This would make the element uncontrollable, so the water ballast has to be contained in known positions within the element. The system usually employed consists of internal water tanks—usually four or six—depending on the size of the element. The tanks are usually of very simple construction. A PVC liner backed by plywood panels with timber walings and a steel column support system is often used. A tank of this form is shown in Figure 14.19. Panels made of steel are used on some longer tunnels as they can be dismantled and reused on later elements. Generally, one of the tunnel walls provides support for one side of the tank. Sometimes, the ballast tanks stretch right

Figure 14.19 Internal water ballast tank. (Photo courtesy of Kent County Council/BAM Nuttall/Carillion/Philip Lane.)

across the tunnel bore, making use of both tunnel walls. However, this blocks access through that bore and requires walkways up and over the tanks, so is usually only used in multibore tunnels, where alternative access is available.

The ballast tanks are provided with pipework and a pumping system to fill and empty the tanks. This system controls the buoyancy of the element during and immediately after placing. The amount of water in each individual tank can be adjusted so that the element floats correctly for transportation.

An alternative to using internal water tanks to control the buoyancy is to use vertical cylinders placed on the tunnel roof. This method is only really practical in shallow water tunnels as the top of the cylinders have to remain above water level at all times. It was used on the Fort Point Channel Tunnel in Boston, which was built in a shallow harbor and in a constrained site that did not lend itself to using conventional pontoons for immersion (Figure 14.20). With this method, water is pumped into the cylinders, which decreases the buoyancy of the element. Each time water is added, the weight increases, and so, the element sinks and displaces more water until an equilibrium point is reached. Further water is added to lower the element further until the next equilibrium point is reached, and so on, until the tunnel element is resting on its temporary support. With an array of cylinders, very accurate control of the element is possible.

Figure 14.20 Immersion cylinders used for the Fort Point Channel Tunnel. (Photo courtesy of D. Wrock.)

Immersion equipment

As well as the equipment that has to be attached to the element to enable it to be positioned, there is also a considerable amount of marine plant required. The main piece of equipment is the immersion rig that takes the weight of the element as it is being placed. In its simplest form, a pontoon is used; this is a floating steel box that can support the weight of the tunnel element in its immersion condition with the required overweight. However, there are several options for how the immersion rig and pontoons can be configured. It can be a single pontoon positioned above the tunnel or a catamaran type rig, where the pontoons float either side of the element, with a beam spanning over the element between them. The element is then suspended from the beam. Two sets of pontoons/catamarans are needed—one at each end of the element; an example can be seen in Figure 14.21.

There is one disadvantage with pontoons on top of the element. The element initially has to be able to support the weight of the pontoons before it becomes negatively buoyant. This means the element has to be lighter to start with, which may require the use of lighter-weight material during construction, or a larger internal air space—both of which are more costly. It is important that the designer knows which method is to be used so that the element can be designed accordingly. If a change from a catamaran to a pontoon system is made too late in the design process, it may not be possible to provide sufficient buoyancy in the element to support the pontoons.

Figure 14.21 Limerick Tunnel immersion pontoons. (Photo courtesy of NRA/DirectRoute.)

The immersion rig has to be positioned accurately when the element is being placed. It is held in position by anchor wires that are attached either to fixed positions on shore, or in deeper water, to marine anchor points that are themselves anchored to the bed. The winch stations are located either on the immersion equipment or the shore, as appropriate. If marine anchor points are being used, then the winch stations will generally be on the immersion equipment, but otherwise, shore-based winches can be used. The arrangement has to be designed to accommodate all the positions of the element while it is being maneuvered into position. As technology improves, it may become possible to keep the immersion rig in position with computer-controlled thrusters, but this has not been attempted yet.

The immersion is controlled from a command position, which can be a small cabin on the immersion pontoons, as shown in Figure 14.21, or located on top of the access shaft. All the positioning and survey information is relayed to the command position to enable them to coordinate and control the operation.

Other equipment is available to place elements. Large marine sheer legs, or floating cranes, can be used to support one or both ends of an element. There are some very large-capacity floating cranes available around the world, but these tend to be used mainly for smaller utility or metro type immersed tunnels. They have also been used in combination with immersion pontoons on larger tunnels where the immersion pontoon is at the primary end and the floating crane holds the secondary end. Floating cranes are expensive to hire and must be able to reach the tunnel site, but can be cost-effective if they

Figure 14.22 Lay barge on the Hampton Roads Tunnel. (Photo courtesy of W. Grantz.)

are readily available and the contractor does not have his own equipment. Another technique, popular in the United States, is to use a lay barge that straddles the tunnel, running the whole length of the element, rather than having separate equipment at each end. These either float or rest on spuds on the sea bed. An example of a floating lay barge can be seen in Figure 14.22. This was used for the construction of the Hampton Roads Tunnel in Virginia. With a roof-mounted cylinder type of water ballasting system, there is no need for any form of barge or pontoon system and the element can be placed directly with positional control by winch cables. Thus, there are a variety of immersion systems that can be used and the specific method and equipment applied will usually be the decision of the contractor.

FLOAT UP

The floating up operation is a critical activity in the construction of a concrete immersed tunnel. It is the time when you find out if the design works, whether it has been constructed accurately, and whether the element will behave as predicted. Of course, nothing is left to chance and the contractor will make a detailed volume and density calculation during construction and know precisely what behavior to expect. Prior to floating, the bulkheads and internal ballast tanks will have been fitted and tested. The water ballast tanks will be full, as they will be filled prior to flooding the casting basin so that the element does not float. Often, the elements are stored for some time on the bottom of the basin. Around the casting basin, various mooring points are usually set up to attach the mooring lines that will control the element when it is floating.

Depending on which placing method is being used, either the support pontoons can be lifted or floated into position on top of the element or the catamaran rig floated in either side of the element. This does not have to be done at this stage as these can be attached to the element after it is afloat, either inside or outside of the dock or casting basin, depending on where there is space and the equipment to fit them. An additional mooring station may be necessary if this is done outside of the basin. The electrical power for the element water ballast pumps can be coupled up and mooring lines attached to the element to control it when it floats.

There are two options for floating the element. It can either be floated on a rising tide or be floated by controlled de-ballasting. In the first method, the pumps are used to start emptying the water ballast tanks and the lifting points on the element are connected to the pontoons that will control the element while it is afloat. Sufficient water is left in the ballast tanks at this stage so that the element remains negatively buoyant and does not try to float.

As the tide subsequently rises, the remaining water is pumped out of the ballast tanks. When the tide rises high enough up the element and gives it sufficient buoyancy, it will float and come under the control of the mooring lines. The design freeboard is small, being around 300–450 mm, so once the tide has risen above this level, the element should float. There is generally a small lag before the element overcomes the bottom suction. If this point is passed and the element does not lift, then efforts are made to loosen it by pumping air through porous pipes cast into the ground under the element. If this does not work, then the water ballast tanks should be refilled so that the element does not lift off the bottom and the reason for the malfunction can be investigated. The element must not be left in an unstable buoyant condition just stuck on the bottom through suction. If it suddenly overcomes the suction, it will rise to the surface in an uncontrollable manner, which would be dangerous for the personnel involved as well as likely causing damage to the marine equipment and, perhaps, the element itself. However, such instances of non-floatation are very rare and tunnel elements are generally floated successfully without incident. The water in the ballast tanks is then finally adjusted to obtain the correct trim in the element. The element is now ready to be transported out of the casting basin to deeper water so that it will not ground when the tide falls.

With controlled de-ballasting, the water in the tanks at one end of the element is emptied so that end of the element floats first. The tanks at the other end of the element are then emptied to bring up the other end of the element. Each method has its advocates, who believe their method gives greater control of the element.

TRANSPORTATION

Transporting the element from where it has been made to the tunnel site can involve a variety of techniques. For concrete elements, this might involve a tow if the distances are large. The shells for a steel tunnel can be transported on semi-submersible platforms to the concreting jetty or towed with a small draft with minimal keel concrete placed inside the element.

Towing a complete tunnel element is not a difficult operation, but it does require careful planning and cooperation with the relevant authorities. A sufficient depth of water must be available over the tow route and, if it is not, then dredging of the approach channel may be required. Another option is to reduce the draught of the element, either by floatation devices or by reducing its weight—for example, by only constructing part of the roof thickness. The consequences of such partial construction on the long-term durability of the tunnel need to be carefully considered. For example, ensuring the same quality of concrete construction is achieved when the roof slab is cast in a floating condition in perhaps a more exposed marine environment.

Before the tow is committed, a number of checks have to be made. Paramount of these are the weather and current forecasts for the tow area. The acceptable conditions for the tow will have been determined during the design stage, when the element was designed to withstand particular wave and current conditions. The forecast for the tow period has to show that these parameters will not be exceeded during the operation. Weather forecasting is still an inexact science and its reliability reduces beyond 48 hours, so a degree of conservatism has to be built in to the go/no-go decision to tow. As well as the contractor and the authorities, the insurers may also be represented during the commitment to tow decisions.

Insurers clearly have a strong interest in such activities; they foot the bill if something goes wrong. If the tow is across the sea or through areas where rough seas are possible, they may insist on extra safeguards for the towing stage. These have included double bulkheads at each end of the element and the installation of additional pumping capacity inside the element. A small leak must not be allowed to grow into a disaster. The pumps must be capable of being brought into operation very quickly as the freeboard of the element is small and could be quickly overcome. Such decisions will, of course, have been made at an early stage to enable the necessary insurance requirements to be satisfied during the design and construction of the element. Requirements for significant additional safety measures over and above what is deemed necessary by an experienced marine contractor are relatively rare, however.

Before commencing the tow, the ballasting of the element is adjusted to give the correct freeboard and an even trim. As the dimensions of the element and the material densities are monitored during construction, the freeboard and trim can be accurately assessed before floatation. If necessary, the trim can be adjusted by placing concrete either internally or externally.

Figure 14.23 shows a tunnel element being towed with a typical tug configuration. Usually, three or four tugs are needed for full control, and here, four tugs of 5,000 HP capacity are being used to tow the 55,000 ton element. The element has the immersion pontoons in position, but this is not always the case if they can be installed at the site. Hydraulic model tests are often used to investigate the forces on an element being towed. If necessary, the element can be temporarily prestressed longitudinally to enable it to accommodate larger waves during the tow. Segmental concrete tunnel elements are, in any case, prestressed to cope with the largest predicted wave.

If long sea tows are required, then sheltered mooring sites have to be identified along the route in case the element has to be moored temporarily to allow a storm to pass. For the Bjørvika Tunnel in Oslo, the concrete tunnel elements were towed approximately 700 km around the Norwegian North Sea coast and several safe sites were identified. Considerable planning goes into the sea tow of an immersed tunnel and, because of that, the risks are identified and mitigated. The result is that, despite many projects requiring sea tows, they have been carried out successfully without a major incident.

Figure 14.23 Concrete element being towed to site. (Photo courtesy of Øresundsbron.)

Transporting steel tunnels generally involves a two-stage tow. First, the steel shell is taken from the fabrication yard to the concreting facility. Then, the completed element is taken to the tunnel site. For the first stage, the element is relatively light. It will normally have some ballast concrete cast inside to act as a keel, but even so, it will only have a draught of about 3 m. In this condition, it can be carried on a semi-submersible barge as an alternative to towing, as illustrated in Figure 14.24. Loading the element onto the barge is a critical operation to ensure that the element is not overstressed and that the combination of barge and element remains stable at all times. Steel elements can also be towed large distances in their launched condition. Here, windage must also be taken into consideration when determining the towing forces—something that does not really apply to the low freeboard concrete elements.

Once the steel element has been concreted, its behavior is then similar to that of a concrete element and the towed journey to the tunnel site is similar. If the tow is a short distance in sheltered water, it may well be sufficient to use just one tug at the head of the element with a small multicat-type craft in attendance to assist with control of the element. Pushboats are also used to move elements in inland waterways.

Figure 14.24 Steel shell element being transported. (Photo courtesy of W. Grantz.)

If an element simply has to be transferred out of a casting basin from one side of a river to the other, then tugs are often not needed. The element can be warped across the waterway using a series of land-based mooring points, or floating mooring pontoons anchored to the river bed. The lines from these mooring points are attached to the immersion equipment and/ or the bollards on the element itself. By carefully sequencing the pulling and relocation of these moorings, the element can be safely and accurately moved across the waterway. One disadvantage of this process is that, at some stage, it will probably be necessary to have a mooring line right across the waterway—either at the front or the rear of the element. This will block the waterway, albeit temporarily and generally less than a day at a time, and can present some difficulty to the navigation authority requiring careful planning in advance.

Figure 14.25 shows a typical warping arrangement as an element is winched across a river. During this operation, the winches that pull in the mooring lines can either be mounted on the lowering pontoons or be shore-based. If there are several mooring points, it is more economical to have the winch drums mounted on the pontoons. Each mooring line generally has a line pull of about 500 kN. The actions of the winchmen in slackening and tightening their mooring lines needs careful planning, and considerable expertise is required to balance the pull on the lines because of the delay, due to line lengthening and shortening as the load changes, and the inertia of the element. The winch cables can be very long; 200–700 m is not unusual as the longer the lines, the more efficient the system.

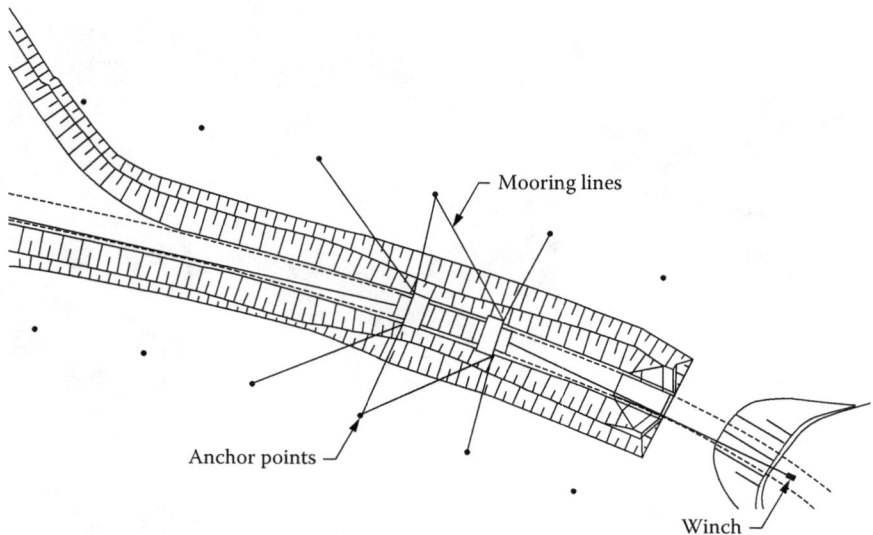

Figure 14.25 Warping a tunnel element across a river.

Figure 14.25 shows temporary anchor points positioned in the waterway to provide temporary anchorage for winch lines. This keeps the length of the winch cables short and confines the corridor in which the marine operation takes place. This may be necessary if the river banks are not close by or there is a requirement not to interfere with marine activities adjacent to the tunnel site. If there are no such constraints, the winch line cables may be much longer. In offshore conditions, other solutions may be needed. For the Øresund Tunnel, the solution developed was to construct a number of small three-legged jack-up platforms that could be placed around each tunnel element immersion position. The tunnel elements were brought in by tugs and transferred from the tugs directly to the anchor platforms, ready for immersion. This can be seen in Figure 14.26, which shows an element being immersed.

Once the element has been delivered to the tunnel site, it is either prepared immediately for immersion or it may be moored, so that it can finally be fitted out for the immersion process. These moorings can be very simple, for example steel or concrete piles to moor the element against. Elements can stay moored for long periods, depending on the element production and placing program. If that is the case, basic services such as lighting and ventilation will be needed and the interior treated as a confined space for safety reasons. Protection from marine fouling and damage from mooring cables may also be needed for the Gina seals.

Figure 14.26 Anchor platforms used for immersing tunnel elements on the Øresund project. (Photo courtesy of Øresundsbron.)

IMMERSION

The immersion process is the crux of the whole immersed tunnel method. The tunnel element has been constructed and transported to the tunnel site, but now it has to be positioned accurately into the trench in the bed of the waterway and joined to its neighbor.

In tidal waters, the state of the tide can influence when the immersion operations take place. If the currents are strong, immersion may take place during the slack water at the turn of the tide. If the elements are being placed in shallow water, then immersion may be carried out on a rising tide so that, if something untoward does occur and the operation has to be halted, the element will not immediately ground on the bottom due to a falling tide.

The element is fitted with all the temporary equipment and is attached to whatever type of immersion rig is being used for placing. The following description is generic and will vary from tunnel to tunnel, depending on the particular circumstances and the immersion equipment used.

Prior to commencing the immersion, all the anchor points to be used are checked for strength; the position of the previous element and the condition of the Gina gasket sealing plate will have been checked; and the state of the trench bottom checked for accuracy and cleanliness by soundings and video equipment.

The element is maneuvered into position, floating above the trench so that it is approximately 10 m away horizontally from the element that it is to be placed against. The element is attached to the lowering system on the immersion pontoons via the lifting lugs attached to the element roof. The protective covering to the Gina seal is removed by divers.

Inside the element, the water ballast tanks are filled until the required immersion weight and trim are achieved. The immersion weight depends on the size of the element but, generally, the element will have a negative buoyancy of about 2000–4000 kN. While the ballast water is being taken in, the lowering system is pretensioned to its maximum lowering force value. On some tunnels, an active ballast system is used, in which the ballast is continually adjusted according to the variation in salinity while the element is afloat or being immersed.

Once the ballasting has been completed, any crew members that have been inside the element controlling the valves should leave the element. During the placing of early immersed tunnel elements, it was once common for crew members to be inside the element throughout the immersion operation. They had access in and out of the element during the process via the access tower. However, the presence of people inside the element during the immersion is a risk that can, and should, be avoided. The risks involved were graphically illustrated during placing of one of the Øresund Tunnel elements, when a bulkhead failed during lowering, causing the element to flood completely and sink. The operation was planned so that no one would be inside the element and no one was injured in the incident, although the immersion crew on the sinking pontoons had a few nervous moments!

When the immersion weight is at the required value, the crew clear, and the hatches sealed, the lowering winches are released and the element lowered at a velocity of about 1 m/min. The element is lowered in a series of vertical and horizontal steps until it is approximately 0.5 m both vertically and horizontally from its final position. The control of the winches and mooring lines is a specialized operation as the element has considerable inertia. The element will not start moving immediately after the winches are activated, and equally, it will not stop immediately after the winches do. So, the final positioning of the element is done in a series of small steps and the winch operators can get a feel for the way the element is behaving.

Checks are also carried out, usually automatically, on the water levels in the sumps at the bulkheads. Inspections are sometimes made by crew entering the element to check the water levels. To minimize the risk to them, the movement of the element is halted while they carry out their checks. If the bulkhead is leaking and water builds up in the sumps, then the immersion is halted and, if necessary, the element is lifted and the bulkhead repaired.

The element is now ready to be landed, either onto temporary supports or on a gravel bed. The final approach alignment is achieved using guiding devices and survey equipment. If temporary supports are being used, the primary end support is first engaged on the previous element, then the temporary support rams are landed on the foundation pads. The negative buoyancy of the element is transferred from the lowering equipment to the temporary support system, but the lowering system maintains a nominal load in the lowering wires. The distance between the elements is measured accurately, often with distance rods operating through the bulkhead of the previous element. Divers attach the pulling device, which will draw the elements together, and carry out an inspection of the Gina seal to check that there is no obstruction as the elements are pulled together. The pulling system is activated and the elements finally close with the forming of the initial seal on the Gina gasket. This is usually checked by divers, but is also known from the tension in the pulling winch, which will reach a maximum design value that has been agreed with the gasket manufacturer.

If the element is being placed on a gravel bed, the procedure is similar. The secondary end of the element is held about 20 cm above the gravel bed while the primary end is lowered to make minimal contact with the bed. The guide system on top of the previous element is engaged. The final movement is achieved by pulling the elements together and lowering the secondary end onto the gravel bed.

When the initial Gina seal has been made, the water between the bulkheads is drained into a sump within the element or into one of the water ballast tanks. Air inlet valves are provided in the bulkhead of the previous element to facilitate this. As water is evacuated, the hydrostatic imbalance on the tunnel element causes it to compress the Gina gasket, the element will slide a few centimeters on its temporary supports or on the gravel bed as this compression occurs, and the element closes up to the previous element. When the Gina seal has compressed, the lifting systems are slackened and the element rests either on its temporary supports or on its permanent gravel bed, as the case may be.

The position of the placed element is checked. If it is not within tolerance, some realignment may be necessary. If the element is on temporary support rams, then the lowering system is reactivated to take the load off the rams and the winching system used to make the adjustment to the position of the secondary end. Similarly, on a gravel bed, the load on the bed at the primary end is minimized and the secondary end lifted clear of the bed to make the required adjustments. Often, hydraulic jacks are located in the immersion joints to achieve the required correction rather than rely on the winching system to make these relatively small adjustments.

When the element is in the correct position, the water ballast inside the element is increased to provide the required safety factor against uplift in this temporary position. The bulkhead doors to the previous element can then be opened and the Gina joint inspected for watertightness from the inside. Once the effectiveness of the Gina seal has been established and the element is in the correct position, then the immersion rig can be released and the immersion process is complete.

Deep tunnels

On deep tunnels, different techniques are needed to place the element. The positional accuracy that can be achieved with a winch wire system in 20 m of water cannot be so easily achieved in deeper water. On the Busan Tunnel, which is in 50 m of water, an innovative system was developed by Strukton Mergor: the external positioning system (EPS). This consisted of two hydraulically adjustable steel portal frame structures that straddled the tunnel element. The EPS units are clamped to the tunnel element and the lowering winches attached to lugs on the top of the EPS frame. When the tunnel element is immersed, the feet of the EPS land on the gravel bed at the same time as the tunnel element. The legs of the EPS frames are then extended to lift the tunnel element slightly and horizontal jacks are used to precisely position the element horizontally. This can be achieved to a 10 mm accuracy. Once the element is in the correct position, the EPS units are then released from the tunnel element and lifted away by the lowering winches. The EPS units are shown in Figure 14.27.

On deeper tunnels, it is desirable to minimize diver operations because of the greater risk at depth. On the Bosphorus Tunnel, diving bells were used and, on Busan, a self-propelled diving bell—more like a small

Figure 14.27 External positioning system (EPS) used on the Busan Tunnel. (Photo courtesy of Strukton Mergor.)

submarine—was used. These techniques, which are borrowed from the off-shore oil and gas industry, will, no doubt, be developed further as tunnels become deeper.

Survey techniques

During immersion, the position of the element has to be known at all times. On relatively shallow tunnels, this is achieved by installing towers on the element and using them as targets for a global positioning system survey. If close to the shore, conventional survey equipment, such as total stations, can be used. In Figure 14.16, a lightweight lattice tower can be seen on the element, with survey targets mounted at the top. The survey tower is often on the opposite corner to the access tower. Targets on these two towers, which remain above water at all times, together with the as-built survey of the element, provide sufficient information to locate the element. Often, towers are used at three corners to increase confidence in the data. This information is relayed directly to the command team, usually in the form of a real-time image of the element compared with its final position. Sophisticated software can be used to give 3D visualizations of the immersion process, as shown in Figure 14.28.

Figure 14.28 Example of 3D real-time survey software. (Courtesy of Strukton Mergor.)

As the immersion process is carried out, other systems are brought into use to obtain the correct final alignment of the tunnel element. These include distance measuring rods, which pass through the element bulkhead and can measure distance from the previous tunnel element with millimeter accuracy. Taut wire systems are also used to check the relative horizontal position between the two tunnel elements. The taut wire spans between the new and previously placed tunnel element and is kept taut as the element closes into its intended position. The angular deviation of the wire that results from misalignment can be measured and corrections made, as necessary.

After the element has been placed and the joint dewatered, further precise surveys of the element position are carried out. These are from the inside of the tunnel, through the bulkhead doors, and through the access shaft using GPS targets. These establish if any realignment of the element is needed. The immersion process is very accurate and the free end of the element is usually within 10 mm of its correct position. If it is further out than this, then the free end may need to be realigned to avoid compromising the internal traffic envelope, particularly if it is a rail tunnel.

BALLAST EXCHANGE

Once the element has been placed, it is important to stabilize it, as soon as possible, by the placing of the foundations and the locking fill. When these are in position, the ballast exchange process can begin. When the element is first placed, negative buoyancy is provided by the internal or roof-based water tanks, external ballast boxes, or water cylinders. In this temporary condition, the factor of safety against uplift is generally about 1.04. This temporary weight has to be replaced by permanent ballast that gives the element the required permanent safety factor against uplift of between 1.06 and 1.10. This permanent ballast in a concrete tunnel normally takes the form of unreinforced concrete, which is placed in the bottom of the bores under the roadway or railway.

If internal water ballast tanks are being used, which is the most usual way of providing the temporary ballast, the removal of them and the placing of the ballast concrete is a patchwork operation. The ballast tanks cannot be removed completely in one go as the safety factor would reduce and the element would be in danger of becoming buoyant. So initially, some ballast concrete has to be placed around the internal ballast tanks. Then, water can be drained from some of the tanks, those tanks dismantled, more ballast concrete placed, and so on, until sufficient permanent concrete ballast has been placed to replace the weight of the temporary ballast. The ballast exchange process is complicated by the need to keep some areas clear of ballast concrete—for example, to leave space to accommodate the installation of drainage pipework and gulleys and to allow the completion of segment joints. This means that the areas available for placing concrete ballast may be quite small.

The sequencing of this ballasting operation has to be considered when devising the layout and capacity of the ballast tanks in the first place to ensure that the replacement operation is as straightforward as possible. During the ballast exchange, it is often permitted to reduce the minimum safety factor against uplift to 1.02 for short periods of time. The partially complete ballast exchange condition can be the governing condition for determining the thickness of the ballast concrete and the size of the ballast tanks. Once the element has been ballasted to the full permanent safety factor, then the internal finishing work can begin.

LESSONS LEARNED

It is, unfortunately, generally true that we learn most lessons when things go wrong. If all goes well, we often might not perceive what risks are being taken and how close to disaster we might be. Modern risk-based procedures, aligned with improvements in health and safety legislation, have made great progress in mitigating risks and reducing accidents in the construction industry. Nevertheless, construction is a hazardous activity, and marine construction particularly so. In this section, we will describe a few occasions when things did go wrong during immersed tunnel construction and the lessons learned from them.

Great emphasis has to be placed on the geometric and material densities during construction so that the bulk density of the completed element is within the design tolerances. If it is too heavy, then it will not float, and if it is too light, then it may be difficult to find room for sufficient permanent ballast to provide the required safety factor against uplift. An associated issue is the placing equipment that is going to be used. The Conwy Tunnel in the U.K. was designed on the basis that a catamaran-type placing rig would be used, which places no additional load on the element while it is floating. However, the contractor decided to use two pontoons to place the element. These pontoons were placed on top of the element, so initially the element had to be sufficiently buoyant to support them while it was being floated to the tunnel site. The two pontoons weighed a total of 7000 kN, which had not been allowed for in the design. Fortunately, the decision was made early on during the construction of the elements and it was possible to save an equivalent amount of weight in the construction. The element had a concrete protective layer on the roof, and by using lightweight concrete for the protection, sufficient weight was saved to compensate for the additional weight of the pontoons. If the method of placing has not been decided at the outset, then the element should be designed to cope with the additional loads of the pontoon method of placing. If they are not subsequently used and the element freeboard is too large, then some permanent ballast concrete can be placed inside the element before floatation.

Positioning of the element should be accurate. The primary end is positively located on the previous element, so the only variable should be the position of the secondary end. This is subject to some variation as the initial position of the secondary end depends, to a large extent, on the accuracy with which the end frames have been set out. The Gina seal bears against the steel end frame on the previous element. If the plane of this end frame is not correct, then the secondary end will be misaligned during the initial placing as the "direction" of the element is determined by the plane of the end frames. If, for example, the end frame is 25 mm out of plane horizontally, then as the aspect ratio of the element will usually be about 5:1, the secondary end will be 125 mm off line. This can be rectified after the initial positioning by simply lifting the secondary end and pulling it across to the correct position. This will, however, result in differential compression in the Gina gasket, which will try to even itself out and push the element off line again. A more positive method is to install jacks in the element joint. Then, if the position of the secondary end has to be corrected, it can be lifted and swung across and the primary end can be locked into that position with the jacks. Once the element has been backfilled, the frictional forces from the backfill will prevent it going off line again and the jacks between the elements can be released so that the long-term thermal movement of the joint is not compromised.

Water density has to be assessed carefully, with consideration given to extreme conditions while the element is in a temporary state. A storm occurred on one project shortly after a concrete tunnel element had been placed. The storm washed a great deal of sediment down the river, which increased the density of the water. The secondary end of the recently placed element started to float. As this happened in the short time between placing and ballasting, the immersion equipment was still on hand and the element was repositioned and stabilized with additional ballast.

A similar event can cause problems for the tunnel foundation. The biggest disadvantage of the sand foundation is that it is placed after the tunnel has been immersed. The foundation operation usually starts immediately after the element has been placed. On one occasion, when this was not done straightaway, a storm again increased the sediment load in the river and this clogged up the entire space under the element. This took several months to rectify as the material was too jammed under the element to be removed by a vacuum pump and a steel plough had to be fabricated, which was drawn back and forth under the element to clear out the sediment. There are also stories of the weight of the element itself being used as a compactor to compress the material under the tunnel to get the tunnel down to the correct level, but this is not recommended.

Failure to clean the trench properly before placing the element can lead to unwanted settlement in the elements. As well as from subsequent level surveys, this can become apparent through shear cracks forming in the tunnel walls, particularly the lighter central walls, as increased load is transferred from the section being affected by the settlement. The causes of the settlement could be insufficient attention to the placing of the sand foundation so that a void is left under the tunnel, or—and this is more likely—the trench is not cleaned properly before immersion and some softer material is left in situ. The consequences can be serious. In extreme cases, it has been necessary to drill holes through the tunnel floor and inject grout under the tunnel to stabilize the foundation and prevent further settlement. As might be imagined, this is not a simple operation. Special drilling equipment is needed to prevent water from entering the tunnel when the floor is penetrated, and isolating valves are needed when the grout is pumped out through the tunnel floor. Pancakes of grout are installed under the affected area of the tunnel.

In some waterways, there is also the possibility of the bed material forming a fluid mud, which can flow down the sides of the trench and under the element. This possibility needs to be considered as part of the geotechnical assessment of the bed material.

We have mentioned earlier the inadvisability of personnel being inside the element while it is being placed. However, simple things can go wrong. The well-publicized failure of a bulkhead on element 13 of the Øresund Tunnel illustrates this—yes, it really was element 13. The buoyancy of the element during immersion depends on the end bulkheads. These are robustly designed and their supporting frames are usually anchored to the roof by large steel brackets and bear against concrete beams on the floor.

Unfortunately, in one instance, some of the hairpin reinforcement bars that anchored the beam to the floor were left out. The result was that, during immersion, as hydrostatic pressure came onto the bulkhead, the capacity of the beam was insufficient and it sheared off, resulting in failure of the bulkhead. The element rapidly filled with water and sank. Any personnel inside would not have had a chance to escape, but no one was injured because the immersion procedure was designed such that no personnel were inside during immersion. The construction site was very well run with a very strict quality control regime, but the incident illustrates that the unexpected does happen and our procedures and processes have to be as robust as possible.

Marine activities are hazardous. There have been incidents, fortunately very rare, of access towers being hit by errant shipping, which emphasizes the need to provide watertight hatches at the bottom of the shaft. It also highlights the need to keep watch for vessels straying out of the designated navigation fairways. There have also been incidents where spuds from temporary marine barges have dropped onto the roof of tunnel elements. Fortunately, tunnel structures are relatively massive and resilient to such events, but it serves to remind us that safety is paramount and considering the unexpected, both in design and construction, is an important part of the engineering and planning process.

Attention to detail is very important in immersed tunnels. What may seem a trivial and unimportant detail can subsequently have consequences out of proportion to the original mistake. A frequent problem is with leaking segment joints in concrete tunnels. As described earlier, segment joints have groutable waterstops running right round the perimeter if the tunnel is to make the joint watertight. The waterstops are fitted with sponge tips and the principle is that these tips are injected with grout to seal any leakage paths in the concrete round the waterstop itself. When they are installed, the waterstops have grout tubes fitted and when the concrete has hardened, grout is injected through these tubes to ensure the seal.

It sounds simple enough to inject the grout in sequence around the tunnel perimeter, checking that grout is issuing from the next vent, thus ensuring that the length of waterstop between the two vents has been grouted, before moving on round the section. However, experience has shown that this operation is not always carried out successfully, resulting in water leaking through the joint. This is not especially serious for the durability of the tunnel and does not affect its overall integrity, but it is not good public relations if a motorist spots a drip falling on his windscreen. It causes a loss of confidence in the use of the tunnel, which can take some time to recover. The solution is relatively straightforward and is to inject grout around the tips of the waterstop to seal the leak. This involves drilling grout injection holes through the concrete to intercept the ends of the waterstop. There are firms which now have considerable experience with these procedures, but to undertake them, the tunnel bore has to be closed, which may not be convenient for the tunnel owner.

Incomplete grouting is not the only hazard to the correct functioning of the waterstop. Waterstops have been nailed to formwork during construction, and mishandled so that the steel part of the waterstop is creased, making it likely that a small void will be created during construction. The construction operatives do not always understand the importance of these waterstops nor the subsequent remedial work required, and so they are not always treated as delicately as they should be.

A similar level of care is needed in the construction of the steel end frames at the ends of the element. To avoid setting up electrical corrosion cells, the end frames are usually isolated from the steel reinforcement. Tests can be, and are, undertaken during construction to check that there is no electrical conductivity, but in practice, subsequent testing sometimes shows that there is such a connection. The consequences are serious, particularly if the element has been placed before the fault is found. The remedy is to install cathodic protection, either sacrificial anodes or an impressed current system. Neither option is simple, given that the tunnel is backfilled and the rock protection placed. In an extreme case, it has been necessary to remove the tunnel rockfill so that divers have access to the end frame. They could then install the system, but it was necessary to drill holes through the roof of the tunnel to make the necessary cable connections. The whole operation was expensive and time-consuming and out of proportion to the original fault of some reinforcement being in contact with the end frame. Again, a small detail can have serious repercussions.

Corrosion of the steel reinforcement can be caused through cracks that are the result of the method and sequence of construction. One early concrete tunnel was constructed as a series of independent 12.8 m long segments that were then stitched together by 1.8 m long infill sections. The tunnel was given an external waterproofing membrane and performed satisfactorily for several years. Over time, however, the flexing of the tunnel due to temperature changes, and the shrinkage of the concrete, led to cracks forming at these construction joints. These cracks were full-thickness cracks, and so led to leakage and corrosion of the reinforcement. Extensive research was needed to develop a solution, which was to install longitudinal prestressing to close the cracks—an expensive operation that led to the closure of the tunnel for several months. The long-term consequences of the method of construction were not understood at the time.

The control of cracking has been a common area where difficulties have occurred. There have been too many projects where contractors have attempted innovative approaches to eliminating early age cracking in lieu of the tried and tested method of concrete cooling. Many contractors have experienced concrete technology experts and some have attempted to achieve control through concrete mix design alone. The occurrence of cracking has led to experts being brought in subsequently to develop and implement cooling and curing programs. An example where innovation was successful is the Øresund Tunnel and the implementation of full section casting with no cooling. This required extensive testing of materials and the analysis of multiple scenarios to ensure the method was successful along with full trialing of the methods.

These are some of the things that have gone wrong. However, although they were inconvenient and, in some cases, expensive to rectify, none was sufficient to cause the tunnel project to be abandoned or delayed unduly. In fact, the Øresund Tunnel was opened early despite one of the elements sinking accidentally. We have learned from these incidents so that the design and construction procedures enable immersed tunnels to be constructed in safety and with confidence.

Chapter 15

Finishing works

Once the elements forming a tunnel have been immersed, joined, and made stable through ballasting and backfilling, the interior finishing works can commence. There are a great many operations that have to be carried out within a tunnel before it can be opened to either road traffic or rail services. Some of these must be carried out sequentially in order to enable access for other operations. This is the case particularly for early finishing operations when access is first obtained into the newly immersed tunnel. Later, it will be possible to undertake multiple finishing operations at one time, but these need to be planned carefully to ensure that access through the tunnel can be maintained. This usually requires the finishing operations to be staggered through the tunnel such that the finishes are installed progressively through the tunnel in a logical order.

A typical sequence for the finishing works inside a tunnel would consist of the following:

1. Install Omega seal to immersion joint.
2. Exchange temporary water ballast for permanent concrete ballast. Initially, only sufficient concrete ballast is needed to give the temporary factor of safety against uplift.
3. Remove bulkhead.
4. Complete the immersion-joint shear keys and joint infill.
5. Cut temporary prestress if it has been provided.
6. Place the remaining concrete ballast to give the permanent factor of safety against uplift.
7. Install fire protection.
8. Apply finishes to walls (paint, cladding, or tiles as applicable).
9. Install cabling and other mechanical and electrical (M&E) installations.
10. Complete road surfacing/lay rail track.

A typical general program for carrying out the finishing works in a tunnel is shown in Figure 15.1. This is just one example of the sequence of

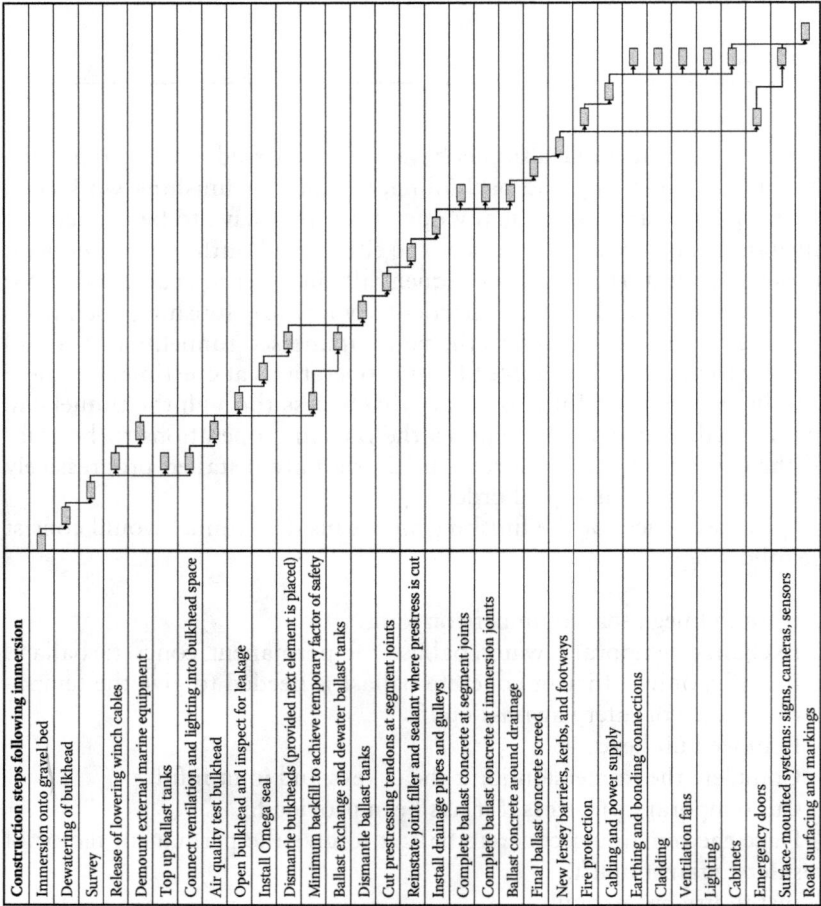

Construction steps following immersion

Immersion onto gravel bed
Dewatering of bulkhead
Survey
Release of lowering winch cables
Demount external marine equipment
Top up ballast tanks
Connect ventilation and lighting into bulkhead space
Air quality test bulkhead
Open bulkhead and inspect for leakage
Install Omega seal
Dismantle bulkheads (provided next element is placed)
Minimum backfill to achieve temporary factor of safety
Ballast exchange and dewater ballast tanks
Dismantle ballast tanks
Cut prestressing tendons at segment joints
Reinstate joint filler and sealant where prestress is cut
Install drainage pipes and gulleys
Complete ballast concrete at segment joints
Complete ballast concrete at immersion joints
Ballast concrete around drainage
Final ballast concrete screed
New Jersey barriers, kerbs, and footways
Fire protection
Cabling and power supply
Earthing and bonding connections
Cladding
Ventilation fans
Lighting
Cabinets
Emergency doors
Surface-mounted systems: signs, cameras, sensors
Road surfacing and markings

Figure 15.1 General finishing works program.

works, but it serves to illustrate the number of separate tasks required and their sequential nature. It frequently takes a year or more to bring a tunnel into operation after it has been placed.

TUNNEL ENTRY PROCEDURES

There are some important safety considerations to be addressed in the early hours, days, and weeks after gaining entry into a newly immersed tunnel. These include

- Maintaining a sufficient number of bulkheads at the exposed end of the tunnel to provide a safe working environment within the tunnel. This point is expanded on later in this chapter.
- An immersed tube should be treated the same as any other tunnel in terms of controlling the entry and exit of personnel, plant, and equipment. Only authorized entry should be permitted, with strict control on the plant and materials allowed into the zone past the bulkheads. Air quality in this area is likely to be suboptimal and should not be compromised further by anything that may generate fumes or reduce oxygen levels.
- Safety equipment should be provided at the main entrance to the tunnel and at the bulkhead entry point. This should include breathing apparatus, buoyancy aids, torches, firefighting equipment, and first aid points with the usual medical and first aid provisions. There is no great history of accidents occurring in this environment, but it should be recognized that it is a higher-risk environment compared to conventional construction sites and should be managed accordingly.
- Air-monitoring equipment should be installed not only in the tunnel generally but also, importantly, in the tunnel elements beyond the bulkhead line.
- Temporary ventilation should be installed as soon as possible. The normal types of temporary ventilation provided in all tunnel construction can be used, although these can only extend as far as the bulkheads. It is not generally necessary to provide a separate system beyond the bulkheads provided air quality is monitored continuously. If it is required beyond the bulkhead, then this must be allowed for in the bulkhead design by making provisions for openings to allow the ventilation ducts to pass through. Any such openings must be capable of being closed off rapidly in the event of leakage occurring beyond the bulkhead.
- Temporary lighting should be installed as soon as possible.

The narrowness of the bulkhead compartment is illustrated in Figure 15.2.

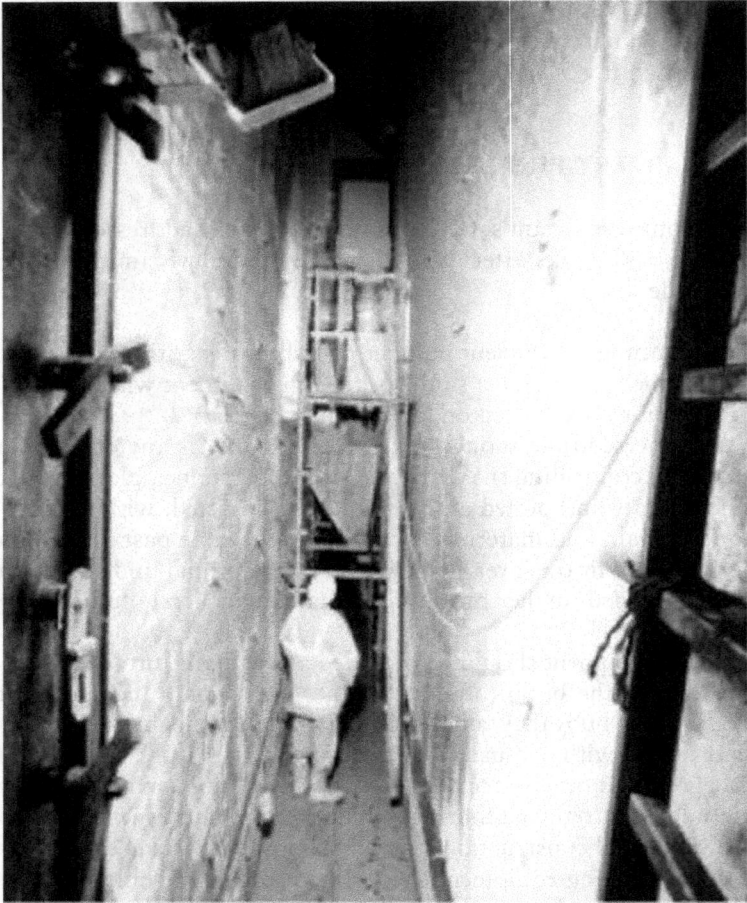

Figure 15.2 Bulkhead compartment. (Courtesy of Kent County Council, BAM Nuttall/Carillion/Philip Lane.)

JOINTS

There are several activities required to finish the construction of the tunnel immersion joints from the inside of the tunnel. These include bulkhead removal, Omega seal installation, shear key formation and casting of infill concrete, cutting of prestressing, and finishing works associated with seismic joints and joints for steel tunnels.

The completion of the joint works is particularly time-consuming due to the sequential activities; the required steps are illustrated further in Table 15.1. Again, this sequence is just a typical example, but it illustrates the complexity of the internal finishing operations. Careful planning is

Table 15.1 Typical joint finishing sequence

	Steps to complete an immersion joint
1	Access bulkhead space after dewatering
2	Install temporary lighting and ventilation
3	Cleaning of joint gap
4	Erect scaffold for Omega
5	Install Omega seal
6	Pressure test Omega
7	Dismantle bulkheads (provided the next element is placed)
8	Corrosion protection to Omega clamping systems
9	Form base slab shear key
9a	Fit formwork over Omega seal
9b	Fix reinforcement
9c	Cast concrete
9d	Install joint gap seals
10	Form joint infill concrete to base
11	Form wall shear keys
12	Complete joint infill concrete to walls
13	Complete joint infill concrete to roof
14	Install remainder of joint seals

required to optimize the construction schedule, and the time to complete all of these activities should not be underestimated. Once the joint finishing works are complete, the general internal finishing works of placing ballast concrete, surfacing, and jointing within the surfacing can be undertaken.

Omega seal installation

It is common practice to install the Omega seal while the bulkheads on either side of the joint in question are still in place. This is partly due to the desire to progress with the internal works as quickly as possible, but this is also a good practice from a safety perspective. It ensures that a double seal against water ingress to the tunnel is in place at all times, whether it is in the form of bulkheads or joint seals. There is a view that if the tunnel element has been backfilled, even only partially, then it is not going to move. Thus, as the watertightness of the seal has already been demonstrated, the bulkhead can be removed first. The authors prefer the safer procedure of installing the Omega seal before removing the bulkhead, even though this makes the installation more complicated.

The access for this operation is difficult as there is only ever between 1 and 2 m width between opposing bulkheads. This is illustrated in Figure 15.2, which shows the joint in the Medway Tunnel in the United Kingdom during construction. The first operations required in the joint space are cleaning and erecting lighting and scaffolding to access the full perimeter of the tunnel. Cleaning can be a significant operation, particularly if the tunnel element has been afloat for a long period of time. Marine fouling will occur, and high-pressure washing and other cleaning methods will be required to get the immersion joint steelwork back to its original condition such that the Omega seal can be clamped to a smooth, clean surface and therefore achieve the necessary degree of watertightness. It is likely there will need to be some minor repairs to paintwork and cleaning of threads to receive the seal clamping bolts. Protection of threads in advance of flooding the casting basin with grease, tape, or caps is advisable to minimize the buildup of corrosion products in the short term while the tunnel elements are afloat. The immersion joint should be temporarily boarded over on the base slab to allow easier access and to protect the components of the immersion joint.

When the joint has been cleaned, the Omega seal can be clamped into position. The seal will need to be brought into the joint as one continuous ring and is therefore heavy and cumbersome to maneuver. Similarly, the clamping bars are heavy components, and it is advisable to install some winches and simple lifting equipment on the scaffolding. The Omega should be lifted to the roof of the tunnel in the first instance and clamped at the corners initially. Although these corners are preformed, and the clamping components are specially designed to ensure that clamping pressure is exerted right into the corners, this is the most difficult part of the operation. If there is any difference in level between the two adjoining tunnel elements, there will be a height difference between the adjacent clamping surfaces and the seal will need careful maneuvering to get a good fit into the corners. When the corners are clamped, the straight lengths between them can be clamped into position.

Testing of the Omega seal to ensure that the clamping has made the seal truly watertight is required. This is usually done as an air or water pressure test in which the space between the Omega seal and the Gina seal is pressurized to a pressure in excess of the water pressure likely to be experienced. However, this may not necessarily be to the full factor of safety used in the design. The test should be carried out over a 24-hour period, and initial pressure drop-off needs to be corrected due to the short-term creep behavior in the rubber materials of the seals. The test should be done while the bulkheads are still in place as any failure of the test would require reclamping of the seal.

The testing facilities comprise simple pipework and valves from the joint space through to the interior of the tunnel. If the test used is a water test, the water will need to be fully evacuated after the test, and the pipework should be positioned to enable this. The testing pipes can be capped off when the test is complete. It is a good idea to do this rather than sealing them permanently, so that the pipework can be accessed later in the life of the tunnel. If a leak occurs at the tunnel joint in the future, this pipework can be used to determine if the leak is coming past the Gina seal. They can then also be used to pump out any water that might collect between the two seals and to permit the introduction of corrosion inhibitors. For these reasons, the end of the test pipe should be as low as possible in the tunnel wall and readily accessible for inspection.

Bulkhead breakout

Breaking out of a tunnel bulkhead can be started when there are a sufficient number of bulkheads in place to offer the necessary level of safety to operatives working in the tunnel. The minimum number of bulkheads that should always be left in place until the tunnel is complete is usually considered to be three. This number comprises the single bulkhead at the free end of the tunnel as far as it has been completed, and the pair of bulkheads at the first immersion joint in from the free end of the tunnel. Controlled entry through these bulkheads can be allowed to perform operations such as ballasting. This access is through the bulkhead door, which is manned to ensure that it can be closed quickly should any leakage occur at the joint or into the last element placed. Provided the last element has been ballasted securely, the risks of movement of that element and any consequential leakage should be minimal, but the retention of the bulkheads is a prudent safety measure that has been adopted worldwide.

The method of breaking out a bulkhead depends on the type being used. Concrete bulkheads are simply demolished with pneumatic tools and the supporting steelwork is dismantled. The steel beams can be recovered and the concrete debris removed. Needless to say, breaking out the concrete in such an enclosed space requires the dust and debris that is created to be managed carefully. Steel bulkheads are probably quicker to remove. If they are of modular form, they can be unbolted and removed for reuse in later elements. The perimeters are often welded to supporting steelwork for watertightness. These welds will need to be burned through, so good ventilation of the area and possibly fume extraction are important. It is also vital during these bulkhead removal operations that the joint seals are adequately protected.

Shear keys and joint infill

There are a variety of shear key solutions available for tunnel immersion joints, which are described in Chapter 9. Whichever form of shear key is used, it can only be installed once the Omega seal is in place. There is no need to wait until the bulkheads are removed before the shear keys are installed. The timing for their installation may depend on the likely settlements that are anticipated, and it is common for some of the settlement to be allowed to occur before the shear keys are formed to reduce the magnitude of forces they have to carry. Therefore, they are more typically constructed some time after the bulkheads are removed, the tunnel has been backfilled, and much of the ballast concrete has been placed. Temporary shear keys used to place a tunnel element can be removed once the tunnel element is securely founded and locked into position by the backfill. If discrete concrete shear keys are to be formed in the walls and base, the remainder of the infill concrete can be placed and the shear key sections left to the appropriate time. This can result in a complex sequence of concrete pours to complete the infill.

It may also be the case, however, that the tunnel designer wants to restrain shear movements continuously and not allow the tunnel elements to settle freely. This may be the case if there is concern about large differential settlements occurring that would compromise the performance of the joint seals, or create difficulty in achieving a smooth road or rail alignment through the tunnel. In this case, the permanent shear keys need to be installed prior to the removal of the temporary keys, and this sequence needs to be considered when designing both temporary and permanent key arrangements to ensure that this transition can be achieved. This might entail the use of demountable shear keys on the internal walls of the tunnel, leaving space for the permanent keys to be placed on the external walls such that the temporary ones can then be removed.

One of the difficulties of any shear key arrangement is getting correct alignment between two adjacent tunnel elements. If tunnel elements are placed on gravel with no vertical shear restraint, independent settlement of the elements can occur, giving rise to a differential level between the elements. Equally differing surface levels on the finished gravel bed between adjacent elements can cause misalignment. If the infill and the shear keys are to be constructed of concrete, this misalignment can mean that reinforcement bar couplers cast into one side of the immersion joint do not align with where the shear key needs to be constructed. Designers should as far as possible allow tolerances in construction details to accommodate a degree of misalignment. The alternative is to undertake remedial work on site by drilling and grouting new reinforcement bars, but this can be costly and time-consuming.

Concrete shear keys and the general concrete joint infill will need simple formwork to support the face of the concrete adjacent to the Omega seal. The formwork must provide a space around the Omega seal that allows the seal to flex as the immersion joint compresses due to longitudinal expansion of the tunnel structure due to thermal variations. This formwork is sacrificial but can be conventional timber construction. The space created to the inside of the Omega seal used to be filled with a fire-resistant insulation material to protect the joint seal from high temperatures in the event of a fire in the tunnel. However, most modern tunnels have fire protection applied to their interior surfaces, and this can overlap the joint gaps at immersion joints and thereby limit temperatures in the void behind the infill concrete. This has been successfully demonstrated by fire testing, and this means the insulating material is not necessarily required.

Concrete infill and concrete shear keys in the roof of the tunnel are difficult to form with conventional concreting techniques. Some tunnels have used spray-applied concrete to the roof, which can provide an effective solution. Other tunnels have used high-flow pumpable mixes. These are generally subject to extensive pretesting to verify the flow characteristics are suitable, and this is now the more usual method rather than spraying.

The quickest method of establishing a shear connection at the joints is to use steel shear keys and provide simple cover plates in front of them. The cover plates provide access for future maintenance and inspection, but they will also need to provide fire protection so that the shear key components are not exposed to high temperatures that could be detrimental to the materials of the key or bearings. Steel shear keys generally need some form of simple bearing installed with provisions to accommodate the longitudinal movements that will occur. This can be an elastomeric pad or similar, but it needs to be a low-maintenance bearing with a long design life.

Cover plates can be used to the walls and roof of the tunnel, but concrete infill over the immersion joint is needed in the base of the tunnel to transfer loadings from the ballast concrete beneath the road or railway.

Cutting prestressing

If the immersed tunnel elements are formed by segmental concrete construction, they will be temporarily prestressed together for the purpose of floating, transporting, and immersing the elements into position. This prestress will need to be released in order to allow the tunnel to settle and articulate in a permanent, backfilled condition. The cutting operation needs to be carried out as quite an early operation once access is available through the tunnel. The work would not be carried out until the tunnel bulkheads are removed as the operation of cutting the prestressing tendons may require reasonably large equipment and the bulkhead door openings would usually not permit access for such equipment.

There may be other items of temporary works obstructing the segment joints where the prestressing tendons need to be cut. For example, ballast tanks are likely to be continuous over large parts of the tunnel elements and will obstruct access to the segment joints in the base slab. Cutting of prestressing will need to wait until the ballast tanks are removed. To do this, temporary water ballast will need to be replaced with permanent concrete ballast. It will be important to plan the sequence of the placing of the permanent concrete ballast so that segment joints can be accessed for cutting the prestress tendons.

Ballast tanks will also most likely obstruct access to the roof slab for cutting of prestressing tendons, and this operation will also need to wait until the tanks have been removed. The sequence in which the tendons are cut will need to be determined by the designer. It is important for the designer to ensure that there will be no unacceptable buildup of stresses around the perimeter of the tunnel structure at the joint, and to ensure that no exaggerated settlement occurs. If, for example, a single segment joint is inaccessible for cutting prestress ducts because of some temporary obstruction, but the remainder of the tunnel is allowed to articulate by releasing the prestressing, then the segments with the prestressing between them would act as a monolithic structure, and it is possible that settlements would not occur evenly as the segment joint is not allowed to rotate. Equally, if one or two tendons are left uncut, they would strain as the joint opens. The detailing can be such that this does not damage the structure, but it is a situation that is best avoided. Although it is important to consider the sequence of tendon cutting, experience shows that this is not generally onerous for the structure or restrictive to the other finishing works in the tunnel, and some flexibility in the sequence is possible.

Prestressing tendons are typically located in the roof and base slab of the tunnel and will be set at a minimum possible distance from the inside surface of the concrete. This minimum depth is needed to locate the tendon and the duct in which it is carried, to be clear of the reinforcement mats at the surface of the structure and to provide space around the prestress duct to ensure that concrete can flow properly during casting. Typically, the centerline of the ducts will be at about 300 mm from the surface at the tunnel segment joints where the cutting needs to be carried out. In order to cut the full diameter of the tendon and duct, a pocket or recess in the structural slab of about 400-mm depth will be required. Individual recesses can be provided at each prestressing tendon or, if the tendons are grouped, a larger recess can be provided to allow the cutting of a group of tendons.

Tendons can be cut using mobile hydraulic cutting devices or can be saw cut. If the tendons are to be saw cut, a large-diameter saw will be required and this would need to be mounted on the inside surface of the tunnel in order to operate safely. The tendons can be cut at the joint gap or to one side of it. If cutting is to be carried out at the joint, it is relatively simple to widen the segment joint gap locally to allow access for a saw blade. If a hydraulic cutting device is to be used, then the joint gap may not be the best location as the gap for access would need to be considerably wider. A separate pocket in the structure just to the side of the joint may be a better approach in this instance. In either case, a local repair will be required to return the structure, or the joint gap and joint sealing materials, to its intended condition.

Safety is an important consideration when using high-level cutting equipment, and appropriate precautions should be taken for protecting the workforce and equipment operators. There is no risk attached to the release of a prestress cable as it is cut. The tendons will be grouted within the ducts, and although the force being carried in a tendon will be released back into the surrounding concrete, this will be transferred through the grout progressively as the individual strands making up the prestress tendon are cut. There should be no sudden release of force that generates any movement or facture or damage to the concrete structure. Therefore, given that safe access is provided and suitable safety equipment is available to the workforce, the operation can be carried out in a safe manner.

Installation of seismic joints

Seismic joints generally feature a restraint mechanism that controls the amount by which the joint may open due to the ground movement occurring in a seismic event. These are most likely to be located in both the roof and the base of the tunnel at the immersion joints, although some tunnels have managed to restrict the restraint mechanisms to the base of the tunnel. The Busan Tunnel also incorporated seismic features at the segment joints along the length of each tunnel element. These were relatively simple postfixed Omega seals that spanned the joints that would be installed in the normal way for the seal.

The seismic restraint devices can be installed once the tunnel elements are immersed and access is available to the immersion joints. Omega seals need to be installed before the seismic restraints are connected; otherwise, they will create a significant obstruction to the Omega seal installation. If there is sufficient concern that seismic restraints must be in place as soon as the elements are immersed, it might be necessary to install the Omega seals with a small amount of support clamping so that the seismic restraints can be coupled as early as possible.

Whether the seismic restraints are formed by prestressing cables or prestressing bars, they will include a length of cable or bar located in a duct on each side of the immersion joint. Protective covers can be removed and the bars slid from within the structure into the joint space. The free ends are then coupled using a proprietary coupling that is appropriate for the particular pretensioning system being used. This is a relatively simple operation, and once it is complete, the remainder of the joint infill concreting works can continue.

Joint finishing for steel tunnels

Steel tunnels historically have had continuous steel and interior concrete at the immersion joints. This requires some welding operations and relatively conventional concreting work. However, this is less common now and steel tunnels are more likely to feature flexible joints with rubber gaskets and internal finishing at the joints that are similar to a concrete tunnel. One Japanese steel tunnel has used a seismic restraint in the form of a wave or bellows plate (see Chapter 10). This would also require internal welding operations.

BALLAST CONCRETE

Ballast concrete is usually placed in a progressive manner through an immersed tunnel and is built up in a number of stages. Some ballast concrete may be placed prior to the floating of the tunnel elements. This can be useful in helping to trim a tunnel element so that it floats with an even freeboard over its length and width, and to control the freeboard more generally to a preferred dimension.

A minimum amount of ballast concrete must be placed in the ballast exchange process described in Chapter 14, whereby temporary water ballast is replaced with permanent concrete ballast. This needs to be sufficient in volume to provide the required temporary factor of safety against uplift, and the depth will have been determined during the design process to achieve this. Given the density ratio between the water ballast and the concrete ballast, the concrete will take up much less space.

The remainder of the concrete ballast to achieve the permanent factor of safety against uplift can be placed at a later date. There may be advantages to leaving this until the bulkheads are removed and better access is available into the tunnel elements. The initial ballast exchange concrete may need to be pumped some distance from the adjacent tunnel element, through the bulkhead and to the far end of the new tunnel element. Depending on the progress of activities such as fixing Omega seals, it might be necessary to pump the distance of two tunnel elements, if other bulkheads are still in place. Therefore, only the minimum amount of concrete required is usually placed in this early operation.

In order to place the general ballast concrete in the base of the tunnel, it will be necessary to remove the ballast tanks. These are usually of modular construction and designed to be easily demountable. It is not generally necessary to remove the ballast tanks through the tunnel bulkheads. In the sequence of finishing operations, the bulkheads are likely to have been removed before the ballast tanks are emptied and dismantled. Usually, tanks are lined so that there should not be any significant residue on the concrete surfaces. The walls will, nevertheless, need to be cleaned before ballast concrete is placed, as the environment in the tunnel elements is often highly saline and water is not just confined to the tanks during the temporary conditions that exist during construction.

For both the ballast exchange concrete and the general ballast concrete, there will be parts of the volume that need to be left until other finishing operations have been carried out. At the immersion joints, ballast concrete needs to be kept clear of the bulkheads so that they can be easily removed and the joint finishing works can be carried out. At segment joints, ballast may need to be omitted initially if there are prestressing tendons to be cut. If the tunnel is for a road, there will probably be a drainage carrier pipe embedded in the ballast concrete, or other utility pipes and ducts. Boxouts will need to be made for these unless they can be fixed at an early stage.

Ballast concrete is generally left low beneath a roadway by around 150–200 mm so that a final topping layer can be laid to a suitable accuracy, on which to lay the road pavement surfacing materials. This final topping layer would usually be finished with a high-tolerance screed. The previous layers and sections of the ballast concrete do not require such accuracy.

As the ballast concrete layer is formed in various parts, there will be a great many joints within it. Although this layer is not structural, it is necessary to ensure that surfaces are clean and provide a degree of mechanical strength across the joint. Although all surfaces do not need to be prepared as construction joints, the detailing of layers should ensure that there are no debonding planes or thin layers of ballast concrete that may crack or deteriorate under highway or rail loading. Similarly, there is no need to make a structural connection between the main tunnel structure and the ballast concrete. Again, it is sufficient to simply make sure that the surface of the tunnel base slab is clean.

It is possible to incorporate precast elements into the ballast concrete. For example, at the Bjørvika Tunnel in Norway, the project client wished to have the ability to quickly remove ballast concrete at the immersion joints. Consideration was given to precasting the sections of ballast concrete that could be placed as general ballast concrete and be covered with the topping layer. Finally, the ballast was designed to be cast in situ but in a series of match-cast debonded sections fitted with lifting eyes so that they could be removed. This facility would assist in removing the ballast quickly if ever there was a need to get to an immersion joint for maintenance or repair work. Care had to be taken with the detailing to ensure that the debonded sections could span the joint gap and allow movements to occur. This required a simple sliding surface to be created to one side of the immersion joint beneath the ballast concrete section so that the immersion joint can open and close without the ballast section causing any restraint, as shown in Figure 15.3.

The ballast concrete will have joints at the tunnel immersion joints and, if the tunnel is a concrete segmental form of construction, at segments joints as well. These joints in the ballast need to accommodate the same degree

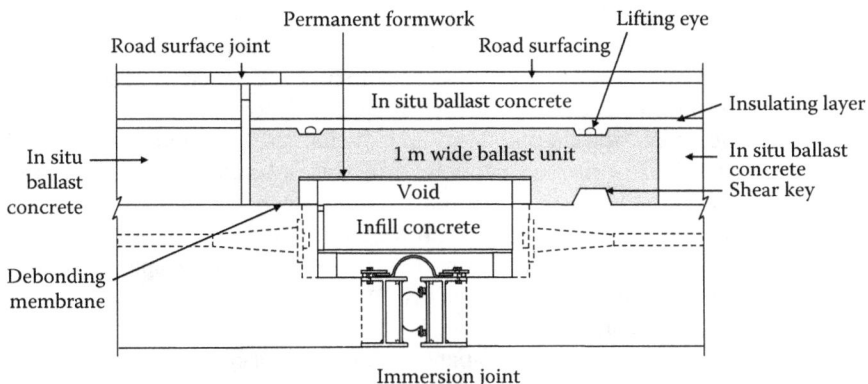

Figure 15.3 Removable ballast concrete section used on the Bjørvika Tunnel.

of movement as the main tunnel structure. In addition, they must provide a barrier to hazardous materials in the tunnel that may be spilled on the road or rail track bed surface and that could percolate down through the joints. The road pavement immediately above the ballast concrete should have a waterproof flexible joint to deal with the tunnel movements. However, there is a small risk that materials such as petrol or diesel could get past the joints. In order to prevent such materials from reaching the important watertight seals (the Gina and Omega seals) in the immersion joint, a suitable watertight seal should be used in the top surface of the ballast concrete. This will be a high-specification material that is resistant to oil-based materials and corrosive alkaline or acid materials.

Some tunnels have incorporated a pressure relief system beneath the ballast layer so that in the event that a segment or an immersion joint suffers some leakage, the water entering the tunnel has a path to find its way into the drainage system rather than forcing its way upward through the ballast concrete joints to the road surface. Opinion is fairly divided as to whether this is of great benefit. There is an argument that there is sufficient longitudinal jointing between ballast and structure that water paths will exist adequately to reduce any pressure buildup beneath the road surfacing.

FIRE PROTECTION

The effect of intense heat produced by fires in tunnels is well documented. Organizations such as the Permanent International Association of Road Congresses (PIARC) and research programs such as Durable and Reliable Tunnel Structures (DARTS), Upgrading Methods for Fire Safety in Existing Tunnels (UPTUN), and Fire in Tunnels (FIT) within Europe have made significant investment in understanding the nature of fires that can occur, the methods of assessing risk for a particular tunnel, the ventilation and suppression systems required to make a tunnel safe in the event of a fire, and the operating procedures to deal with an emergency of this nature. These issues are common to all tunnels, regardless of the form of construction; so it is not the authors' intention to cover this in great detail, although some operational aspects are described in Chapter 16. The risk that fire presents to structural failure is critical to immersed tunnels, and this subject must be carefully considered and dealt with by the tunnel's owner, designer, and constructor.

Intense heat arising from fires at or above 100 MW in magnitude will have two primary effects on the tunnel concrete. The first is surface spalling of the concrete that is directly exposed to the fire. Water trapped within the concrete matrix will turn to steam and expand. If the rate of pressure rise exceeds the speed at which the pressure can dissipate through the concrete matrix, the pressure can give rise to cracking and spall the surface of the concrete. With continuously applied heat, the spalling will become progressive and gradually reduce the thickness of the structural concrete. This progressive spalling (sometimes referred to as raveling) would clearly be seriously detrimental to the structure. Spalling can also be explosive in nature, which will inhibit or even prevent the access of emergency services to the seat of the fire. The second effect is that high heat will cause concrete and steel materials to reduce in strength. Concrete typically will start to exhibit loss of strength at around 400°C when the calcium hydroxide in the cement starts to dehydrate, and the tensile strength of steel will begin to reduce once temperatures go beyond 300°C.

Spalling can be reduced by adding polypropylene fibers to the concrete mix, but these are not effective in reducing the temperatures in the concrete or the reinforcement. Therefore, it is usual for modern immersed tunnels to have fire protection material applied to the inner surfaces above a roadway or rail track, and often also to the external walls, to protect the concrete from damage should a fire break out in a tunnel. These are the watertight elements of the tunnel that are resisting water and ground pressures from the outside. The resulting bending in the structure means the exposed faces on the inside of the tunnel are in flexural tension at the midspan points of the walls and slabs, and any loss of tension steel effectiveness could lead to structural collapse. These surfaces should therefore be protected. It is often discussed as to whether the full height of the external walls should be treated with fire protection, as the heat from a fire may be lower at the base of a wall. Unfortunately, it is very difficult to demonstrate this and little research has been done. Some work has been carried out by The Netherlands Organization for Applied Scientific Research (TNO), and these findings can be applied to tunnels that are represented by the testing. However, given the safety critical nature of fire protection, many owners specify that all exposed surfaces of the wall should receive treatment. This is a prudent approach unless it can be demonstrated by analysis that temperatures arising from a fire will remain at an acceptable level.

It is necessary to decide the performance criteria that any particular project owner would like to apply. Typically, the two main criteria are

1. To prevent any loss of strength of the tunnel structure reinforced concrete. The exposed surface of the roof slab over the road and the external walls will have zones where the internal surface is in tension and, as noted previously, loss of strength of the tension steel could result in structural collapse.
2. To prevent spalling of the tunnel concrete. Spalling behavior can result in a significant loss of section in the tunnel structural elements. Notwithstanding the effects this might have on structural capacity, it will result in a major repair operation to bring the structure back to a serviceable condition, which could take many months.

A number of specifically developed passive fire protection materials are available that have insulating properties and are able to retain their integrity under extreme heat. By coating the surface of the structure with these materials, the temperature at the face of the structure and at the depth of the reinforcement can be limited to ensure that no loss of strength occurs. The practice of limiting the surface temperature of the concrete will, in most instances, also have the effect of eliminating the risk of spalling.

Design criteria

There are a number of fire curves that are used to replicate the heat generation of a fire, as shown in Figure 15.4. The heat development curves generally used are as follows:

- Rijkswaterstaat (RWS) curve
- International Organization for Standardization (ISO) 834/EN 1363-1
- Hydrocarbon Eurocode 1 curve (EN 1363-2)
- HCM modified hydrocarbon curve
- RABT curve (German DIN standards)

The most widely used of these for immersed tunnels is the RWS fire curve for maximum temperature and rate of rise in temperature; this represents a

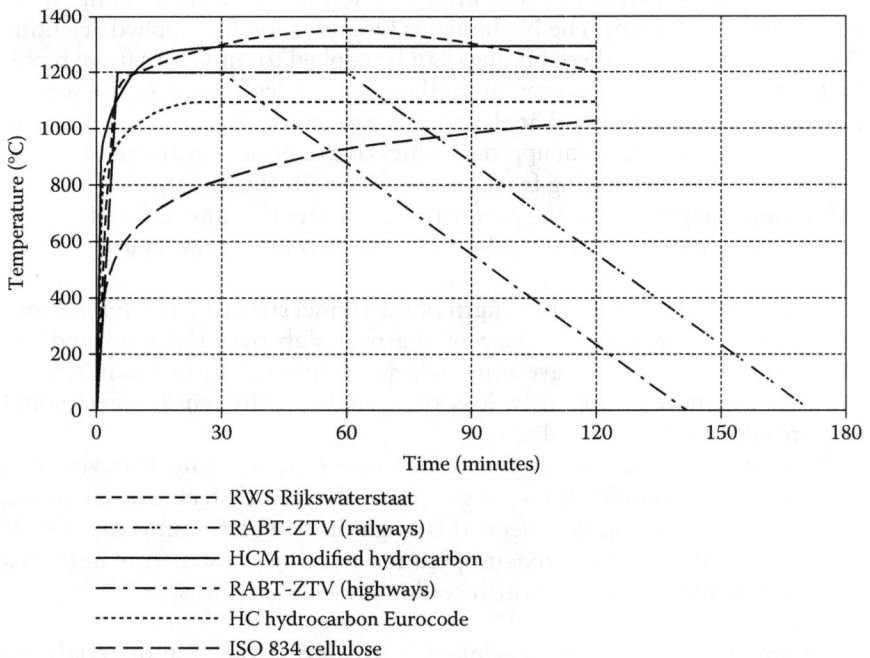

Figure 15.4 Design fire curves.

300-MW fire lasting 120 minutes and reaching a maximum temperature of 1350°C. The modified hydrocarbon curve is also often used for the maximum total heat generated. The hydrocarbon Eurocode 1 curve is also sometimes applied for 4 hours. A combination of curves can be applied in design to cover different aspects and to arrive at a conservative, safe solution.

The normal specifications used for the purposes of selecting the thickness and type of material required are as follows:

- Maximum temperature at the surface of concrete: 380°C
- Maximum temperature at outer reinforcement: 250°C

This will protect the structure from any loss of strength. The criteria may or may not be sufficient to prevent spalling as this is a function of the hardened concrete mix properties rather than temperature alone. This is therefore generally determined through furnace testing. Testing should be carried out on concrete panels that are representative of the final structure, that is, using the actual concrete mix and quantities of reinforcement, which are eccentrically prestressed to achieve the level of tension on the surface that would be expected in the tunnel. Accredited testing laboratories should be used with furnace facilities capable of applying the required temperature curves in all respects: speed of heat gain, maximum temperatures, and durations. Test panels need to be fitted with arrays of thermocouples to monitor surface and internal temperature development. Special features such as the details and geometry at immersion joints should be tested to ensure that the temperatures at joint seals remain within their permissible range. Segment joints should also be tested to check that the protection to the cast-in waterstops is adequate. In some instances, existing test data may be good enough to assess suitability, provided the test data is fully representative of the concrete mix that is being considered.

Application

There are two methods of applying fire protection material to a tunnel. It can be applied either by a spray method or as a board material. Both offer equally good levels of protection and have been tested and used on many immersed tunnels. The spray-applied material can be either cement based or vermiculite based. It is applied with the same equipment as sprayed concrete, and the technique has developed such that a consistent thickness and quality of material can be achieved.

The board systems available can be postfixed to the structure but are often applied to the tunnel roof by lining the shutter with the board material ahead of placing the roof concrete. The boards are then effectively cast in. One perceived disadvantage with this method is that defects in the concrete slab might not be identified as the soffit will never be visible. However, these risks are very small and considered by most to be negligible. The boards would always need to be postfixed to the walls, however, to ensure that defects can be detected. Also, there tend to be more openings and recesses in the walls, so postfixing is more suitable. Many contractors prefer the board system as it is a cleaner process and is more flexible for working around other finishing operations in the tunnel, but both systems are good solutions. Both can be finished with suitable coatings, and both have addressed the issue of adhesion to the concrete substrate. Spray-applied materials are sprayed onto a mesh that is fixed to the concrete by dowels. Board materials are always physically connected to the concrete with fixings detailed to resist high fire temperatures. The decision as to which material should be selected will, as always, come down to cost and the preference of the owner and contractor.

Details of fire protection at the tunnel immersion joints and segment joints need to be developed to ensure that heat cannot pass along the joint interface and cause any detrimental effect to the watertight seals. This will normally require a surface sealant at the inside face of the structure, set into a recess at the joint gap, and an overlapping detail of the fire protection material over the joint gap to provide double protection. The overlapping detail is particularly important at the immersion joints, where quite large joint openings can occur as the tunnel settles and contracts with seasonal ambient temperature changes. It is less of a concern at segment joints where movements are comparatively small. For the Øresund Tunnel project, furnace testing of these details was carried out and the need for overlapping detail at the immersion, joint was verified. At the segment joint, a simple saw cut in the spray-applied protection material was sufficient.

The surface sealant needs to have fire-resistant properties and a high expansion and contraction capacity. Materials originally developed for firestops in building construction have been developed further to give them sufficient resistance to the high temperatures that can be experienced in a tunnel fire. These are typically spongelike materials impregnated with fire-resistant elements and reinforced with plys to ensure that the shape is retained under high expansion. They are not fully resistant and may deteriorate in a fire but will protect the joint gap opening for a specified period of time, either 2 or 4 hours depending on the fire criteria, and can subsequently be replaced.

CLADDING, KERBS, AND CRASH BARRIERS

The finish on the interior walls of a road tunnel serves two purposes. Aesthetically, it provides a suitable environment for drivers, and it also provides the level of reflectance required for tunnel lighting. Practice varies from country to country, but there are three main ways of finishing the tunnel walls in a road tunnel: they can be painted, tiled, or fitted with a cladding system. Rail tunnels tend to not require this as the inside of a rail tunnel is not particularly visible and lighting levels are not critical.

The simplest finish to the concrete tunnel wall is to paint it, and this has been done on several European tunnels. The United States almost exclusively uses ceramic tiles on the walls of its tunnels. Although they may have a slightly dated appearance, they are highly versatile and easy to replace if damaged. They offer some level of fire resistance, but not up to the high temperatures discussed earlier in this chapter, so combination with fire protection or a secondary reinforced concrete panel needs to be considered. Cladding is typically provided to a height of 3–4 m above road level, starting above the verge or above the level of any concrete safety barrier adjacent to the carriageway. The installation of such a system is shown in Figure 15.5. Requirements for height and extent may be set out in applicable tunnel design standards. Various cladding systems are available, including ceramic-coated steel and powder-coated aluminium. It is possible to combine the cladding with a fire protection system. Ultimately, it does not matter what material is used provided it gives the desired reflectance, is low maintenance and durable, and poses no risk of components loosening due to wind pressure effects or deteriorating through fatigue due to the repeated passage of vehicles.

Practice also varies considerably around the world on the type of crash barriers that are installed adjacent to the carriageway. In many European countries, the New Jersey barrier profile is used with a simple hard strip

Figure 15.5 Cladding fixing in the Øresund Tunnel. (Courtesy of Øresundsbron.)

adjacent to the main carriageway. This is also normal in the United States. In the United Kingdom and Ireland, the tunnel standards suggest a raised verge only, with a standard precast kerb, and allow the tunnel wall to serve as the crash surface provided it is continuous and smooth. Local highway standards are likely to dictate the solution in any particular country.

DRAINAGE

Road tunnels will require a gravity drainage system to collect water or liquid arising from washing, firefighting, or spillages arising in the tunnels. This is typically a gulley pot collection system feeding into a carrier pipe. Combined kerb and gulley systems are also used extensively, and combined crash barrier (New Jersey profile) and drainage gulleys are also used. The system needs to be trapped to prevent the spread of flame in the event that a spillage of flammable material enters the system and is ignited. Inlet points are typically at 25–30 m spacing, but their frequency may increase locally to the low point of the tunnel where gradients are flatter and it is necessary to introduce additional inlets to clear water from the roadway in a suitably short time. Similarly, the spacing may be closer at the tunnel portals to remove rainwater runoff from the roadway as quickly as possible and prevent it from running along the road down the tunnel.

Carrier pipes are set into the ballast concrete. Often, the minimum depth of ballast is set to accommodate the drainage system components, as it is expensive to recess the base slab to accommodate the pipes and gulleys. The exception is at the low point of a tunnel where it is not normally possible to fit the drainage sump into the depth of ballast concrete alone and a special feature is required in the structure.

Rail tunnels have a far lesser risk of hazardous hydrocarbons arising in a tunnel and finding their way onto the drainage system. Usually, surface drainage channels formed in the track bed concrete will suffice, channeling liquid down to the low point sump. If a carrier pipe is preferred, then it will be of relatively small diameter and so is easy to accommodate.

RAIL TRACKS AND ROAD SURFACING

The final finishing of the road surface in an immersed tunnel is fairly conventional. Typically, a two-layer wearing course is laid to form a road surface, with the first layer acting as a regulating course to smooth out any irregularities. Headroom in the tunnel is limited, and this has an impact on the choice of plant and equipment, but otherwise this is quite a conventional process and normal road paving machines can be used. Barriers, kerbs, and footways can similarly be installed in a conventional manner.

Figure 15.6 Fixing booted track system in the Øresund Tunnel. (Courtesy of Øresundsbron.)

The installation of rail tracks requires long lengths of tunnel to be available, so other finishing operations either need to be completed beforehand or will have to wait until after the track is laid. Low-height trackforms are used, often booted block systems that are fixed in position to line and level and have concrete placed around and beneath them, as shown in Figure 15.6.

M&E INSTALLATION

The installation of all the M&E equipment described in Chapter 7 is generally the last, and most complex, part of finishing operations. The vehicles needed to fix the lights and fans are road-going vehicles; often, mobile platforms are used, sometimes even old open-topped double-decker buses, and these need a smooth, uninterrupted surface to work on. This means that the ballast concrete and any regulating layer will have to be completed first. The M&E contractor really requires uninterrupted access if the work is to be carried out efficiently. The bituminous road surfacing will not be placed at this point as it is susceptible to damage from oil and diesel spills. Care must also be taken to avoid damaging any wall coating during these installations.

When the equipment has been installed, it will have to be tested and commissioned prior to opening the tunnel. There is nothing specific to immersed tunnels in this regard, and the procedures are similar to any other form of road or rail tunnel.

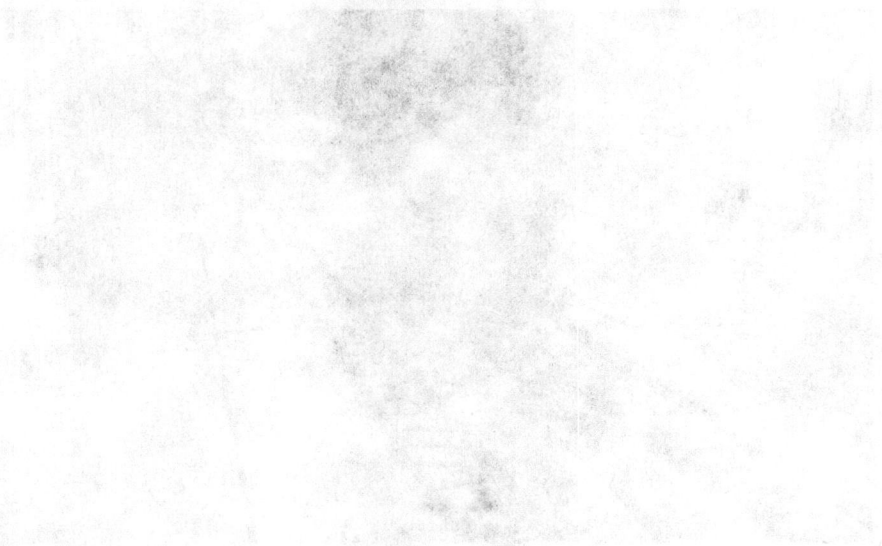

Chapter 16

Operation and maintenance

The operation of an immersed tunnel is similar to the operation of any other type of road or rail tunnel. It involves operating the tunnel on a daily basis, planning and carrying out routine maintenance, and dealing with emergencies. It is essential that the eventual operator of the tunnel is involved in the early stage of its planning so that enough early focus is put on the operational aspects. The operator, in conjunction with the tunnel owner, requires high availability of the tunnel to traffic, high safety standards, and low cost. To achieve the right balance of these three interrelated factors, the design and operation have to be coordinated at an early stage. The way the tunnel is going to be operated will have a significant impact on the design of the tunnel and the facilities that it has to accommodate. This is particularly so for an immersed tunnel where, as has been mentioned many times, internal space is at a premium. In general, road tunnels have more operational and maintenance requirements than railway tunnels. This chapter will concentrate on the operation and maintenance in regard to road tunnels, but many aspects will be equally applicable to rail tunnels.

OPERATION

Operating procedures

In any tunnel, the operating procedures that have to be followed by control room personnel must be fully documented and must cover both normal and emergency operating conditions. This will be a comprehensive set of documents detailing responsibility and actions for all normal and emergency situations in the tunnel. It will cover the duties of the tunnel manager and the control room operators, procedures for rapid response in the event of a fault or emergency, and access and safety procedures, as well as detailing the normal operating procedures for monitoring traffic and the environment in the tunnel and dealing with events and incidents in the tunnel.

Most traffic incidents in a tunnel are minor, involving breakdowns or traffic shunts. These require only the attendance of a breakdown vehicle, and traffic signs can be used to temporarily close the affected lanes while the incident is cleared. A major tunnel incident will require the response of the emergency services, fire, police, and ambulance. Some major tunnels may have their own firefighting capability, but this is not common. In the early stage of developing the procedures for a major incident liaison is essential with all the emergency services to agree what facilities need to be built into the tunnel to enable them to deal with incidents effectively. This is the case especially for the fire service in terms of access, water supply, and control points for the ventilation system, deluge, and fire suppression systems. They are the people who have to enter the tunnel to deal with the fire and are entitled to have the facilities they need to do so as safely and effectively as possible. Each country will have its own procedures, but the control and command structure must be clearly set out and understood by all parties.

It is important to know who has responsibility for which action, for example, who takes overall control, who operates the tunnel ventilation system and the like, should it be the tunnel operator or the emergency services, and if so, which service? This needs to be clearly set out in the documentation and must be agreed on by all parties early in the planning of the project, as it forms the basis for design. There can be a degree of automated response to an incident; for example, control of ventilation fans for specific scenarios can be pre-programmed for a quick response.

The major incident plan must consider what to do with the traffic. For example, is there an alternative route to bypass the tunnel? The initial response to a major incident in a tunnel is to stop all traffic so that no one else enters the tunnel. But the details of these responses are common to all tunnels and therefore will not be considered further here, as we are concentrating on immersed tunnels. Another problem, common to nearly all tunnels, is that there is not sufficient time to commission the equipment and train the personnel. Tunnel opening dates are often set some time in advance, particularly if it is to be a royal or presidential occasion. Delays then occur during construction, leaving insufficient time for training. This training is of paramount importance to the safe operation of the tunnel, so reduced commissioning and training periods should be resisted if at all possible.

The operation of tunnels is a continuously evolving practice as technology improves and procedures are refined to account for actual incidents. Those planning a tunnel should seek the opinion of tunnel operators, who may have access to national tunnel operators' forums, and national and international organizations such as The World Road Association (PIARC) and the Organization for Economic Cooperation and Development (OECD) for the latest recommendations on operating standards.

The requirements for rail tunnels are similar, although they are less complicated than road tunnels. Fewer systems are installed, simplifying the response to an incident. There may be a need for an interface between a local control center and the systemwide rail operations center. Indeed, there may not be a need for a local control center at all if incident response can be satisfactorily dealt with by the systemwide operations center.

Control building

A key aspect, particularly for a road tunnel, is whether or not the tunnel is going to be continuously manned. This will influence what monitoring and control systems are in put in place in the tunnel and approaches, and where the monitoring systems are located. For a manned tunnel, it is common to have a control building adjacent to the tunnel where all the monitoring and control systems are routed, and there will be personnel stationed there to respond to any incident in the tunnel. For an unmanned tunnel, the monitoring will be done remotely, possibly by a local authority or the police. In these circumstances, the tunnel systems must be such that the initial response to an incident can be instigated remotely, for example, by the automatic deployment of variable message traffic signs to divert traffic away from the tunnel, or the automatic control of the tunnel ventilation and other fire/life safety systems.

The control building is often situated on top of the cut and cover portal section. This is convenient for routing the cables into the tunnel as well as ensuring that the personnel have ready access to the tunnel for maintenance or emergency. If the tunnel is provided with a separate central maintenance or escape bore, then there can be direct access from the control building. The building will house the electrical transformers and switchgear, standby generation and battery backup systems, as well as the tunnel monitoring and control systems. In addition to what is needed for the tunnel systems, the building may have to provide room for stores, workshop facilities, and the messing requirements for the operation and maintenance staff. Around the building, there must be sufficient hardstanding for parking and vehicles delivering heavy electrical equipment and access for emergency vehicles. Consideration might be given to providing a helicopter landing pad if this would be part of the emergency services response. A typical layout of a service and control building on top of the tunnel is shown in Chapter 8, Figure 8.7. If the tunnel is tolled, then the service building may be situated a little way away from the tunnel at the toll plaza. Here, it can be combined with the facilities required for toll collection.

Hazardous loads

The passage of hazardous loads through the tunnel must be carefully considered at the outset of design. They are dealt with in different ways by different jurisdictions; some will allow free passage, some prohibit them completely, and some allow passage under controlled conditions, for example, by forming convoys that are escorted through the tunnel. If they are to be escorted through, then a marshalling area will be needed close to the tunnel entrance where the hazardous loads can wait for their escort. This can be combined with the toll plaza if there is one.

These aspects will have an effect on the size of the service building that has to be provided. Even if the tunnel is considered to be unmanned, there will have to be a small service building to house the plant rooms and provide a temporary control center for the tunnel in the event of an emergency and provide facilities for visiting maintenance personnel.

Spillages

Another important consideration is the treatment of spillages of hazardous liquids in the tunnel. They cannot be allowed simply to drain away into the tunnel drainage system and then be pumped to the public drainage network with the other tunnel water. Interceptors have to be provided in the tunnel sumps and the hazardous liquid separated and stored before being taken away in special tankers for disposal. The extent of the separators needed may be influenced by whether hazardous loads can use the tunnel freely. The arrangements for this affect the dimensions of the drainage sumps in the tunnel, which are usually very constrained anyway by the lack of space generally under the carriageway in an immersed tunnel. The design of the drainage sumps in an immersed tunnel is a key factor in the early design to ensure that sufficient space is provided for any separators or special pumping equipment. The storage sumps are often located at one of the portals.

The tunnel operator must have clear procedures in place for dealing with a spillage incident, which must be consistent with the decisions made during the design development. The operator must be able to identify a spillage as soon as it occurs, assess the nature of the risks that are presented, and make the appropriate decision on how to remove the liquid, either by direct pumping to a tanker at the low-point sump or by controlled pumping out of the tunnel up to the portal sump and potentially onto the oil separators and silt traps at the discharge point.

Overheight vehicles

Overheight vehicles are a serious risk to any tunnel and particularly so to an immersed tunnel. As the rectangular tunnel shape matches the rectangular traffic envelope, the tunnel height can be minimized, leaving little room above the notional design headroom. In a circular bored type of tunnel, there is more room above the carriageway and the problem is not so acute. An overheight vehicle would be physically stopped by the portal roof as it tried to enter the tunnel, but this in itself would cause a potentially dangerous traffic incident. A marginally overheight vehicle could get into the tunnel below the portal roof but still be high enough to damage the lighting or ventilation equipment as it drove through the tunnel. This would present a danger to the following traffic, as well as disrupting tunnel operation while repairs are carried out. Such an incident could be caused by a flapping tarpaulin on top of a high vehicle.

To prevent this type of occurrence, overheight vehicle detection equipment is installed on the tunnel approaches. This is linked to variable message signs that warn the driver of the offending vehicle and also to the control room to warn the operators. Having approached the tunnel, there must be provision for the overheight vehicle to turn around and drive away from the tunnel. As most tunnels are on dual carriageway roads, this is usually facilitated by an access road passing over the tunnel portal enabling the vehicle to rejoin the opposite carriageway. This is usually the same road that gives access to the control room above the portal, so it has to be designed with this use in mind.

Lighting

The key environmental systems inside the tunnel are the lighting and ventilation. In most tunnels, both these systems are automatically controlled by sensors in the tunnel and approaches. These should be routed to wherever the tunnel is monitored so that personnel can check the status of the systems at any time and are advised of any system warnings or malfunctions.

Traffic control

The volume of traffic must be considered in developing the traffic operation system for the tunnel. Will it need speed and lane control indicators, or even traffic stop signals if it is likely to become very congested? These can be important in multilane road tunnels, which is where immersed tunnels gain a cost advantage over bored tunnels. Will the tunnel be run under a contraflow system with two-way traffic in one bore during maintenance operations? Again, this will require suitable lane control systems to be installed in the tunnel and its approaches.

The installed systems must enable the tunnel operator to implement a traffic diversion scheme swiftly and efficiently, whether it is for a planned maintenance closure, where every available minute of the closure may be vital for completing repairs and replacing components, or for managing an incident such as a vehicle breakdown or a road traffic accident. The operator should have written procedures for a variety of scenarios that may arise that allow him to switch sign displays quickly and provide manpower on the ground for setting up barriers, markings, and lighting as necessary. It should also provide all information necessary for contacting and liaising with traffic police and local authorities as appropriate.

MAINTENANCE

Planning

Any tunnel is a valuable piece of infrastructure, and if it is to be serviceable for a long period of time, then maintenance will be needed throughout its lifetime. Maintenance of a tunnel will involve the following operations, though it should be noted that maintenance demands are greater for road tunnels than rail tunnels:

- Planned maintenance involving tunnel cleaning, maintaining equipment in accordance with the manufacturer's schedules, inspection, and performance testing of equipment. This routine regular maintenance is aimed at preventing tunnel closures due to failure of equipment.
- Technical inspections of equipment and the fabric of the tunnel. These inspections are usually undertaken every 2 or 3 years during routine maintenance closures and the aim is to identify what work needs to be done, say, over the next 5 years so that it can be scheduled accordingly.
- Corrective maintenance of equipment to keep the tunnel open on a day-to-day basis. This would generally be small-scale maintenance such as replacing failed sensors.
- Major refurbishment of either the tunnel fabric or equipment that has been identified by the technical inspections. This might typically be expected at 40- to 50-year intervals.

The maintenance costs can be significantly affected by decisions made during the design and construction phases. It is important, therefore, that people with sufficient experience in tunnel maintenance are involved in those stages. The amount by which they can influence the maintenance costs lessens as the design and construction process progresses.

It is common for an operation and maintenance manual to be prepared before the tunnel opening. This will list all the equipment installed in the tunnel, together with the manufacturer's recommendations for maintenance and lists of spare parts. In this respect, it is important that manufacturers give an undertaking to maintain spares for the equipment for a certain period of time, at least 10 years. If not, the tunnel owner should consider laying in and storing parts, which might become scarce during the lifetime of the equipment.

The manual will set out the planned maintenance regime for all the equipment giving details of what has to be inspected and the inspection intervals. Over time, these inspections form a valuable database for the tunnel, which is useful in informing decisions on future maintenance schedules as well determining what works and does not work for future tunnels.

Implementation

Maintenance work needs to be planned to minimize its impact on the operation of the tunnel. It should wherever possible be carried out during the periods of lowest use of the tunnel. This will usually be at night with the tunnel having to be back in full use for the next morning's rush hour. Maintenance personnel will need protection, both from traffic using the tunnel and from the vehicles being used for the maintenance. In a rail tunnel, the line will be closed completely to trains during maintenance, but that is not necessarily so for a road tunnel. Road traffic will often continue to use the tunnel, albeit in a very restricted way, during the maintenance work. The environment inside a road tunnel is harsh, noise levels are high, the traffic produces very turbulent air, and vehicle loads can be a danger to workers if they overhang the sides of the vehicle. It is better to completely separate the traffic from the tunnel maintenance workers. Rather than just close off one lane, which may not give sufficient protection to workers, it is preferable, and indeed typical, to close one bore of a road tunnel completely during maintenance and run the other bore in contraflow at reduced speeds.

A traffic management system will have to be implemented to close one bore and this will require a crossover to be located somewhere near the portal so that the traffic can cross to the other carriageway. A gantry sign may be needed on the tunnel approaches to warn traffic and get it into the correct lanes. There should also ideally be some form of physical barrier at the portal to stop traffic entering the closed bore. It is surprising what some drivers can do in error and every precaution should be taken to protect the workforce.

Structure

The main concrete or steel structure of the tunnel will have been designed for the 100 year or so lifetime of the tunnel and can be expected to perform satisfactorily over this period. There are relatively few regular maintenance activities required for immersed tunnel structures but there are specific monitoring and inspection facilities that can enhance an owner/operator's knowledge of how a tunnel and some of its hidden features are performing.

The durability of a concrete structure can be affected by saline intrusion into the concrete through a number of mechanisms. These are discussed in greater detail in Chapter 11. Chloride ingress depassivates the concrete, leading to corrosion of the embedded reinforcement. This should have been accounted for in the original design, but visual inspections of the structure are necessary to identify leaks that could be indicative of coming problems. As well as visual inspections, the risk of corrosion is assessed by embedded probes fixed at different depths in the concrete to monitor the progress of chlorides through the concrete cover zone, and hence predict the onset of corrosion. This provides a tool for planning future maintenance activities and budgets. Similarly on tunnels with external steel membranes, it is common to install a corrosion monitoring system to check the corrosion rate of the external steelwork. This can be done by reference electrodes that can be monitored from inside the tunnel or retractable coupons that again can be pulled in through the tunnel wall for inspection. Such items should be checked annually, at least initially until long-term corrosion rates have been established.

Water leakage is a commonly perceived risk to durability. In an immersion joint, it is unlikely that both the Gina and the Omega seals will leak. This would have been identified during the initial construction phase. However, if they did leak, the water would be evident inside the tunnel or would be noticed as an additional water flow in the tunnel drainage system. Vent tubes are installed between the Omega and the Gina seals, and these can be inspected during tunnel cleaning to check that no water is building up between the two seals. Inspection pipes can also be used to check if water is building up under the ballast concrete. An array of such inspection pipes was installed in the Bjørvika Tunnel for this purpose. In addition, inspection chambers were installed at each side of the immersion joints to identify any water inflow arising over the length of each tunnel element. Water is intercepted and a sump provided so that the water can be evacuated and the refill time measured to arrive at an inflow rate. This can be seen in Figure 16.1.

Segment joints should be checked for leakage during tunnel cleaning operations, although if they do leak it will be apparent to users of the

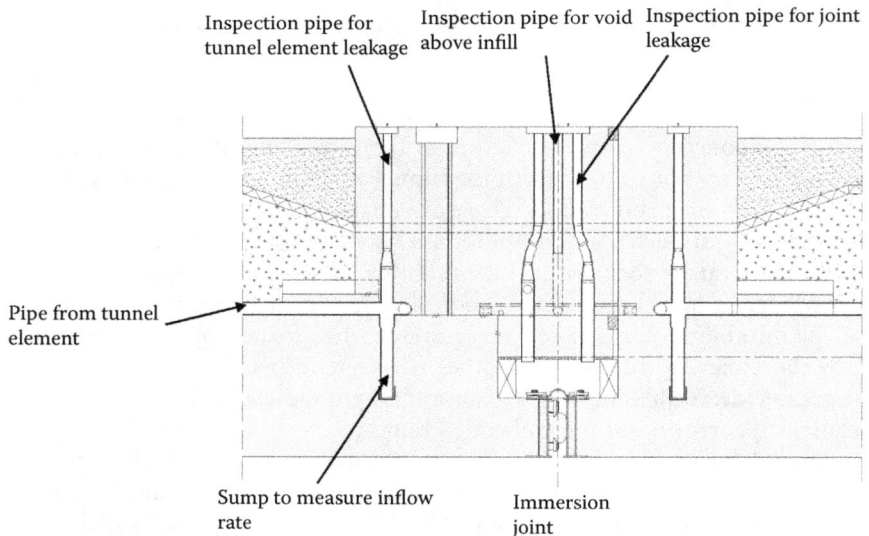

Figure 16.1 Inspection facilities for immersion joint leakage.

tunnel. If a leak does occur here, it is common to grout the leaking joint during the next wall cleaning cycle. Modern flexible expandable grouts are usually effective in sealing any leaks in segment joints. Re-treatment may be necessary after a period of time due to seasonal opening and closing of joints over the life of the tunnel, which will eventually cause the sealing grout to debond and become less effective.

Reference electrodes are usually attached to the immersion joint end frames to check that they are not corroding and that the corrosion protection system is working correctly. End frame corrosion can set in if the end frame is in contact with the reinforcement, and although this should have been checked during construction, incidences do come to light during service. They can lead to major maintenance intervention as some sort of cathodic protection system has to be installed.

The most regular form of tunnel maintenance is cleaning of the tunnel. The tunnel walls will need cleaning approximately every 3 months, varying slightly with the use of the tunnel. This is carried out by specialized lorry-mounted washing equipment that jets the walls and roof. Inside an immersed tunnel, the concrete walls may be painted or clad with special panels or tiles. Care must be taken during washing that it is not too vigorous thereby damaging the paint coating or indeed the protective coating of any other equipment in the tunnel. Modern paint coatings that are used in tunnels are designed to be resistant to this sort of treatment. Equally, the luminaires are resistant to the pressures used during this cleaning operation. In the United Kingdom, IP65 standard is usually defined as the requirement.

The tunnel drainage system will need to be inspected and cleaned out at regular intervals. This will include the oil separators that are installed as part of the low-point sumps. Monitoring of the drainage system is important as, if there has been no rain, the system should be dry. Running water is indicative that there is either a leak or that water is flowing down the tunnel approaches into the tunnel. The performance of the drainage system gives an early warning of potential problems. Similarly, the frequency with which the tunnel sump pumps operate can give an indication that they are pumping more water than expected.

The civil maintenance activities for an immersed tunnel can be broken down by category into the main tunnel structure, the joints within the structure and between the tunnel elements, and the finishes that are applied internally. The typical activities that may be required are shown in Table 16.1. Most authorities will have established processes for undertaking periodic inspections of road and rail tunnels and the table will need to be adapted for these and for the specific features of each tunnel. Rail tunnels may have a different inspection and cleaning regime to roads but the activities apply equally.

The routine inspection program would be carried out during the planned regular cleaning and maintenance closures every 3 months. This includes

Table 16.1 Typical civil works maintenance schedule

Element	Component	Maintenance Activity	Frequency
Structure	Internal surface	Inspection for seepage, staining, spalling	Routine
		Mapping of cracks—check for changes	General
		Mapping of cracks—remapping	Principal
		Testing for carbonation and chloride ingress	Principal
		Reapplication of silane	As required
	Shear keys	Inspection for signs of distress at bearing surfaces and corrosion of steel components	Principal
		Inspection for deterioration of bearings (if used)	Principal
	General	Corrosion cell monitoring	General
		Performance of sacrificial anodes	General
		Activate impressed current or alternative CP system	As required
	Settlement	Level survey through tunnel and approaches	General
Immersion joints	Omega seals	Inspection of seal and clamping system condition if access is available. Replacement is not an anticipated maintenance activity	Principal
	End frame steel	Test performance of sacrificial anodes, if installed	General
	Void spaces	Check for water between Gina seal and Omega seal and if present check if it is saline or fresh	General
		Check for water on inside of Omega seal, remove as necessary, check for presence of corrosive bacteria, and introduce anticorrosive agents (if used)	General
		Renew internal sacrificial anodes to joint steelwork (if used and accessible)	As required
	General	Check tunnel interior perimeter for evidence of leakage	Routine
		Monitor joint movement—opening, rotation, differential settlement	General
	Structure seal	Inspect if visible on walls and roof, for signs of deterioration or leakage	Routine
	Ballast seal	Inspect during road surface renewal	As required
	Road surface seal	Inspection for deterioration and leakage	Routine

(Continued)

Table 16.1 Typical civil works maintenance schedule (*Continued*)

Element	Component	Maintenance Activity	Frequency
Segment joints	Structure seal	Inspection for signs of deterioration or leakage—walls and soffit	Routine
	Ballast seal	Inspect during road surface renewal	As required
	Road surface seal	Inspection for signs of deterioration or leakage	Routine
Internal finishes	Fire protection boards	Inspect for deterioration of surface color, integrity of fixings, damage to boards, and joint overlap details	General
	Spray-applied fire protection	Inspect for delamination, spalling, dampness, deterioration of surface coating, and integrity of joint overlap details	General
	Cladding panels	Inspect security and integrity of fixings, corrosion of support frame, and deterioration or discoloration of surface coating	Routine
	Tiles	Inspect for delamination, cracked tiles, staining/leakage	Routine
	Paint	Inspect for deterioration of coating layer or surface color	Routine
	Finished surfaces	Test luminance levels are satisfactory	Principal
	Cleaning	Overall clean of interior structure, cladding, etc.	Routine

all general inspection items that may indicate leakage. The general inspection, which is carried out every 1–2 years, should pick up the more detailed inspection of the tunnel movement and specific checks for possible leakage. The principal inspection, which is typically every 5 or 6 years, is the full inspection of all elements of the structure that would include all routine and general inspection items. The frequency of some activities may be increased in the early months and years of a tunnel's life until such time as observations and behavior are established as stable. The immersion joint seals—the Gina and the Omega gaskets and their clamping systems—and the cast-in grout injectable waterstops at segment joints are inaccessible and require no routine maintenance.

Electrical and mechanical systems

The maintenance of the electrical and mechanical systems will be the same as for any other type of tunnel. Major maintenance must be anticipated and planned well in advance. This will include such items as the ventilation

Table 16.2 Typical replacement periods and maintenance frequency for mechanical and electrical equipment

Component	Life expectancy	Maintenance frequency
Electrical		
Transformers	30 years	12 months
Cables	40 years	
Switchgear	20 years	12 months
UPS sets	15 years	3 months
Luminaires	20 years	6 months
Control sensors	15–20 years	3–12 months
CCTV cameras	15 years	1 month
Control equipment	20 years	1–3 months
Stainless steel fixings	100 years	3 years
Galvanized steel fixings	15 years	3 years
Mechanical		
Longitudinal jet fans	20 years	5 years
Axial fans	30 years	4000 hours to minor
		25,000 hours to major
Electrostatic precipitators	10 years (estimated)	3 months
Pumps	15 years	1 month
Fire hydrants	30 years	6 months
Fire main pipework	30 years	6 months
Sprinkler systems	20 years	3 months
Standby generators	20 years	3 months
UPS sets	15 years	6 months

fans and road surfacing. Table 16.2 shows the life expectancy of some of the major items likely to be installed in a tunnel. It also gives typical maintenance intervals for the equipment. Although life expectancies are quoted for items such as control equipment, they may well become obsolete and need to be replaced by more up-to-date equipment long before then.

RAIL TUNNELS

A rail tunnel does not have as much equipment to maintain, as the ventilation, lighting, and drainage are all less than in a road tunnel. But the maintenance of the immersed tunnel fabric, such as joints and end frames, and monitoring for corrosion will be the same as for a road tunnel. Track maintenance is a separate issue, but the requirements are not onerous, as it is typical to have a concrete track bed rather than a ballasted system because it requires less headroom and is low maintenance.

The implementation of a well-thought-out operation and maintenance system will ensure that the tunnel can operate as intended safely and efficiently for many years. As with most pieces of infrastructure, taking short cuts or skimping on maintenance will, in the long term, be detrimental to the operation of the tunnel.

Chapter 17

Contract forms

Many words have been written on the forms of contracts that are used in civil engineering. Without going into the detail of each particular form, in this chapter we discuss some of the overriding principles to be remembered when they are applied to immersed tunnel projects. Using these, it is possible to develop a contract for an immersed tunnel that is fair and reasonable and can be implemented successfully. Major disputes are rare in immersed tunnel contracts, which is a testament to the knowledge and experience of the people and firms involved in the industry.

The form of contract used should be appropriate to the country of the project. The participants will be used to that form and can understand how it is interpreted in daily use. Introducing a novel contract into a complicated infrastructure scheme is not recommended and can lead to difficulty and dispute. Most countries will have standard forms of contract that they use, and if not, then the International Federation of Consulting Engineers (FIDIC) form of contract is usually acceptable. It is not impossible to develop a particular set of conditions of contract, and this has been done successfully on some immersed tunnel contracts. The client, however, has to be prepared to invest significantly in developing such conditions if they are to be successful.

A main objective of any form of contract should be to assign risk and responsibility in the most fair and appropriate manner. Both parties, client and contractor, should know what they are expected to do and what the obligations of each to the other are. The guiding principle should be that the responsibility is placed with the party best suited to deal with it. If a party has no control over a risk or cannot protect itself against it, then it should not be required to take the responsibility for that risk. If asked to do so, it may come at a high price for the project sponsor.

FORM OF CONTRACT

The construction of an immersed tunnel requires close integration of permanent and temporary works. Each contractor will have their own preferences for building and placing the tunnel elements. The design of tunnel elements requires certain assumptions to be made about the temporary works that are going to be used to transport and place the tunnel elements. This lends itself to the design and build form of contract. The preferences and expertise of the contractor can be built in to the design of the tunnel at the outset. It allows contractors to maximize on their experience and expertise to develop innovative solutions in conjunction with their chosen designer. This is particularly relevant if there are decisions to be made about the number of tunnel elements and the size and location of an element fabrication facility. For example, there will be a lot of abortive design work if the tunnel elements are designed to be 150 m long, and a contractor finds a very convenient and cheap casting facility that will only accommodate 120-m-long elements. Design and build avoids the contractor being subject to these unnecessary constraints.

Integrating design and construction creates opportunities for optimization that will not occur if the design is completed before the contractor becomes involved. This can involve faster construction, better quality, and improved durability as well as reduction in construction cost.

One area of difficulty that does cause some issues in developing a design-and-build contract is the extent of a client's specifications. All too often, clients specify how something is to be done as opposed to what is to be achieved. To get the best out of design and build, the contractor and their design team must be allowed as much flexibility as possible to develop solutions that achieve the client's objectives. The client, and in particular their engineering advisors, have to consider very carefully whether something is really necessary or whether they would just prefer it. All engineers will have their own preferred details and methods that have worked for them before, but the specification of these must be resisted. Thus, the specification should be based on performance requirements. It should say as little as possible about how the contractor is to achieve those requirements. There will of course be some aspects where the client will have some particular requirement, for example, the architectural treatment at the portals, but these should be kept to a minimum.

There are many items that are specific to an immersed tunnel, such as tunnel joints, the definition of watertightness, the methods of achieving watertightness, and the durability of the structure. The details of these have been developed over a number of years and a number of projects. There are usually minimum requirements for these, which enable the client to obtain an immersed tunnel that is serviceable and durable for its intended design

life. It also protects an inexperienced contractor from overlooking important aspects of design and construction. It is therefore appropriate that
the requirements for such specific items are set out in contract documents.
There is no common immersed tunnel specification, and with relatively few
immersed tunnels constructed, it is important that lessons learned are fed
back into future projects. The employers' requirements for both construction and design and the construction specifications are the most appropriate vehicle for doing this. However, these minimum requirements should
be written into the contract in a way that does not stifle innovation. There
are also only a small number of consultants worldwide who get involved in
early contract preparation, so the specifications tend to be carried forward
and developed from project to project, which is a healthy situation. This is
leading naturally to a more standardized specification for immersed tunnels, but as yet, no one has endeavored to formalize a document for the
industry.

This approach leads to a question of what aspects of the design should
be included in a design-and-build contract. The contractor will produce a
design during the tender, but what status does or should this have in the
contract? As ever there are extremes. The contractor's design can be treated
merely as supporting information to the price. In this case, it will have no
contractual status once the contract is signed. The contractor will be free
to change their design provided it complies with the employer's specification. They could tender one design and then construct another. This begs
the question of why the design submission was required by the client in the
first place. The other extreme is that the entire tender design submission is
bound into the contract. Again, this is unsatisfactory, as perhaps not all the
preferred tenderer's proposals are acceptable or desirable to the client, and
therefore should not become contractual obligations.

The compromise approach adopted on some contracts has been
to require detailed technical proposals at the tender stage but not to
automatically treat them as part of the contract. In the Øresund Tunnel
contract, the client retained the option to decide what parts of the tender
were to be bound into the contract, that is, which parts of the contractor's
offer were fundamental to his acceptance of the tender. This enabled the
client to hold the contractor to those parts of the tender that he considered desirable, but it left flexibility over the remainder. This approach was
taken a stage further on the Medway Tunnel contract in the UK, where
both the client and the contractor agreed on what each of them wanted
and incorporated it into the contract. The client included what was fundamental to his acceptance of the tender, and the contractor included what
was fundamental to his offer. In both contracts, this approach worked
well, giving both sides the degree of assurance they required over what
they were buying and selling.

Design and build is not universally appropriate for immersed tunnels. In a highly constrained site, there may be a client preference to fix many aspects of the project and prepare a conventional engineer's design. This will, for example, enable the client to cover aspects of the tunnel that may have to be agreed on with third parties or for which they have a specific preference. In doing so, the client should still identify which parts of the design can be optimized by the contractor to suit their proposed construction methods.

The particular preference of a client for procurement by a specific method, or national cultures for business, may dictate a preference for developing a full engineer's design before procuring a construction contract. In this instance, it is important to ensure that the designer is highly experienced in this field and has a full grasp of the design intricacies, the construction methods, and the variety of equipment that is utilized, so that the design is buildable and offers contractors as much flexibility as possible.

Maintenance

Whether or not the owner has their own maintenance organization, it can be useful to involve the design-and-build contractor in the maintenance of a tunnel. This will put emphasis on the quality of construction and the components installed. There is always a trade-off between equipment cost and maintenance; generally, the cheaper the equipment, the more it costs to maintain. With the contractor involved in the maintenance, they will consider the whole-life cost rather than the initial cost.

CONTRACT DOCUMENTS

The documents forming a typical design-and-build contract will normally consist of the following:

- Form of contract
- Employer's requirements
- Design requirements
- Project application document (detailed implementation of specific codes and requirements)
- Payment mechanism
- Reference conditions
- Materials specification
- Workmanship specification
- Definition drawings
- Illustrative design

The last two will be replaced by detailed design drawings if the contract is not design-and-build. The form of contract will normally be decided by the client, but it is important that the team preparing all these documents understands the philosophy and approach of the client so that all the other documents are drafted in a compatible manner.

The employer's requirements should include details of any arrangements entered into by the client. This could include land availability and/or any legal agreements entered into with landowners or third parties, environmental agreements, and agreements in principle with the port authority for temporary closures of the waterway and diversions of navigable channels. It will also include decisions made by the fire life safety and operating committee during the early planning of the project. Any agreements, or use or availability of casting basins or dry docks safeguarded by the project owner for use by the contractor would also be included.

The design requirements should have clear requirements for design checking. Because of the unusual nature of the construction, there should always be an independent third-party check by an appropriately experienced company, or by the client's engineer.

Based on experience from previous tunnel contracts, the following points should be taken into account when preparing contract documents:

- The documents should be clear and leave no room for interpretation. Tunnel contracts are often multinational undertakings, and cultural differences can lead to different interpretations if the documents are not sufficiently precise.
- On the employer's drawings, there will be a need for background information that may not be mandatory in the contract. The contract needs to be clear on what information is mandatory and what is subject to the contractor's choice.
- The status of an illustrative design must be clear. In a design-and-build contract, it will have no status other than as an illustration of a solution that satisfies the employer's requirements. It is often used, however, as a reference in terms of what is expected when determining if a contractor's proposal is acceptable. When setting out the illustrative design, the employer's team must take care to illustrate the design but not to give definitive information on dimensions or materials if these are subject to the choice of the contractor. The contractor has to take full responsibility for the design and build part of the works even if they choose to build precisely what is on the illustrative design.
- The approach to drafting the specification documents must be consistent and not overprescriptive for minor construction items when these are best left to the discretion of the contractor.

Number of contracts

The number of separate contracts used on an immersed tunnel project will vary from project to project. On a small project, it might be appropriate to have just a single contract that covers the tunnel and its installations and approaches. On more complex projects, several separate contracts might be used, for example

- Dredging and reclamation.
- Tunnel and approaches.
- Shore-to-shore installations: If the tunnel is only part of a crossing, for example, if there is also a bridge or causeway, then some crossing-wide contracts might be applicable. These might include mechanical and electrical and communications contracts.
- Railway installations: If the tunnel is part of a railway line, it is more sensible to have system-wide contracts for railway installations such as track, power, and signaling.

A clear definition of the interfaces between contracts is required. What is to be handed over, what access is to be given, or what the interface is, together with dates and conditions, should be clearly set out. Cooperation agreements will be needed if contractors are working in shared areas of a site, and a process for early planning meetings between contractors and the employer is needed to introduce flexibility of working and to prevent disputes from arising.

GROUND CONDITIONS

The main risk in an immersed tunnel project is the ground. In general, the site of the tunnel, and hence the ground, is chosen by the client, so the contractor has little choice in the matter. Is there a risk of encountering ground conditions that differ from preconstruction ground investigations? Is there a risk of encountering buried obstacles? These could be of archaeological importance and could delay construction, or there might be unexploded ordnance, which would be a danger to construction, or a hard rock intrusion into the generally soft river- or seabed. Similar situations could occur on the land works in built-up areas with utility services in the vicinity of the tunnel approaches. These are not always as well recorded as they should be.

All the preconstruction information obtained by the client must be made available to the contractor, but it must be clear who is responsible in the event that something unexpected occurs. The responsibility for dealing with all these is often placed on contractors, so it will be their responsibility to assess the available information and decide if it is sufficient or if they need to carry out further investigation to enable them to mitigate their risks.

One method of dealing with this is to invite two tender options with the risk of unforeseen ground conditions being with either the client or the contractor. This enables the client to assess the premium the contractor wants in order to accept the risk. The client can then decide if they want to pay this premium or take the risk themselves. This approach does have its drawbacks. If the risk is with the contractor and there is a major unforeseen condition, the contractor may well claim additional monies anyway, so the client could end up paying twice.

Another way of apportioning the risk for ground conditions that has proved to be successful on immersed tunnel contracts is to introduce reference conditions. These reference conditions set out for each geotechnical parameter the range for which the contractor is responsible. If the parameters fall outside this range, then the client bears the responsibility. This approach is also valid for weather conditions as well as for ground conditions. The maximum and minimum values of each parameter are set out. If the range is set too narrow, then in practical terms all the risk is with the client. Equally, if the range is too wide, then all the risk is with the contractor. By setting appropriate limits, the client can establish a contract that has a reasonable risk allocation between the two parties.

DISPUTE RESOLUTION

Whatever the form of the contract being used, disagreements can occur. Clients often believe that contractors' claims are unjustified and are to be resisted. Contractors believe that their claims are justified and should be pursued. Whatever the merits of a particular case, it is important that the disagreement is settled as quickly as possible so that all parties can concentrate on finishing the project as efficiently as possible. To this end, the formation of a disputes resolution board can be very effective. This is a panel of, for example, three experts who consider and give an opinion on disputes. One is usually selected by the client, one is selected by the contractor, and one is chosen independently or by joint agreement.

The dispute resolution board members should be involved from the outset of the project and be updated regularly on its progress. Ideally, they should visit the site three or four times a year and meet all the parties. This should happen on a regular basis whether or not there is a dispute. Thus, if a dispute occurs, they can consider it quickly as they are fully informed on what is happening in the project and do not require extensive briefings. Their decision is implemented by the parties until and unless it is overturned by subsequent arbitration. This provides a mechanism for the speedy resolution of an issue that then enables the project to continue and not get bogged down in a protracted argument. Given that the panel members, who are experts in the field, have reached their decision, this would give a strong steer on what is likely to happen if either party took the dispute further, say to arbitration, and more often than not their decision is accepted by the parties with no further disagreement. Often, the existence of the board is sufficient to encourage the parties to reach an agreement without even referring the dispute to the board.

INSURANCE

On major projects such as immersed tunnels, consideration should always be given to taking out project insurance. Traditionally, insurance on a project has been split between two types of policy: (1) professional indemnity insurance, which covers the negligence of consultants for design, and (2) contractors all risk, which covers the contractor for damage and accidents during construction. Each party takes out their own insurance. This results in duplication and gaps in the cover. If there is an incident, it is often difficult to allocate responsibility to one party and the legal processes involved in reaching a decision are time-consuming and costly. It also places emphasis on finding out who is responsible rather than solving the problem.

Project insurance overcomes these issues. It is a policy that covers the whole of the project delivery team (client, consultants, contractors, and their supply chains) for losses incurred on the project. It covers all aspects such as unforeseen conditions, accidental damage, design errors, and the cost of remedying defects. It can also extend to loss of revenue due to delays in the project. It also includes the marine activities that are particularly important in an immersed tunnel contract. With separate insurances, there is often dispute about who is responsible in the event of a marine incident.

Project insurance has the advantage that it encourages cooperation and teamwork, which can be vital to the success of a project if there is any major incident. This was demonstrated by the incident of the sinking of tunnel element 13 in the Øresund contract. The contract had project insurance and the cooperation of the owner, contractor, and designer in responding to the incident made the project the priority rather than who could be blamed. The result was that the element was recovered, the delay due to the accident was pulled back, and the tunnel was completed earlier than it was scheduled to be the day before the accident.

PAYMENT

Payment terms are important in a construction contract; indeed, they are important in any contract. If the design-and-build form of contract is used, then payment based on a bill of quantities is not appropriate as the quantities are not known at the outset. A milestone payment system is better in these circumstances, and it also requires less administrative effort in determining the value of interim payments. However, there are many types of milestone payment mechanisms. Unless there is an initial mobilization payment, the fewer the number of interim milestones, the greater the need for contractors to fund their costs from their own resources. For an immersed tunnel contract, there may be a greater need for early payments as the cost of initial temporary works can be high, for constructing casting basins, for example, or completing large-scale temporary earthworks at the tunnel approaches. Payment milestones should consider the large temporary works items as well as the permanent works construction.

From the client's perspective, they need to ensure that payments are consistent with the value of the work completed, so that if the work is halted for some reason they have not paid out more than the value of construction to date. A simplified set of milestones related to the major items of the work, probably some 40–50 in number, is usually sufficient. Generally, milestone payments are "all or nothing," that is, if a milestone has not been reached then the contractor receives no payment for that milestone. However, this does have the disadvantage that the contractor may be unduly penalized for just missing a milestone. Some contracts include valuing the percentage completion of milestones on a monthly basis and paying the contractor accordingly. This procedure is not overly onerous on administration as it is relatively quick and simple to agree on the percentage completion of, say, 35 milestones based on monthly progress. Others would argue that the all-or-nothing milestone system concentrates the mind of the contractor better.

PRIVATELY FINANCED CONSTRUCTION

Immersed tunnels are often part of privately financed schemes. The concession contract will usually contain a design-and-construct contract for the tunnel, with the base specification having been developed by the client. There will be the addition of operation and maintenance requirements over the lifetime of the concession, which will impose obligations to maintain equipment and renew life expired items. The requirements for the condition of the tunnel when it is handed back to the client at the end of the concession will also be defined.

The tunnel is not expected to be new when it is handed back, but equally, it should not be life expired. The handback conditions must be set out explicitly. Phrases like "in good condition" should not be used. The required residual life of each item is normally set out in a schedule. A detailed inspection of the structure and equipment should be carried out to determine compliance. Table 17.1 sets out some typical residual lives that might be expected for various parts of the tunnel, based on an initial 100-year design life.

Table 17.1 Residual life

Item	Residual life
Reinforced concrete	Design life less the concession period
Structural steelwork (excluding steel shell/ membrane)	30 years
Road surface	10 years
Luminaires	Renewed within last 6 months of concession
Mechanical equipment	5 years
Equipment with life expectancy less than 5 years	Renewed within last 6 months of concession

The mechanism for payment to the concessionaire may also affect the operation and maintenance of the tunnel. It might be through the collection of tolls, or the concessionaire may receive payment based on performance criteria such as lane availability. In either case, it will be important to keep the tunnel operating and the traffic flowing. This may involve obligations for vehicle recovery and liaison with the local police and emergency services.

Concession contracts also often involve the concessionaire inheriting and operating an existing tunnel while building a new crossing. The New Tyne Tunnel in the United Kingdom and the 2nd Midtown Tunnel in the United States are good examples. It is difficult to establish the condition of an existing tunnel and predict the amount of maintenance that will be required. The residual life may be low and there may be defects that are not yet apparent that may expose the concessionaire to an unreasonable maintenance liability. There is a need for a fair mechanism to deal with this, because if the whole liability is put on the concessionaire, they will probably not accept it.

A reasonable approach in this instance may be to assess the residual life that might be expected of any historic structure provided it has had a reasonable level of ongoing maintenance throughout its life, and strength and condition assessments show it is not in need of major structural repair or replacement. This might typically be on the order of 10–15 years.

Chapter 18

The future

Immersed tunnels will undoubtedly continue to be used as a method to cross waterways where they offer a cost advantage over other forms of crossing. They will also be used in challenging conditions or in circumstances where physical constraints prevent other solutions.

The method of construction offers a viable solution for tunneling in soft marine ground conditions, in highly active seismic regions, in both inland and offshore environments, in constrained urban sites, and where there are environmentally sensitive conditions. In many circumstances, the immersed tunnel will be a cost-effective solution that is competitive with bridge and bored tunnel crossings of waterways. There are some obstacles to overcome with regard to perception, for example, the immersed tunnel often gets some unfair press when it comes to environmental impact. The process of education in this respect will take some time and will always be necessary because of the infrequent occurrence of immersed tunnel projects. However, projects such as the Fehmarnbelt crossing between Denmark and Germany, where a fair and full evaluation between bridge and tunnel, taking into account temporary and whole life impacts, has been carried out and shown the tunnel to be favorable, will help the industry's understanding enormously. Hopefully projects such as this, together with this book and the ongoing work of the International Tunnelling Association (ITA), will help prospective project sponsors understand immersed tunnels better. To illustrate the current extent of immersed tunnel development, some notable achievements are given in Table 18.1.

GOING LONGER

It can be seen that a number of these records are not that recent. In the future, challenges will become greater as engineers seek to push the boundaries of technology; in particular, it is likely that deeper and longer tunnels will be built and construction methods improved to speed up the rate

463

Table 18.1 Current immersed tunnel record holders

Criteria	Tunnel	Value	Year constructed
Longest rail immersed tunnel	BART, San Francisco, United States	5.825 km	1970
Longest road immersed tunnel	Øresund Tunnel, Denmark	4.0 km	2000
Longest tunnel element	Guldborgsund Tunnel, Denmark	230 m	1988
Largest tunnel element (by displacement)	Øresund Tunnel, Denmark	55,000 tonnes	2000
Widest tunnel element	Drecht Tunnel, Netherlands	49 m	1977
Deepest tunnel	Bosphorus Tunnel, Istanbul, Turkey	55 m water depth	2008
Longest tow	Bjørvika Tunnel, Norway	700 km	2010
Most lanes of traffic (road and rail)	Drecht Tunnel, Netherlands	8 road traffic lanes	1977
Fastest concrete element production	Øresund Tunnel, Denmark	8 weeks for a pair of 175-m-long elements	2000

of tunnel element production, and hence reduce the overall construction period. Tunnel element lengths may also increase to facilitate this.

The first immersed tunnels were relatively short, around 1 km long, and the great majority are still below 2 km. In structural terms, there is no limit to the length of an immersed tunnel; all that is needed is to join more and more tunnel elements together. It is not quite that simple, however, and a major limiting factor is the ventilation system. Rail tunnels need less ventilation than road tunnels, so it is no surprise that the longest immersed tunnel built so far is the 5.825 km BART rail tunnel in San Francisco. This is a twin-bore single steel shell design built between 1966 and 1969.

For a road tunnel, developing a strategy for the safe day-to-day operation of the tunnel as well as dealing with emergency conditions is the most critical aspect of the tunnel design. The operational strategy must include a ventilation system capable of dealing with a long tunnel and ensure that safe means of escape and emergency access is provided. Longer road tunnels are becoming possible as vehicle emissions improve. The 4 km long Øresund Tunnel opened in 2000 was a landmark increase in immersed tunnel length for roads. Several other tunnels around 4 km long are in construction or planned, but even longer tunnels are envisaged. The BART tunnel will soon be overtaken as the longest tunnel by the Hong Kong Zhuhai Macao Bridge Tunnel at 6 km, forming part of the 28 km link between Hong Kong and Macao. Also, at the time of the writing of this book, the Fehmarnbelt project is undergoing the planning process, and this 19 km long immersed tunnel will become the longest and largest immersed tunnel to be built. Twenty years ago, it could hardly be imagined that a 19 km immersed tunnel would be constructed in challenging offshore conditions. Undoubtedly, that project will see a step change in the techniques developed for construction, similar to those seen for the nearby Øresund Tunnel in the late 1990s, and will serve as a platform for further development of the technology.

Longitudinal ventilation designs are being developed to cope with longer tunnels. The limiting length depends very much on the traffic density, and longer tunnels are possible for more lightly trafficked roads. Transverse ventilation systems can cope with longer tunnels but the sizes of the ducts required to move the air in and out can become excessive. Intermediate ventilation islands offer a solution, but these are large constructions that form obstructions in the waterway and are best avoided if at all possible. Electrostatic precipitators that clean the air at intervals along the tunnel are also being developed, although they currently require a lot of space and use a lot of energy. With these techniques, and the reduction in vehicle emissions, longer immersed tunnels will be technically possible.

With long road tunnels, however, human behavior also has to be considered. Driving a long distance in a tunnel can be a monotonous experience that could result in tiredness, slower reaction speeds, and hence, more accidents. It is necessary to stimulate drivers' senses to maintain their awareness throughout the journey. This can be achieved with a mixture of sound, lighting, and images throughout the tunnel. The environment needs to be made attractive and enjoyable and promote a feeling of security as some people have phobias about entering long tunnels, particularly undersea ones. The combination of measures needs to stimulate and reassure the driver. The Laerdal Tunnel in Norway, which is a 25 km long rock tunnel, has a special lighting scheme in the intermediate rock caverns to stimulate drivers' senses. Immersed tunnels do not have the opportunity to widen the tunnel unless special elements are constructed for that purpose, but combinations of lighting and artwork could provide similar stimuli, although the use of art that reminds drivers that they are driving underwater is probably best avoided. This is an area that will need more study as longer tunnels are contemplated.

These long tunnels also present logistical problems during construction simply because of their size. Multiple casting facilities are needed, the dredging quantities are vast, and there are considerable environmental challenges in dealing with such large volumes of dredged materials. The construction program needs careful consideration and multiple workfronts considered to complete the tunnel within a reasonable timescale. Even the quantities of concrete and steel required, and the workforce needed, can stretch the capacity of the region in which the tunnel is being built. All these need careful consideration during the feasibility and planning stages, but as techniques and capabilities improve, we can expect the boundaries of the immersed tunnel method to be extended.

Provided the operational safety systems and procedures continue to develop and make tunnels safer and safer places, there is no limit on the length of an immersed tunnel project, in the same way there is no limit to the length of a bored tunnel. Operating procedures and systems are already at the point where it can be demonstrated that with the correct approach the tunnel environment is as safe, if not safer than the open road. Fire safety is key to the thinking of planners and engineers for all tunnel projects, and this has advanced significantly to the point where smoke can be controlled and fires suppressed and managed sufficiently such that the risk of injury or loss of life to drivers and passengers involved in an incident can be controlled to acceptable levels.

GOING DEEPER

Immersed tunnels are going deeper than ever before. The Bosphorus Tunnel in Istanbul has been built at a record depth of 55 m below mean water level to the underside of the tunnel structure. The Busan–Geoje link is similar at a depth of 48 m below mean water level. The Busan and the Bosphorus tunnels have seen developments in the technology used to control and immerse tunnel elements at these great depths. Design and detailing of the structure and the joints between the tunnel elements needs to take the high hydrostatic loads into account, but we are by no means at the limits of design today, and it is possible to go substantially deeper. This may require special development of equipment for dredging and forming foundation layers for the tunnels, but the larger infrastructure projects will support this development. The construction techniques for placing tunnel elements have already developed in their sophistication to be able to place tunnel elements at depth, borrowing technology from the offshore industry.

With this in mind, there is nothing stopping the development of tunnel crossings that exceed the Fehmarnbelt in scale, and the previously fanciful notions of long sea crossings such as the Bering Strait have come into the realms of possibility. Old proposals to cross the English Channel or, more recently, the Irish Sea, could become a reality. Whether bored or immersed tube tunnels, such projects would require massive investment, and this is more likely to be the impediment to their development rather than any limit on the ability of the engineers to design or construct such crossings. Projects such as the Spain to Morocco crossing are being studied; this is intended to be a bored tunnel and would be a major advance in terms of a sea crossing. This would be an immensely challenging project. An immersed tunnel would be no less safe for such a crossing, and given the step change that will be seen with the Fehmarnbelt Tunnel, immersed tubes could been seen to be used for such crossings in the future.

MATERIALS

What materials would be needed for such future mega projects? The trend for concrete tunnels has been seen in recent years throughout the world, and this is likely to remain the material of choice for conventional river and sea crossings around the world. The decline of the steel shipbuilding industry in the Western world means that concrete construction may always be seen as the more flexible option. However, for deeper tunnels, it is more likely that steel could continue to be used because of its inherent ductility and the ability to create stronger composite structures to resist high water pressures.

Studies undertaken for the Bosphorus Tunnel showed the benefits of considering composite technology and modifying the structure shape to deal with high hydrostatic loads. Figure 18.1 shows some of the options that were considered, and the capacity to go to great water depths was demonstrated by using circular and curved tunnel cross sections.

Improvements can be made to various components forming an immersed tunnel. For many years, debate has gone on about the best method to ensure the durability of the structures. As described in Chapter 11, techniques such as cathodic protection are currently most often used, but the selection of alternative materials will solve this problem in the future. The reliance on steel to reinforce our structures and to form the end frames and clamping components for the immersion joint seals will always create a problem of durability. There is no reason why joint frame and clamping components cannot be manufactured using high-strength plastics. This would require the right project to invest in a new approach combined with research and development with universities and manufacturers, and perhaps with sponsorship and funding from scheme promoters. This would be a significant advance if it could be achieved as ongoing maintenance, and inspection would be greatly eased.

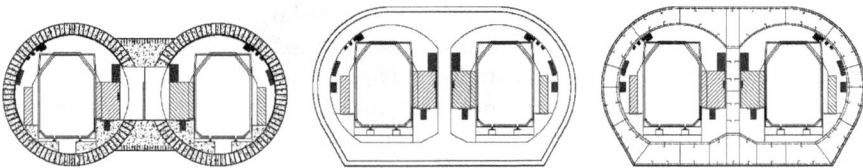

Figure 18.1 Conceptual designs for the Bosphorus Tunnel.

Reinforced concrete technology will continue to move on slowly, but it is already a reliable, long-lived material. Engineers working on immersed tunnel projects already embrace the best of concrete construction practice and will continue to do so.

Construction techniques will no doubt continue to develop. Currently, large tunnel elements of around 200 m are the longest that are constructed. There is no reason why this cannot be extended provided the marine plant can handle the elements and the conditions do not exert overly onerous loadings on the elements. However, for long deep tunnels, perhaps using composite construction, it could be possible to create and place tunnel elements of much greater length. Retaining sufficient flexibility in the structure may be key to achieving this, such that deformations can occur due to waves and currents in temporary conditions without overstressing the structure.

SUBMERGED FLOATING TUNNELS

Submerged floating tunnels (SFTs; also known as the Archimedes bridge) are no longer a distant dream, and considerable effort has been put into their development since the 1990s. Although this book does not attempt to deal with this technology in detail, many aspects of the design approach and construction methods will be very similar to that of the immersed tunnel. This type of construction could be utilized where the depth of water is extreme such that construction of a bored tunnel or immersed tunnel is not feasible, and the waterway of such width that it cannot be spanned by a long-span bridge structure. The Norwegian Fjords are a classic example of conditions where water depths are hundreds of meters and waterway widths may be several kilometers. The first scheme that nearly came to fruition was the Hogsfjord crossing in Norway, which was on the verge of entering the construction procurement process when it was withdrawn from the government construction program. Studies have also been carried out for crossings of deep lakes such as Lake Washington in Washington state, Lugano Lake in Switzerland, and sea straits such as the Strait of Messina, and various locations in Japan.

At the time of the writing of this book, construction of the first prototype is about to begin in China at Qiandao Lake in Zhejiang Province. Once built, this will serve as a pedestrian tunnel in a popular area of natural beauty, but will also act as a prototype to assist larger-scale developments such as the crossing being considered of the Jintang Strait in the Zhoushan archipelago. The tunnel structure for this prototype is planned to have an aluminum outer shell, a steel inner shell, and it is infilled with concrete. It will be 100 m in length and 3.5 m in internal diameter; it is a buoyant structure to be anchored in place by cables. The realization of this prototype would be a massive step forward for this technology. The first full-scale submerged floating highway tunnel that is now likely to be built is currently being planned in Norway, across the Sognefjord. Various combinations of floating bridges and submerged floating tunnels are being considered.

This method of construction can follow one of the following two principles, as shown in Figure 18.2:

1. The tunnel structure is positively buoyant, that is, it floats and is held down by tethers secured to the seabed so that it remains below the water surface.
2. The tunnel is negatively buoyant, that is, it sinks and is supported from above by floating pontoons or from below on trestles, which hold the tunnel at the required level below the water surface.

Tethering systems for buoyant tunnel structures would comprise steel cables anchored to the seabed into ballasted caisson structures or to piled anchorages and connected to the tunnel structure. Similar types of systems could be used for suspending the tunnel structure from pontoons, although most conceptual schemes that have been developed have adopted a fixed

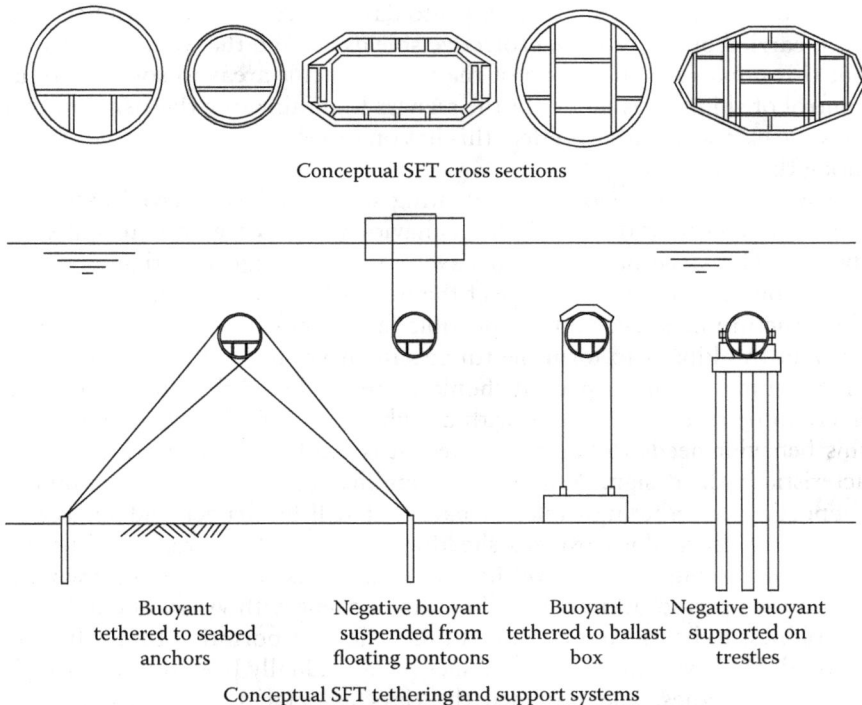

Conceptual SFT cross sections

| Buoyant tethered to seabed anchors | Negative buoyant suspended from floating pontoons | Buoyant tethered to ballast box | Negative buoyant supported on trestles |

Conceptual SFT tethering and support systems

Figure 18.2 Submerged floating tunnel construction types.

rigid framework beneath each floating pontoon to which the tunnel is connected. Slender cable systems will be susceptible to dynamic effects, and this has been the subject of much study by universities in the late 1990s and early 2000s. Support structures for negatively buoyant tunnels would most likely comprise reinforced concrete trestles with either piled or caisson foundations and crossheads on which the tunnel sections would rest and be fixed in position, effectively forming an underwater bridge.

The tunnel structure may be conventional steel and/or concrete, similar to an immersed tunnel structure, and the method of construction is likely to be very similar to that of an immersed tunnel with elements that are sealed by bulkheads being connected underwater while suspended. Dewatering of joints will allow the hydrostatic forces to compress joint seals and create a watertight structure. Therefore, the joints between the elements making up the tunnel are likely to follow the same design principles to the immersed tunnel.

One of the subject areas that receives a great deal of debate is the risk to the structure from impact by shipping or submarines or from anchors trailing from ships passing overhead. The tunnel only needs to be held at a depth of around 50 m below the water surface for shipping to pass above, and to reduce the effects of surface waves and currents on the tunnel structure, but this still leaves it susceptible to damage from impact. Studies have been carried out to provide protective shields around the tunnel structure, but this remains perhaps one of the most difficult areas to solve. Careful control of shipping is no doubt the answer to minimizing the risk, but this must be balanced with the effect this has on vessel speeds and journey times along the fjords.

The dynamic behavior of the floating structure has received extensive study to determine the oscillating behavior under wave and current conditions. The structure needs to have a natural frequency that prevents excitation due to the movement of the body of water, and the detailing of the structure must account for possible fatigue effects due to oscillations. Current thinking is to form the tunnels in curves to improve the behavior of the tunnel in this respect. It should be remembered that the tunnels are likely to be relatively massive with a high natural frequency, so, although this behavior needs to be investigated, it is unlikely to be a critical characteristic in the design. Perhaps more significant could be the effect on the supporting or tethering cables or bars that will be slender and subject to oscillating effects due to vortex shedding under steady current conditions.

The race to build the first of these structures has already been drawn out over a long period of time. It will require a client with vision and a degree of boldness to be the first to build a full-scale transportation tunnel by this method. However, the technical issues have gradually been solved through successive studies, and provided the risks associated with accidental or environmental effects can be managed down sufficiently, it is only a matter of time before the first submerged floating tunnel is constructed. Whether the form of construction becomes at all prevalent will then depend on how competitive the solution can be with other forms of construction, and only time will tell as to whether we will see many of these types of structure in the future. More information on this form of construction can be obtained from the ITA.

LOCATION

So where will tunnels be built in the future? The rise of Asian economies in the early twenty-first century means that there will be an increase in the number of projects constructed in the Asia region. This is already being seen in China, which has overtaken the Netherlands in terms of the number of projects in planning and that will soon be under construction. However, the infrastructure of Europe and the Americas will require constant renewal and expansion and immersed tunnels will continue to be required. Refurbishment of existing immersed tunnels will become a more frequent challenge to the engineering community and for the owners of the aging assets. This has already been seen in the United States with seismic retrofits of existing tunnels, and in both the United States and Europe, where tunnels reaching ages of 40–50 years have needed modification or repair to bring them up to current standards and enable them to carry modern road traffic. Projects such as the Second Midtown Tunnel in Virginia, where three existing immersed tunnels beneath the Elizabeth River carrying traffic between the towns of Norfolk and Portsmouth, combined with a new tunnel to dual the existing Midtown Tunnel and double the road-carrying capacity, are likely to be a common type of project in the developed world from now on.

In summary, there is a bright future for the immersed tunnel. Although fewer than 200 tunnels have been constructed to date, the technology is well developed. There are tried and tested methods that are carried from project to project, but there is ample space for innovation and for extending the possibilities for immersed tunnel crossings.

Figure 18.3 The end. (Courtesy of W. Grantz.)

Glossary

This glossary has been developed by the International Tunnelling Association, Working Group 11—Immersed and Submerged Floating Tunnels, and is reproduced here with their kind permission. It was produced to facilitate international communication on the subject of immersed tunnels in the English language. The terms defined relate to typical design and construction practice for steel and concrete immersed tunnels.

Access shaft Temporary access shafts are commonly provided to allow entry of personnel and occasionally equipment to the interior of an immersed tunnel while floating or submerged. The shafts are usually removed when alternate access is available, such as along the tunnel. The access shaft may be attached to the temporary end bulkhead or may be attached over a temporary hole in the structure, which have to be made watertight later.

Backfill Material placed around the sides and over the top of the tunnel within the excavated trench after the tunnel is installed in the trench. The material is usually granular, rocky, or excavated.

Ballast

  Permanent ballast Nonstructural solid material placed inside or outside an immersed tunnel to increase its effective weight permanently. Material placed outside should either be attached to the tunnel or retained, thereby preventing accidental falling off or loss of the material. Backfill that may be scoured or accidentally dredged away is not ballast.

  Temporary ballast Material used to temporarily increase the effective weight of the tunnel or a tunnel element during the fabrication and installation phases until replaced by backfill or permanent ballast. The material may be solid or liquid.

Ballast tank Temporary tanks constructed within a tunnel element for the purpose of filling with temporary water ballast to increase the weight of the tunnel to aid trimming of the element while floating, to provide overweight for immersion of the element, and for temporarily holding the tunnel element in position on the bottom of the dredged trench until sufficient ballast weight is provided by permanent concrete ballast and backfill to prevent uplift.

Binocular section Term used to describe an element consisting of two adjacent circular steel tunnels, usually each of two lanes, combined into a common structure.

Bore Term borrowed from mined tunneling to describe a cell.

Box (shape) An indication that the overall cross section of the tunnel is approximately rectangular.

Bulkhead An upright watertight partition used to generate compartments, usually totally closing off the inside of a cell. Temporary bulkheads are provided at the ends of tunnel elements to keep water out (make them watertight) during the floating and installation stages.

Buoyancy The resultant upward force on a body partially or fully immersed in a liquid caused by the pressure of the liquid acting on the body. The magnitude of the force is equal to the weight of liquid displaced.

> **Positive or negative buoyancy** Jargon expressions for the amount by which buoyancy exceeds the weight of a body when totally immersed in a liquid. Positive buoyancy indicates that the body tends to float (buoyancy > weight), while negative buoyancy indicates that it tends to sink (buoyancy < weight).

Casting basin A place where elements for immersed tunnels can be fabricated in the dry, and which can be flooded to allow the elements to be floated out and taken away. Generally used for concrete tunnels.

Cell Continuous space within the cross section of an element, bounded by walls, floor, and ceiling. A cross section may contain many cells, hence multiple-cell box, where, for example, separate cells may be used for each traffic direction, emergency egress, utilities, supply air, and exhaust air.

Chamfer Corners of box section tunnels are often chamfered (beveled, with the corners missing) to remove unnecessary space where they serve no useful purpose, or thickened to reduce moments and shears (haunches) and to allow dragging anchors to pass more easily over the tunnel.

Cill Usually, the highest point of the floor on which or against which the gate rests, and over which elements must pass during removal from the fabrication facility.

Closure joint See "Joint, closure or final joint."

Concrete cooling The process of pumping cold water through a network of embedded pipes in the walls and/or slabs of a concrete immersed tunnel element for the purpose of controlling temperature rise during curing and eliminating early-age cracking due to restraint to expansion and shrinkage at early ages.

Concrete tunnel Term applied to a tunnel not designed to leave the fabrication facility until the external concrete structure is essentially complete. Steel plate, if used, is usually limited to acting as a waterproofing membrane. (See also "Steel tunnel.")

Segmental concrete tunnel The concrete tunnel elements are formed with a series of match-cast segments that are discontinuous and able to rotate relative to each other. Segments are temporarily prestressed together during floating and immersion of the elements. The concrete is watertight.

Monolithic concrete tunnel The concrete tunnel elements are constructed as a continuous concrete structure using construction joints with continuous reinforcement. An external waterproofing membrane is usually applied.

Dam plate Term used in the United States for the temporary end bulkhead.

Draft The depth below the still-water surface of the deepest part of a floating body.

Dredging The operation of excavating the trench. It is usually carried out in two stages: first, bulk dredging, and second, trimming the excavation shortly before placing an element. Compensation dredging refers to additional dredging of a waterway to make up for loss of water depth elsewhere in the cross section.

Dry dock Usually, a man-made area that can be dewatered for the repair of ships. A dry dock may also be a semi-submersible floating structure. Immersed tunnel elements are sometimes fabricated or repaired in dry docks. The term is also sometimes applied to a graving dock or a casting basin.

Duct Term used to describe a cell, particularly for supply or exhaust ventilation, or for utilities.

Element A length of tunnel that is floated and immersed as a single rigid unit. The rigidity may be temporary and later released.

End frame The steel frame cast into the end of a tunnel element around the full perimeter of the tunnel on which the Gina seal is mounted for the immersion joint between the tunnel elements. Typically comprises an I-section cast into the concrete with shear connectors and a secondary counterplate welded between the exposed flanges to a close tolerance on which the Gina seal is mounted and secured using clamping bars.

End of tunnel element

Primary or inboard end The end of the tunnel element that is to be connected first. This end will face either the previously immersed and adjoining element or the terminal structure. This end is usually the end equipped with the immersion gasket.

Secondary or outboard end The other end of the tunnel element.

Fabrication   The stage of construction of a tunnel element before it can float. The fabrication facility may be a casting basin, graving dock, dry dock, shipyard, or a greenfield site. The construction of a tunnel element may need to be completed at an outfitting dock.

Factor of safety (with regard to uplift)   The ratio of the weight of a tunnel, or a portion thereof, to the buoyancy. Different required factors of safety may be specified depending on whether backfill is included or removable items are excluded, and depending on the stage of construction. Water density must be specified, since buoyancy will vary with changes in water density.

Fitting out   Also known as outfitting, this term refers to work that is carried out while the element is afloat. It may consist, for example, of completing any remaining necessary construction of the element prior to immersion, the addition of ballast, the installation or removal of temporary equipment such as navigation lights, survey beacons, and access shafts, and adjusting the trim of the floating element. Some of the work may be necessary before transportation (towing), but the remainder must be completed after towing. Additional construction applies mainly to steel tunnels, where much of the internal structural concrete may not be completed until the element is close to its final destination. Some of the work may not be carded out until the element is supported by the immersion equipment.

Freeboard   The height above the still-water surface of the highest part of a floating body.

Full-section casting   A construction method for concrete segmental tunnels whereby the full cross section of the tunnel, over the length of a tunnel segment, is constructed in one continuous concreting operation. The method removes the need for construction joints around the tunnel perimeter and, therefore, eliminates the risk of early-age cracking due to restraint at the construction joints.

Gasket   A device that acts as a seal between two contacting surfaces.

Gina gasket   A proprietary form of gasket used to seal immersion joints, particularly on concrete tunnels. It consists of a full-bodied rubber section able to transfer large compression forces, and a soft nose able to provide an initial seal under low compression. For binocular sections, each circular tunnel usually has its own gasket around the perimeter, whereas most other forms of tunnel use a single gasket around the external perimeter. The gasket provides a temporary seal and compression contact face during immersion installation, remains in place, and may provide a permanent seal at flexible joints.

Omega gasket or seal   This seal, shaped like the Greek letter Omega (Ω), is installed across flexible immersion joints from within a tunnel after immersion and joining. It may form a secondary permanent seal or become the primary seal. It is bolted to the internal faces on each side of the joint. It may be replaced in a similar manner on an as-needed basis. Because of its shape, it can sustain fairly large longitudinal and transverse movements at the joint.

Soft-nosed gasket See "Gina gasket."

Temporary immersion gasket This is usually an extruded rubber section that acts as a seal when it is compressed. After completion of the permanent joint, the seal is no longer needed. This type of gasket is commonly used in the United States.

Gate Usually either hinged to a wall or floating, this structure is used to close off the fabrication facility from the adjacent water to allow dewatering of the facility.

Gravel (bed) foundation See "Screeded foundation."

Graving dock An area that can be dewatered to form a casting basin.

Greenfield site An area above water level converted to enable the construction of tunnel elements, usually steel shell tunnels. The elements may be side or end launched into the water when capable of floating, or may be incrementally launched.

Grouted foundation bed A foundation formed by filling the space between the underside of an element and the preexcavated trench bottom with grout. Until this operation is complete, the elements require temporary support.

Haunch A thickening of a wall or slab to increase locally the bending strength and shear capacity of the section.

Immersion The phase of construction covering the period between the element floating on the surface and installed on its foundation or temporary supports at bed level.

Incremental construction A method of construction whereby a short section of an element is constructed, and then jacked along to enable the adjacent section to be cast against the previous section.

Installation This phase of construction covers preparation for immersion, immersion, foundation preparation, backfilling, and completion of the interior works.

Jet fan A ducted propeller, usually mounted adjacent to or above the traffic, that helps to maintain air velocity within that cell.

Joint

Closure or final joint A joint where the last element has to be inserted rather than appended to the end of the previous element; a marginal gap will exist at the secondary end. This short length of tunnel will need to be cast in place and is known as the closure or final joint.

Construction joint A horizontal or vertical connection between monolithic parts of a structure, which is used to facilitate construction. A waterstop is commonly placed in such a joint.

Earthquake joint   An immersion joint of special design to accommodate large differential movements in any direction due to a seismic event. It is also applied to a semirigid or flexible joint strengthened to carry seismic loads and across which stressed or unstressed prestressing components may be installed.

Expansion joint   Also known as segment or dilatation joint, this term refers to a special moveable watertight joint between the segments of a tunnel element.

Immersion joint   The watertight joint that is dewatered when an element is installed at the seabed. It may remain flexible or can be made rigid, as is usual with steel tunnels. An immersion gasket or soft-nosed gasket is usually used, and an Omega seal is installed later.

Keel clearance   The least vertical distance between the deepest part of a floating body and the bed beneath.

Keel concrete   Concrete, often ballast, placed in the lowest portion of an element.

Lifting lugs   Temporary lifting points from which an element is suspended during immersion, usually removed after an element is set on its foundation.

Locking fill   Backfill, usually granular, placed carefully around the lower part of a tunnel to hold it in position.

Outfitting   See "Fitting out."

Portal   The structure or the end face of the structure at the two ends of the tunnel at the interface of the covered and open sections.

Prestress, temporary   Used mainly in concrete tunnels to temporarily lock a flexible joint, to modify stresses until immersion, or to provide additional strength during transportation and installation.

Pulling jack   Device attached to the primary end of a tunnel element being immersed, which connects to the secondary end of the previous tunnel element and enables the tunnel element to be pulled up to the previous element and create the initial seal of the Gina gasket.

Pumped sand foundation   See "Sand bedding."

Rigging   A system of lines, winches, and hoists used to control the position of an element, both horizontally and vertically, especially during immersion. Lines may be attached indirectly to the shore, anchors, pontoons, derrick barges, or other lowering equipment.

Rock protection or armor   The provision of larger stone or rock to prevent erosion or dredging of required backfill or bed. The term is also applied to systems for protecting a tunnel against potential collisions and dragging anchors.

Roof protection   Protection provided to the waterproofing membrane on the roof against accidental damage. This term is also applied to combinations of backfill and rock protection placed above the roof to protect against sinking or grounding vessels.

Sand bedding   A foundation formed by filling the space between the under-side of an element and the preexcavated trench bottom with sand. The sand is placed hydraulically with the sand-flow or sand-jetting method. Until this operation is complete, elements require temporary support. A small gap may exist at the underside of the element after this opera-tion, so that the temporary supports must be released or deactivated to lower the element onto the foundation.

Sand flow   A method of sand bedding whereby the sand–water mix is transported through a pipe system with fixed outlets in the soffit of the element. The mix is usually discharged through one outlet at a time. As the velocity of the mix decreases after leaving the outlet, sand is depos-ited by gravity to form a firm pancake-shaped mound almost touching the underside of the tunnel, with a small depression beneath the out-let. While pancake dimensions vary, an area of 100 m² would not be unusual. The sand–water mix may be supplied either externally through inlets in the roof or walls or internally through nonreturn valves.

Sand jetting   A method of sand bedding whereby the sand–water mix is transported through a jet pipe that can be moved anywhere in the void between the underside of the tunnel and the trench bottom. As the velocity of the mix decreases after leaving the jet, sand is deposited by gravity. This work can only be done from the outside.

Screeded foundation   Following trench excavation and before immersing an element, a gravel foundation is prepared by screeding to close tol-erances and onto which elements are placed directly without further adjustment. Temporary supports at bed level are not required.

Segment   A monolithic section of a tunnel element separated from other segments only by vertical joints. For concrete tunnels, a segment is typically the length of a single concrete placing operation. Some tunnel elements, particularly in the Netherlands, consist of a number of dis-crete segments held rigidly together during installation by temporary prestress and joined by expansion joints.

Shear dowel   A device to transfer shear across a joint. Shear dowels are sometimes used in concrete tunnel elements across immersion, closure, or expansion joints to provide continuity of alignment. Such shear dowels must permit relative longitudinal movement. For immersion and closure joints, the dowels would be embedded in cast in place con-crete at the joint face after immersion.

Shear key   A device to transfer shear across a joint, usually a moveable immersion joint. In concrete tunnels, the shear key components may form integral parts of the structure of each element. The keys are usually placed in the space between the end bulkheads adjacent to the immer-sion gaskets so that they can be inspected and repaired if necessary.

Sill   Spelling used in the United States for "Cill."

Snorkel   See "Access shaft."

Squat The additional draft of a floating body moving relative to the water in which it floats, as compared with the draft when stationary. It is caused by a reduction in water pressure below the body because of directional changes in flow around the body. When keel clearances are marginal, squat may cause elements under tow to touch the bottom. Similarly, an element below a passing vessel may experience uplift due to squat or propeller wash. This may need to be taken into account in selecting safety factors against uplift during installation.

Steel shell tunnel

    Single-shell tunnel Term applied to a tunnel consisting of elements where an outer structural steel membrane (the shell) is constructed first, very much in the manner of a ship. The steel plate also acts as a waterproofing membrane. Elements are usually designed to leave a greenfield site before the structural concrete is placed, though this may not be the case when other types of fabrication facility are used. Depending on floating stability requirements, keel concrete may or may not be placed prior to launching. In this condition, draft is usually less than 3 m, making long tows relatively easy while afloat. Nevertheless, transport on barges is not uncommon. The shell plate acts as an external form plate for the structural-reinforced concrete with which it is designed to act compositely. While stability and strength requirements may require some of the structural concrete to be placed before transportation, it is typical for this concrete to be completed during outfitting, close to the final location. Ballast may be located inside, but more often outside, on top.

    Double-shell tunnel An outer steel plate, usually octagonal in shape, is added to a single shell tunnel element to act as an external form plate for tremie concrete placed as permanent ballast. The tremie concrete protects the inner shell plate from corrosion, while the outer form plate is left as sacrificial. The behavior of the inner shell plate, and the compositely acting reinforced concrete within it, is similar to a single shell tunnel element, except that the stiffening elements are usually placed outside the inner shell plate.

Submerged floating tunnel (SFT) A tunnel through water that is not in direct contact with the bed. It may be either positively or negatively buoyant, and may be suspended from the surface or supported from or tied down to the bed.

Submersion The part of the installation activity that takes an element from being afloat to sitting on the bed.

Sump Sumps (reservoirs) are provided at the portals and at low points (nadirs) to contain quantities of runoff and leakage water compatible with the storage requirements of the pumps provided. Oil–water separators are usually required, and sumps within a tunnel most often discharge to portal sumps.

Suspended slab Slab provided to span across a cell above a space, such as ceiling and roadway slabs when ventilation ducts above or below the roadway are used.

Terminal structure The nonimmersed structure abutting the first and last immersed tunnel elements.

Towing or transportation Several phases of construction may involve towing. The fast tow is only a short distance from the fabrication facility to the location where outfitting for the main tow is to be carried out, if needed. The main tow is to the location where outfitting for immersion is to be done, usually close to the immersion point. The final tow is to the immersion point.

Trench The space left below bed level after excavation for the immersed tunnel and for its foundation is complete.

Tube Roadway, track, and service cells are each often referred to as tubes. Also used in the jargon expression "immersed tube tunnel," meaning immersed tunnel (both circular and box shaped), it was perhaps originally intended to imply an immersed tunnel with a circular cross section.

Unit Term sometimes used to refer to an element.

Ventilation

Longitudinal ventilation A system in which fresh air is supplied at one end of the traffic tunnel and polluted air is expelled at the other.

Semi-transverse ventilation A system in which a separate ventilation duct is used for the supply of fresh air through many supply vents along the tunnel. The polluted air is discharged through the end of the traffic tunnel. Also used to describe a system where fresh air is supplied from the end of the tunnel and polluted air is drawn out over the length of the tunnel by exhaust fans.

Transverse ventilation A system in which separate supply and exhaust duct systems are used, so that fresh air is distributed and polluted air is collected over the length of the tunnel by supply and exhaust fans.

Warp Cables used to move elements. Warping is the act of moving elements using warps, usually out of a narrow channel or dock.

Waterproofing membrane A skin provided external to the tunnel to improve the watertightness of concrete. The membrane may be of steel or other more flexible materials. It may be sprayed on or applied to the exterior surface, or the concrete may be placed onto or against it. Most types of flexible membrane require protection against damage by backfill.

Waterstop Special components embedded in concrete construction joints to reduce the permeability of the joint. Waterstops may be flexible for expansion joints.

Watertightness A measure of the capability of a tunnel to resist the penetration of water (leakage).

References

Although reference material has not been cross-referenced throughout the book, the authors felt it would be useful to include a list of background documents that have been used in the preparation of the book, other than national and internationally recognized design codes and standards.

J. Baber, & L. Narvestad. Maintenance considerations for immersed tunnels, with particular reference to the Bjørvika Tunnel. In Proceedings of the Fifth Symposium on Strait Crossings, Trondheim, Norway, 2009.

J. Baber, T. A. M. Salet, & L. K. Lundberg. Øresund Tunnel—Control of early age cracking. In Proceedings of the IABSE Colloqium on Tunnel Structures, Stockholm, Sweden, pp. 175–180, 1998. ISBN 3-85748-096-3.

P. B. Bamforth. *Early-Age Thermal Crack Control in Concrete*. CIRIA, 2007.

G. O. Barratt, & R. Hamlin. The development and construction of offshore cooling water culverts for South Humber Bank Power Station. In Proceedings of the ICE—Water Maritime and Energy, Vol. 130, No. 1, pp. 11–23, Paper 11435, March 1, 1998.

G. O. Barratt, & M. J. V. Sheridan. Construction of the cooling water tunnels using immersed tube techniques. In Proceedings of the Institution of Civil Engineers—Civil Engineering, Vol. 108, No. 5, pp. 63–72, 1995. ISBN 13: 9780727720191, ISBN 10: 0727720198.

H. G. Blaauw, & E. J. van de Kaa. Erosion of bottom and sloping banks caused by the screw race of manoeuvring ships. Paper presented at the 7th International Harbour Congress, Antwerp. Delft Hydraulics Laboratory, the Netherlands, 1978.

J. Busby, & C. Marshall. Design and construction of the Øresund Tunnel. In Proceedings of the Institution of Civil Engineers—Civil Engineering, Vol. 138, pp. 157–166, Paper 12233, November 2000. Thomas Telford Ltd, London.

F. T. Christensen. Ice ride-up and pile-up on shores and coastal structures. *Journal of Coastal Research*, Vol. 10, No. 3, p. 681, 1994.

CIRIA, CUR, & CETMEF. *The Rock Manual: The Use of Rock in Hydraulic Engineering*. 2nd ed., C683. CIRIA, London, 2007. ISBN: 978-0-86017-683-1, ISBN: 0-86017-683-5.

Comité Euro-International du Béton. *Concrete Structures Under Impact and Impulsive Loading*. Synthesis Report, CEB Bulletin No. 187, 1988.

J. M. Duncan, & A. L. Buchignani. *An Engineering Manual for Settlement Studies*. Department of Civil Engineering, University of California, Berkeley, CA, 1976.

C. R. Ford (ed.). *Immersed Tunnel Techniques 2*. In Proceedings of the International Conference organized by the Institution of Civil Engineers in Association with the Institution of Engineering in Ireland and in Cork, Ireland, Vol. 2, April 23–24, 1997. ISBN 13: 9780727726049, ISBN 10: 0727726048. Thomas Telford Ltd, London.

A. Glerum. Developments in immersed tunnelling in Holland. *Tunnelling and Underground Space Technology*, Vol. 10, No. 4, pp. 455–462, 1995.

A. Glerum. Motorway tunnels built by the immersed tube method. *Rijkswaterstaat Communication No. 25*. The Netherlands Government Publishing Office, 1976. ISBN 9012011884.

W. Grantz. Immersed tunnel settlements: Part 1—Nature of settlements. *Tunnelling and Underground Space Technology*, Vol. 16, pp. 195–201, 2001.

W. Grantz. Steel shell immersed tunnels—40 years of experience. *Tunnelling and Underground Space Technology*, Vol. 12, No. 1, pp. 23–31, 1997.

A. Gursoy. Immersed steel tube tunnels: An American experience. *Tunnelling and Underground Space Technology*, Vol. 10, No. 4, pp. 439–453, 1995.

C. J. A. Hakkaart, & Wim't Hart. Dutch high speed railway immersed tunnel projects. In *(Re)Claiming the Underground Space*, Vol. 1, pp. 291–296, published by A. A. Balkema, Lisse, the Netherlands. Proceedings of the ITA World Tunnelling Congress in Amsterdam, the Netherlands, 2003.

M. Hamada, et al. Earthquake observation on two submerged tunnels at Tokyo Port. Paper presented at the Soil Dynamics and Earthquake Engineering Conference, Southampton, England, pp. 723–735, 1992.

Y. M. A. Hashash. Seismic behavior of underground structures and site response. Proceedings of the 1st International Conference on Seismic Retrofitting, Tabriz, Iran. Iranian National Retrofitting Centre, 2008.

Y. M. A. Hashash, J. J. Hook, B. Schmidt, & J. I.-C. Yao. Seismic design and analysis of underground structures. *Tunnelling and Underground Space Technology*, Vol. 16, No. 4, pp. 247–293, 2001.

J. I. Hawkins. Report upon the experimental brick cylinders laid down in the River Thames at Rotherhithe in the year 1811, for the purpose of ascertaining the practicability of forming a tunnel under the river. Held by Institution of Civil Engineers.

Institution of Civil Engineers. *Immersed Tunnel Techniques: Proceedings of the conference organized by the Institution of Civil Engineers*, Manchester, April 11–13, 1989. ISBN 0-7277-1512-7. Thomas Telford Ltd, London, 1990.

International Tunnelling Association. *Immersed Tunnels: A Better Way to Cross Waterways*. Tribune—special edition, May 1999.

International Tunnelling Association, Immersed and Floating Tunnels Working Group. *State-of-the-Art Report*. 1st ed. Pergamon Press, Oxford, 1993.

International Tunnelling Association, Immersed and Floating Tunnels Working Group. *State-of-the-Art Report*. 2nd ed. Pergamon Press, Oxford, 1997.

ITA Working Group 11 for Immersed and Floating Tunnels. *An Owner's Guide to Immersed Tunnels*. J. Baber (ed.). International Tunnelling and Underground Space Association, Lausanne, Switzerland, 2011. ISBN 978-2-97000624-5-5.

W. Janssen, & S. Lykke. The fixed link across the Øresund: Tunnel section under the Drogden. *Tunnelling and Underground Space Technology*, Vol. 12, No. 1, pp. 5–14, 1997.

W. Janssen, P. de Haas, & Y.-H. Yoon. Busan–Geoje Link: Immersed tunnel opening new horizons. *Tunnelling and Underground Space Technology*, Vol. 21, No. 3–4, p. 333, 2006.

O. P. Jensen, T. H. Olsen, C. W. Kim, J. W. Heo, & B. H. Yang. Busan–Geoje Project South Korea: Construction of immersed tunnel in off-shore wave conditions. *Tunnelling and Underground Space Technology*, Vol. 21, No. 3–4, p. 333, 2006.

V. K. Kanjlia, Central Board of Irrigation and Power (India), ITA (India), ITA, et al. Underground facilities for better environment and safety. In Proceedings of the ITA World Tunnel Congress, Agra, India, September 22–24, 2008.

T. Kasper, J. S. Steenfel, L. M. Pedersen, P. G. Jackson, & W. M. G. Heijmans. Stability of an immersed tunnel in off-shore conditions under deep water wave impact. *Coastal Engineering*, Vol. 55, No. 9, p. 753, August 2008.

H. Kimura, I. Kojima, & H. Moritaka. *Development of Sandwich-Structure Submerged Tunnel Tube Construction Method.* Nippon Steel Technical Report No. 86, July 2002.

O. Kiyomiya. Earthquake-resistant design features of immersed tunnels in Japan. *Tunnelling and Underground Space Technology*, Vol. 10, No. 4, pp. 463–475, 1995.

O. Kiyomiya. Flexible joints between elements for large deformation. In *(Re) Claiming the Underground Space*, Vol. 1, pp. 329–334, published by A. A. Balkema. Proceedings of the ITA World Tunnelling Congress in Amsterdam, the Netherlands, 2003.

O. Kiyomiya, Y. Higashijima, Y. Ohgoshi, & H. Yokota. New type flexible joint for the Yumeshima immersed tunnel. *Tunnelling and Underground Space Technology*, Vol. 21, No. 3–4, p. 333, 2006.

J. Krokeborg (ed.). Strait crossings. In Proceedings of the 2nd Symposium on Strait Crossings, Trondheim, Norway, June 10–13, 1990, 2007. ISBN 13: 9789061911180, ISBN 10: 9061911184.

S. Lotysz. Immersed tunnel technology: A brief history of its development. *Civil and Environmental Engineering Reports*, No. 4, pp. 97–110, 2010.

S. Lotysz. Who immersed the tunnel: The early 19th century pioneers of modern technology. In ICOHTEC (International Committee for the History of Technology) 33rd Symposium, Leicester, UK, August 15–20, 2006.

G. D. Mainwaring, Y. K. Lam, & L. W. Weng. The planning, design and construction of the Tuas cable tunnel and future power transmission cable tunnels in Singapore. In Proceedings of the Rapid Excavation and Tunnelling Conference, Santiago, Chile, 2001.

C.-P. Müller, J. Kampmann, M. Tonnesen, & I. B. Kroon. Design of the Busan Geoje immersed tunnel for accidental events and extreme loads. In Proceedings of the ITA World Tunnel Congress, Publication No. 2059, September 22–27, 2008.

R. Narayanan, H. Bowerman, & F. Naji. *Application of Guidelines for Steel-Concrete-Steel Sandwich Construction: 1: Immersed Tube Tunnels.* Publication No. 132, Steel Construction Institute, UK, 1997. ISBN 9781859420133.

Port and Harbour Research Institute, Ministry of Transport, Japan (ed.). *Handbook on Liquefaction Remediation of Reclaimed Land.* Taylor & Francis, UK, 1997.

Proceedings of the Delta Tunnelling Symposium, Vols. 1 and 2, November 16–17, 1978. Royal Institution of Engineers, the Netherlands.

Proceedings of the Øresund Link Immersed Tunnel Conference, Copenhagen, Denmark, April 5–7, 2000. Øresunds-konsortiet, Øresund Tunnel Contractors I/S, and International Tunnelling Association.

M. J. Prosser. *Propeller Induced Scour.* Fluid Engineering Centre, BHRA, Cranfield, 1986. ISBN: 0900337184.

S. Quanke, C. Yue, Y. Li, & J. C. W. M. (Hans) de Wit. China's Hong Kong Zhuhai Macao Bridge (HZMB) link—Stretching the limits of immersed tunnelling. In Proceedings of ITA 37th General Assembly and World Tunnel Congress, Helsinki, 2011.

N. Rasmussen. Concrete immersed tunnels—40 years of experience. *Tunnelling and Underground Space Technology,* Vol. 12, No. 1, pp. 33–46, 1997.

J. C. Robertson (ed.). *The Mechanic's Magazine, Museum, Register, Journal and Gazette,* Vol. 47, 1845.

K. Stiksma (ed.). *Tunnels in the Netherlands—Underground Transport Connections.* Rijkswaterstaat, the Netherlands, 1987. ISBN 90 6618 591 0.

U.S. Department of Transport Federal Highway Administration. Technical manual for the design and construction of road tunnels—Civil elements. Publication No. FHWA-NHI-10-034, December 2009.

G. M. J. Williams, & K. W. Innes. Structural aspects of submerged tube tunnel construction. *The Structural Engineer,* Vol. 50, No. 8, Issue 2, 1972.

N. Yoshida (ed.). *Remedial Measures against Soil Liquefaction: From Investigation and Design to Implementation.* Taylor & Francis, UK, 1998.

T. L. Youd, I. M. Idriss, et al. Liquefaction resistance of soils: Summary report from the 1996 NCEER and 1998 NCEER/NSF workshops on evaluation of liquefaction resistance of soils. *Journal of Geotechnical and Geoenvironmental Engineering,* Vol. 127, No. 10, pp. 817–833, 2001.

Index

For Product Safety Concerns and Information please contact our EU
representative GPSR@taylorandfrancis.com
Taylor & Francis Verlag GmbH, Kaufingerstraße 24, 80331 München, Germany

www.ingramcontent.com/pod-product-compliance
Lightning Source LLC
Chambersburg PA
CBHW060421220326
41598CB00021BA/2251

*9 7 8 1 1 3 8 0 7 6 1 8 1 *